Globalization and New Geographies of Conservation

GLOBALIZATION & NEW GEOGRAPHIES OF CONSERVATION

Edited by Karl S. Zimmerer

The University of Chicago Press   Chicago & London

Karl S. Zimmerer is professor in and chair of the Department of Geography and a member of the Institute for Environmental Studies at the University of Wisconsin–Madison. He is the author of *Changing Fortunes: Biodiversity and Peasant Livelihood in the Peruvian Andes* and coeditor of *Nature's Geography: New Lessons for Conservation in Developing Countries* and *Political Ecology: An Integrative Approach to Geography and Environment-Development Studies.*

The University of Chicago Press, Chicago 60637
The University of Chicago Press, Ltd., London
© 2006 by The University of Chicago
All rights reserved. Published 2006
Printed in the United States of America

15 14 13 12 11 10 09 08 07 06   1 2 3 4 5

ISBN: 0-226-98343-9 (cloth)
ISBN: 0-226-98344-7 (paper)

Library of Congress Cataloging-in-Publication Data

Globalization and new geographies of conservation / edited by Karl S.
Zimmerer.
   p.   cm.
  Includes bibliographical references and index.
  ISBN 0-226-98343-9 (cloth : alk. paper) — ISBN 0-226-98344-7 (pbk. : alk.
paper)
  1. Conservation of natural resources. 2. Natural resources—Management.
  3. Globalization. I. Zimmerer, Karl S.

  S936.G56   2006
  333.72—dc22

                               2005031370

Dedicated to the memory of my brother, Alexandre Zimmerer (1959–2002), who passed away shortly before the conference that launched this book. I hope it will contribute in some meaningful way to the livelihoods and landscapes of his family, their community, and tropical mountains and coasts like the ones in Jamaica that he loved.

Contents

Acknowledgments

It is a pleasure to be able to acknowledge the many people who have made this book possible. I owe an especially large debt of gratitude to the many presenters, participants, and commentators at the conference "Spaces of Hope? Conservation, Environment, and Development amid Global Change," which was held in Madison, Wisconsin, in April 2002. Richard Schroeder, who delivered the keynote address, deserves special mention, as do Thomas Bassett, Susan Paulson, Marcus Lane, and Eduardo Brondizio, whose presentations and participation offered vital contributions. The late Fred Buttel, who attended the conference, shared his astoundingly incisive and constructive comments with me the next week, and they helped to shape the project at its inception. The ideas, debates and discussions, and papers and commentaries that emerged from the "Spaces of Hope?" conference formed a general point of departure for the project that became this book. The authors of the book's chapters exercised a particularly high level of interest and commitment, and even greater patience, as we undertook a follow-up process of research and writing that lasted more than two years.

I am especially grateful to Bradford Barham, who has codirected the Environment-and-Development Advanced Research Circle (EDARC) and who contributed a vital intellectual force to this project and the follow-up. I also appreciate the fine work of Ryan Galt and Margaret Buck, who designed and created several maps and helped out with so many of the conference logistics and organizational details. Not least, I gratefully acknowledge the financial backers of the conference: the International Studies Institute, which supports EDARC, and the Nave Fund of the Latin American, Caribbean, and Iberian Studies Program (LACIS), both of the University of Wisconsin–Madison. Several colleagues contributed indispensably to the editing and production of the book. Eric Carter contributed his superb editorial skills at various stages and offered many substantive improvements as well, even while finishing his

PhD at UW–Madison. Onno Brouwer, Marieka Brouwer, and Richard Worthington of the UW–Madison Cartography Laboratory offered excellent graphics design and crafting.

In the final stretch of preparing this volume, I benefited from the amazing intellectual home of the Agrarian Studies Program at Yale University during the 2004–5 academic year. I have tried to extend the program's unique interdisciplinary framings of people, politics, and landscape into these pages. Finally, Christie Henry and the University of Chicago Press have been ideal counterparts in bringing this work to fruition.

1 Geographical Perspectives on Globalization and Environmental Issues: The Inner-Connections of Conservation, Agriculture, and Livelihoods

KARL S. ZIMMERER

Environmental issues have undergone a process of becoming more "global" during recent decades. Most science and policy circles no longer view the environment as a resource or management issue of mainly regional or national importance. Rather, the world of nature is now seen predominantly as a life-sustaining ecosystem of global dimensions. The global scale is noticeably well suited to the properties of many contemporary environmental issues, such as the biodiversity crisis, water resource availability and quality, and global climate change, which have parameters of plainly global proportions. Awareness of a global perspective on the environment thus stems from how we increasingly understand, envision, and attempt to address these issues (Buttel, Hawkins, and Power 1990; Buttel and Taylor 1992; Cosgrove 2001; Mooney 1998). While this global casting of science, policy, and public awareness and debate on environmental issues has been vastly expanded since the 1970s, it is built upon the foundations of earlier environmentalism and the environmental sciences that stem from the periods before and after the Second World War (Frank 1997; Grove 1996; Guha 2000).

In addition to the rise of a global perspective on the environment there has been a substantial growth of environmental globalization. The term *environmental globalization* refers to the increased role of globally organized management institutions, knowledge systems and monitoring, and coordinated strategies aimed at resource, energy, and conservation issues. Here the sense of *globalization* is the functional integration of internationally dispersed activities or phenomena that indicates the general scaling-up of phenomena that occur over greater distances worldwide in ever shorter amounts of time (Johnston 2000). It resonates with the idea of "processes, procedures, and technologies—economic, cultural, and political—underpinning the current 'time-space' compression which produces a sense of immediacy and simultaneity about the world" (Brah, Hickman, and Mac an Ghaill 1999, 3). In the case of environmental globalization, "functional integration" is manifest in the expanded role of global organizations and agreements. At the same time, the sense of immediacy and simultaneity is apparent in the daily plethora of updated environmental data in which globalization creates the thickening and multiplication of information flows about the state of nature worldwide.

The meaning of globalization that is based on integration and immediacy-simultaneity is increasingly a part of *both* the heightened awareness of the broad human dimensions of environmental problems (biodiversity crisis, water resource issues, global climate change) *and* political strategies for environmental protection that are conceived and crafted institutionally at the global scale (Jasanoff 2004; Mol 2001; Solbrig, Paarlberg, and de Castri 2001; Speth 2003). These political strategies include various global treaties, multilateral agreements, and coordinated management organizations, as well as the environmental activism of globally networked groups. With this increased globalization also comes a growing recognition of the powerful *informal linkages* between a wide gamut of contemporary environmental issues and various other forms of globalization. Increasingly environmental issues are functionally integrated through such informal yet highly influential linkages to economic, political, cultural, technological, and demographic changes that are themselves forms of globalization (Brah, Hickman, and Mac an Ghaill 1999; Johnston 2000; Johnston, Taylor, and Watts 2002).

The view of globalization as an active part of environmental issues through both the adoption of a global perspective and the integration effects of globalizing processes leads to the production of a new kind of Janus-faced juxtaposition of tensions and uncertainty—rather than seeing globalization as the root cause of blanket outcomes of "good" or "bad" (Harvey 1996; Mol 2001; Roberts and Thanos 2003; Speth 2003). Environmental and social consequences have often been negative, often to extreme extents, as a result of globalization changes. Predominant forms of economic and political globalization are strongly anti-environmental, for example, in the destructive clearing of tropical forests—legal or illegal, for crops, pasture, or forest products—that is propelled by global markets, development policies, and the general export of negative environmental impacts to less regulated nations (Chase 2002; Gwynne and Kay 2004). This widespread anti-environmental globalization is currently driven and guided through the powerful influences of a cohesive organizing principle or paradigm of Globalization, which suggests an uppercase *G*. This Globalization has taken shape since the 1970s and 1980s due to the worldwide consolidation and growth of neoliberalism as a predominant economic and political model that has continued to evolve and expand notwithstanding challenges. Neoliberal-led Globalization is driven through the forces of new political, economic, and technological changes that take shape through neoliberal policies and that often have a heavy impact on environments and resources (Bray and Bray 2002; Harvey 1996, 2000; Huber and Solt 2004; Peck 2004). It is this Globalization that is targeted in protests against trade agreements such as the World Trade Organization (WTO) and "free trade" that lacks adequate safeguards for the environment, human rights, and work conditions. The broader debates on globalization (and Globalization) are deeply seated and sorely needed, as much as they are often superheated (Ellwood 2001; Stiglitz 2002).

Yet globalization in the broad sense, which emphasizes the current variety of processes, also carries the possibility for environmental improvement and suggests the potential for direct benefits to sustainability. Potential for the positive environmental impacts of globalization are evidenced, for example, in the growth of global markets in "green products," worldwide environmental standards and regulations for industry, and sustainability policies for rural development (Roberts 1998; Roberts and Thanos 2003). The Janus-faced tension between the latter potentially "green" type of environmental globalization and the various other global-scale forces is often not recognized. This tension needs to serve as a centerpiece of debates on linkages between environment and development, and on such specific areas as conservation and agriculture, in the context of globalization. Indeed, this tension is the centerpiece of this book and is highlighted in the material covered in this chapter. Focusing on the interface of conservation and agricultural livelihoods, my collaborators and I have set out to examine several of the main tensions in this arena of environmental issues that are related to globalization (both in the broad sense of globalization and, more narrowly, as the Globalization described above).

Here in chapter 1, then, I present the background to a pair of basic premises that underlie this volume's view of the newly unfolding issues that combine the global conservation of nature and landscapes, the sustainability of global environments, and the affairs of worldwide development. The general discussion and case studies that follow are designed to provide a number of new understandings aimed at concerned persons—ranging from students, researchers, and teachers to practitioners, activists, and policymakers.

The first premise is that the realm of environmental conservation, in particular the worldwide expansion of protected areas, is a specific component of globalization that increasingly entwines nature and society in various important ways. The recent and ongoing growth of conservation areas, in terms of both official designations (territorial, legal, political) and the discourses and rhetoric surrounding them, is demonstrated to offer a number of similarities and some key differences with respect to other facets of globalization related to the environment (Hall 2000; McCarthy and Prudham 2004; Neumann 2005; Perreault 2005; Perreault and Martin 2005; Rudel 2002; Wells and Brandon 1993; Western and Wright 1994; Zerner 2000). Yet we recognize that the worldwide expansion of conservation has been absent from the general analysis of environmental facets of globalization (Mol 2001; Speth 2003). This absence is perhaps unsurprising, given the large extent, complexity, and interdisciplinary nature of global conservation and protected areas in particular. Notwithstanding this omission, the undertaking of conservation at the global scale tends to receive a high degree of publicity and scientific interest and to invoke a series of vigorous (and vociferous!) debates. Given growing interest and global dimensions, the expansion of protected areas is primed to illuminate the complex and often contradictory forces of globalization.

On the one hand, then, the design and management of conservation areas are typically upheld as integral to environmental globalization. As evidence of this importance, the global expansion of designated conservation areas is typically presented as a showcase of international and national environmentalism worldwide and is regularly highlighted in such influential and representative accounts as the *World Resources Report* (Christen et al. 1998; IUCN 1992; WRI 1990, 1992, 1994, to the present). On the other hand, the establishment of these conservation areas is often hotly contested due to disputes over the design, location, and coverage of territories; debates over the efficiency of techniques for managing plant and nonhuman animal populations; uncertain or unfavorable impacts on livelihood activities and resource access among persons in or near the demarcated area; and, in some cases, the forced displacement of local people who often are economically poor (e.g., agricultural smallholders, peasant farmers, and livestock herders) and socially disadvantaged (e.g., ethnic minorities, indigenous peoples; Brechin 2003; Naughton-Treves 1997; Neumann 1995, 1998, 2005; Stevens 1997; Stonich 2001; M. D. Turner 1999b; Zerner 2000; Zimmerer and Carter 2002).

Several lines of broad-based evidence support the first premise. These lines of evidence, which are outlined here and discussed more fully in the section below titled Overview of the Main Premises, serve to demonstrate that the expansion of conservation areas worldwide is a facet of environmental globalization.

1. Various organizations with international or global reach have effectively promoted the expansion of protected areas at the global scale. These organizations range from units of the United Nations and World Bank to global nongovernmental organizations (NGOs) that specialize in establishing protected areas and other conservation goals. The activities of these global conservation organizations are highly varied, for they encompass financing of protected-area establishment and management plans, coordi-

nation of conservation management activities within and among protected areas, scientific advising, information gathering, public relations, and fund-raising.

2. Awareness and realization of the importance of conserving global biodiversity—the "biodiversity phase" of modern environmentalism that is clearly connected to economic globalization (Brechin 2003; Brechin et al. 2002; Escobar 1995; Goldman 1998; Guha 2000; Worster 1990)—has fueled the designation of many protected areas worldwide, in both developed and developing countries. Biodiversity conservation activities in the latter countries are funded primarily by global and international organizations. For example, biodiversity conservation (including protected areas) in Latin America is estimated to depend on global and international organizations for more than 90 percent of total funding (Castro and Locker 2000).

3. Globalization plays a major role in the politics of environmental conservation, because global political and economic processes influence how protected areas are understood and discussed, granted power, prioritized in policy agendas, and inscribed as territory. Globalization shapes emerging conservation discourses, for it invokes the idea of improvement through integration that fits neatly with the global framework of conservation. These globalization discourses on protected-area conservation are especially influential and popular among the more economically and politically powerful groups (the "elite") of developing countries (Bassett and Zuéli 2000; Christen et al. 1998). References to globalization discourses, as well as to global support networks, are utilized in many situations within the anti-Globalization politics of poorer and socially disadvantaged persons and groups that reside in proximity to conservation areas (Forsyth 2001; Haenn 2002; Peet and Watts 2004; Zerner 2000).

The second main premise of our book is that globalization has forced conservation to interface increasingly with agriculture and other types of livelihoods and resource use (livestock raising, forest extraction, fishing, hunting). Rapid growth of this interface of conservation is driven by numerous factors that include the territorial expansion of protected areas and the management emphasis of these broadened conservation efforts. As discussed further in this chapter's final section (Overview of the Main Premises), the worldwide coverage of designated protected areas is estimated to have expanded more than tenfold in area during the past three decades (Zimmerer, Galt, and Buck 2004). Measuring less than 1.0 million km^2 in 1970, and estimated at 5.2 million km^2 in 1985, the area of publicly designated protected areas grew to more than 12.2 million km^2 by 1997 (IUCN and WCMC 1998). These estimates, which must be treated as coarse approximations, suggest that the most significant expansion of protected-area coverage took place over the course of the past couple of decades (Speth 2003).[1] This growth of protected areas has continued from 1997 to the present. In addition to its territorial, legal, and political ramifications, protected-area expansion is discursive and rhetorical through its many deployments in global environmental debates and discussions. The term *inner-connections*—to describe the varied and deeply embedded interactions of this vastly expanded conservation coverage with the realms of agriculture, livelihoods, and land use—is meant to encompass various sorts of relationships between protected areas and resource use that range from synergistic coexistence to open conflict to managerial incorporation.

Incorporation of agriculture and resource use into conservation programs, whether tending toward cooperation or antagonism, is an important characteristic of the expansion of global conservation and protected areas (Wiseman and Hopkins 2001; Zimmerer 2005). Thus, the second premise of our book is extended to focus also on the land use aspects of these enlarged conservation efforts that, throughout the world, frequently impinge on the activities of farmers, livestock-raising pastoralists, and other resource users. The characteristics of their livelihood activities and the variation of their social, economic, and political power are keys to the interface with globalization. Even at first glance it is easy to acknowledge the scope of the increased incorporation of land use in conservation. In support of our argument is the large extent of designated conservation coverage that belongs to the globally recognized categories calling for sustainable utilization through agriculture and resource use. In 1997 the global coverage of management for sustainability activities amounted to more than 7,125,000 km^2, an area equivalent to about three-fourths the size of Canada.[2] Equally or more persuasive than this quantitative measure, since it sums the often inaccurate estimates of the global conservation databases, is the rhetorical significance of presenting and publicizing these numbers. Regardless of estimating inconsistencies, these numbers play a vital role of image creation in the world of global con-

---

1. Like certain other facets of environmental globalization (such as global treaties on trade in endangered species), such trends were built on important developments that took place in preceding decades as well (Frank 1997).

2. Another means of evaluating the increased global coverage of management for sustainability activities is through using globally standardized categories (discussed further in the third section of this chapter; see also Zimmerer, Galt, and Buck 2004). A distinction can be drawn between the global coverage of those official protected areas whose primary purpose is designated as promotion of the sustainability of agriculture and resource use versus those intended primary or exclusively for strict conservation. Using this distinction, and mindful as before of the coarse approximations that are characteristic of the global conservation data sets, the former categories (i.e., resource and land use such as agriculture) are seen to rise from 55.8 percent of the global total in 1985 to 58.2 percent in 1997.

servation. One primary image of the expansion of protected areas is the incorporation of resource and land use, such as agriculture, into global conservation.

The increased interface of conservation areas with agriculture and resource use is an integral part of a "third wave" of conservation that gained prominence in the late 1980s and early 1990s (on this historicization see Guha 2000; see also Brandon, Redford, and Sanderson 1998; Redford and Padoch 1992; Schwartzman, Moreira, and Nepstad 2000; Western and Wright 1994). The term *third wave of conservation* refers to the shift to sustainability as one of the defining goals of conservation worldwide. The goal of sustainability—or the still more general term *sustainable development*—has been granted a high level of priority similar to that given strict preservation in at least certain conservation circles (Cavallaro and Dansero 1998; Sneddon 2000). This incorporation of agriculture and resource use is particularly evident in so-called developing countries; a general estimate is that developing countries contain approximately two-thirds of the total global coverage of protected areas, more than 7.5 million km². Of this developing-world coverage, approximately 4.5 million km², or an area that covers more than twice the size of Mexico, is found in areas that are designated specifically for agriculture and resource use under global conservation guidelines. The interface of conservation areas with persons involved in agriculture and resource use is also far more widespread in developing countries due to the existence of more dense rural populations whose livelihoods depend on farming, livestock raising, and other forms of resource extraction (Solbrig, Paarlberg, and de Castri 2001). These characteristics are a primary reason for the focus of our book on developing countries' experiences with conservation and globalization.

The pair of premises described above may be used to highlight how I have designed this book to address geographical aspects of the recent intensification of the inner-connections between global conservation and livelihoods. In this context *geographical* refers to (a) human-environment interactions that are important to the conservation-globalization nexus and (b) the changing spatial organization of activities related to conservation-globalization issues. These geographical aspects are of major interest to interdisciplinary environmental studies in general; more particularly, the thematic foundation of each point is major emphasis of the modern discipline of geography and its extension into interdisciplinary environmental studies (Batterbury, Forsyth, and Thomson 1997; Forsyth 2003; Neumann 2005; Peet and Watts 2004; Robbins 2004; Stott and Sullivan 2000; Zimmerer and Bassett 2003). As a result, our book is intended for the wide audience comprising *both* those specializing in human-environment interaction and spatial aspects of the conservation-globalization nexus *and* the general reader with interest in environmental issues and development-with-sustainability concerns. In the book we focus on geographical aspects of both protected areas per se and the increased interface of conservation with agriculture and resource use. These topics are of broad, timely relevance to the fields of environmental studies and land resources; resource management and development studies; conservation biology; regional, land use, and resource planning; environmental, human, and physical geography; ecological anthropology; and environmental sociology.

In establishing a geographical perspective on conservation and globalization this book advances the series of four arguments that are introduced below. Following the brief overview here, I introduce each of the main arguments in more detail in the next section (Expanding Dimensions of Conservation and Globalization). By way of introduction at this stage I refer to the central idea or ideas of the four main arguments. To do so, I make reference to a group of central concepts—namely, conservation territories, spatiality, scale, sustainability, and decentralization—for which the definitions and usages of key terms are offered.

First, this book argues that the entwining of conservation and globalization has led to a new range and unprecedented number of spatial arrangements whose environmental management goals and prescribed activities may vary from strict nature protection to sustainable utilization. These spatial arrangements are etched in *conservation territories,* which refer to the designated spaces of nature protection and resource management (Daniels and Bassett 2002; Dinerstein 1995; Neumann 2004; Zimmerer 2000; Zimmerer and Carter 2002). The proliferation of these territories worldwide has depended on the extension of existing types of management units as well as the rapid evolution of novel management units, such as community conservation areas, watershed-based projects, and the management or buffer zones of biosphere reserves (Agrawal and Gibson 1999; Ravenga et al. 1998; Schroeder 1999; Wells and Brandon 1993). Technology changes, especially advances in remote sensing (RS), geographical information systems (GIS), and global positioning systems (GPS), are central to this proliferation of conservation territories (St. Martin 2001; B. L. Turner et al. 2001; M. D. Turner 2003). These geospatial technologies provide much-needed tools for global conservation managers, as well as an emphasis on spatial categorization that is characteristic of the role of the global environmental sciences in the new conservation territories.

My collaborators and I further argue that globalization has produced new spaces for conservation that go beyond the mere extension and modifications of conventional conservation territories. These other spatial arrangements increasingly include environmental networks (both contiguous and discontinuous), in which landscape connectivity and matrix quality are paramount, as well as planning process-based overlapping or multifaceted spatial units that are associated with the diversity of coexisting management goals and institutional processes (Morehouse 1996; Soulé and Terborgh 1999; Terborgh 2002). In order to describe this range of types of spatial arrangements, and the processes of how and why they form, we rely on the idea of *spatiality,* which refers to a concept that helps us to see environmental spaces and configurations in which "physical extent" is "fused with social intent" (Johnston 2000, 782; see also Amin 2002). A conservation territory can be seen as taking shape through the spatiality that becomes inscribed as a result of various forces, particularly scientific ideas about protecting nature, the social power and governance authority that empowers these ideas, and the various other conditions that shape area formation such as economics (e.g., resource and settlement location) and politics (e.g., jurisdictional differences). The idea of spatiality does suggest how both physical space and biologically defined area are central to a conservation territory, since these biogeophysical parameters are never mere background, as they interact with the human activities that entail relations of power and difference (Escobar 1999; Paulson and Gezon 2004; Peet and Watts 2004; Zimmerer and Bassett 2003). Our volume offers a number of examples of the resulting social-environmental networks that offer promise for conservation concerns.

Second, my collaborators and I argue that the interaction of conservation and globalization is leading to new *scales* of importance to environmental management. The idea of scale, and the process of scaling, refers to the spatial patterning of environmental processes and human-environment interaction (such as the scale of shade-grown organic coffee habitats, agrobiodiversity flows, and pastoralists' use of range resources, which are examples of topics examined in detail in the chapters). Diverse sorts of environmental processes and human-environment interaction are expected to produce differing sorts of scaling or spatial manifestations. This viewpoint is vividly phrased as "The landscape 'level' is dead," which means to say that there is no one or pregiven "landscape" but rather that this scale depends on the nature of the particular objects and interaction processes that are of interest (Allen 1998; see also Marston 2000; Swyngedouw 1997; M. D. Turner 1999a; Wiens 1989; Zimmerer 2000; and Zimmerer and Bassett 2003). Since more than one type of process or interaction is typically involved in an environmental issue, the spatial manifestation of scaling is different from the idea of a fixed unit of areal coverage. Our book argues that certain scales and networks of sustainable environmental techniques—for example, the regional scale of existing land-use coordination and urban-rural networks of management—do not necessarily correspond to the prescriptions of currently predominant policies that are framed according to social units such as individual communities or broad economic sectors such as agriculture. By identifying these other key scales and networks, global organizations can interact more effectively with local and national counterparts.

Third, our book argues that the interaction of conservation and globalization is resulting in an unprecedented degree of importance of boundary or border issues, often transnational in scope. The spatial and environmental dynamics of transnational border issues are increasingly counted among the chief challenges of global conservation (Wells and Brandon 1992; Western and Wright 1994). The challenges of border issues to global conservation are a consequence, in the first place, of the expanded activities of global organizations and institutions that transcend the boundaries of single countries. As a result, global conservation is often based on transnational arrangements of environmental management and territorial regulation that cross the boundaries of two or more national states (Duffy 2001). Such transnational linkages are considered key to the current and future prospects of resource management for such environmental priorities as biodiversity conservation and water resources. Border issues are also central to the many cases of local agriculture and resource use that may not correspond entirely, or correspond barely at all in certain cases, to the spatial designations of management (Agrawal and Gibson 1999). As a result, our book argues that border issues deserve special attention as resource management and conservation policies become increasingly informed by spatial concepts.

Fourth, the book argues that decentralization is one of the major expressions of the influence of globalization on conservation. *Decentralization* is defined broadly here as the trend toward an emphasis on incorporating the local level of environmental governance, in which local communities, municipalities, villages, households, and conservation units are employed as key units of environmental projects and programs (Agrawal 2005; Ribot and Larson 2005). These local units are empowered with authority over environmental decision making, although in practice thus far the reality of centralized control often falls short of the stated decentralization goals. Still, the decentralization programs that have been put in place for conservation management often affect the distribution of funds from donor groups or governments and the determination of resource access. At the same time, decentralization is generally a common feature of globalization, as most global organizations and institutions prefer to interact in more direct ways with local entities, ranging from interlinked businesses and small producers to NGOs (Dicken 1992; Johnston, Taylor, and Watts 2002).[3] Our book takes the position that environmental governance at the local level is necessary for the

---

3. An example in economic globalization is the local-level production contracting of global seed and agribusiness companies; local contracting arrangements and contract technicians have replaced national agricultural agencies in many places worldwide (Watts and Goodman 1997).

success of a majority of initiatives for conservation and environmental management. At the same time, however, we are aware that a potential pitfall of governance at this level is the influence and entrenchment of sharp social inequalities that operate in such local milieus as villages and communities (Agrawal and Gibson 1999; Berkes 2004; Brechin 2003; Brosius, Tsing, and Zerner 1998; Holling and Meffe 1996; Kellert et al. 2000; Ribot 1999).

My collaborators and I assert that such power inequalities are prone to redoubling as a result of the so-called power effects of the expansion of conservation territories and social networks (e.g., migrant groups, women's groups) of such projects as ecotourism and sustainable development initiatives (Brechin 2003; Murdoch and Marsden 1995; Peluso and Watts 2001; Sundberg 1998; Zerner 2000). Either the dynamics of these new power effects of decentralization may enable effective environmental management and democratization, or, alternatively, they may compromise or undermine democratic governance—and, in some cases, such change may threaten the success of conservation. Furthermore, our book argues that the influences of national-level conditions and concerns are still centrally important to conservation and, more specifically, to the organization of resource spaces and environmental governance (Christen et al. 1998; Duffy 2001; Scott 1998; Swyngedouw 1997). Notwithstanding the attention-grabbing trends of environmental globalization that diverge from the national scale (e.g., village-level decentralization, transnational projects among multiple adjoining countries), the influences of national-level governance are both persistent and, in many cases, strategically reformulated. The importance of national sovereignty has reemerged, for example, in the face of such environmental issues as biodiversity-related intellectual property rights and oceanic mineral resources (Jasanoff 2004). As a consequence, the role of the state is a much-needed component to the understanding of current conservation and globalization (Scott 1998).

Expanding Dimensions of Conservation, Globalization, and Livelihood Issues

This book brings together a group of twelve studies that have been carefully chosen to address the expanding geographical dimensions of conservation and globalization with special reference to agriculture and local livelihoods. The studies originated as invited presentations at the conference "Globalization and the New Geographies of Conservation," sponsored by the Environment-and-Development Advanced Research Circle (EDARC). It was held at the University of Wisconsin–Madison on 19–20 April 2002. The content of each chapter subsequently evolved through follow-up research and writing in order to build a shared focus on the main premises of the book—namely, the expanded globalization of conservation and the growing interface of global conservation with livelihoods, agriculture, and resource use. The central arguments of our volume, briefly introduced above, are developed more fully following the outline of the book. We also describe the rationales for the regional coverage, which is shown in figure 1.1, and research methodologies of the chapters in each section.

Spatialities in Global Conservation and Sustainability Projects
The first section of our book is focused on the spatial arrangement and dynamics of conservation projects organized for the purpose of environmental sustainability. The case studies in this section cover a range of spatialities from agricultural field-level arrangements of certified organic coffee production (chap. 2) to monitoring designs for sustainable forestry as part of certification projects (chap. 3) and spatial diffusion and networking of ecodevelopment projects for beekeeping and honey production in tropical regions (chap. 4). The chapters explore the tensions between the "local" basis of spatial designs for conservation and sustainability that not coincidentally have become closely tied to the globalization of these efforts (Braun and Castree 1998). Most of the volume's other chapters contain a similar emphasis on spatialities, both conservation territories and networks. Also meriting mention is that the examples of conservation and sustainability projects in the case studies are typical of a broad swath of efforts that have been undertaken during recent decades through global agencies and institutions (the World Bank, Fair Trade organizations, international environmental NGOs). The location of the case studies in Mexico and Brazil is fitting since the largest share of international funding for global biodiversity conservation, including the "green market" sustainability-type projects that are presented here, has been destined for the countries of Latin America (Castro and Locker 2000).

In chapter 2 ("Certifying Biodiversity: Conservation Networks, Landscape Connectivity, and Certified Agriculture in Southern Mexico"), Tad Mutersbaugh examines the field-level spatialities of organic coffee production in Mexico, which is a leader in this rapidly growing strategy for sustainable agricultural production for global markets. One key element of organic coffee production worldwide is the process of "green market" certification, whereby local farmers and farmer organizations participate in monitoring and evaluation routines in order to inform global consumers and global certifying agencies or institutions. Green-market environmental certification is generally rife with the Janus-faced tensions of sustainability prospects that are juxtaposed with signs of possible threats to livelihoods stemming from the same or other aspects of globalization. Indeed, many cases of green-market certification are closely related to neoliberal-style Globalization. A vivid example of green-market certification under neoliberalism is the paradox of organic farming in California (Guthman 2004). Development of the strong ties of

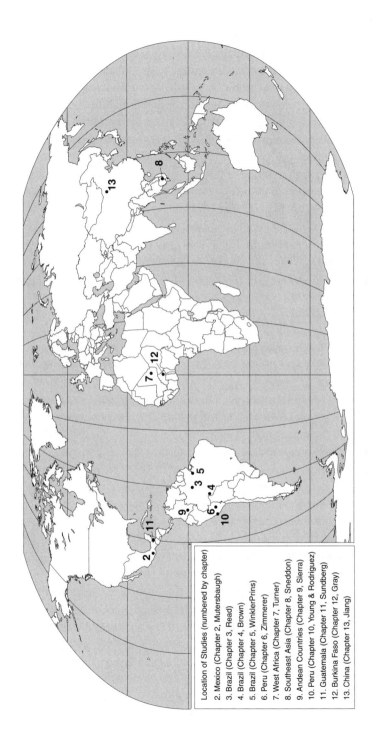

FIGURE 1.1. Locations of case studies (labeled according to chapter number)

Location of Studies (numbered by chapter)

2. Mexico (Chapter 2, Mutersbaugh)
3. Brazil (Chapter 3, Read)
4. Brazil (Chapter 4, Brown)
5. Brazil (Chapter 5, WinklerPrins)
6. Peru (Chapter 6, Zimmerer)
7. West Africa (Chapter 7, Turner)
8. Southeast Asia (Chapter 8, Sneddon)
9. Andean Countries (Chapter 9, Sierra)
10. Peru (Chapter 10, Young & Rodriguez)
11. Guatemala (Chapter 11, Sundberg)
12. Burkina Faso (Chapter 12, Gray)
13. China (Chapter 13, Jiang)

organic certification to agribusiness firms, including leaders in the neoliberal Globalization of fruit and vegetable growing, is an important thematic message to this volume notwithstanding the seemingly distant setting.

Certification of organic coffee in Oaxaca, where Mutersbaugh has conducted extensive field studies among small-scale or peasant farmers (Mutersbaugh 2002), many of whom are Mixtec and Zapotec people, is designed to verify the use of certain agronomic techniques and to ensure the overall biodiversity of plants and animals, both in fields and in surrounding habitats. Indeed, the centrality of organic coffee production as a conservation and globalization issue is tightly tied to the ecological and socioeconomic significance of these agroecosystems of shade-grown coffee (Hecht 2004; Rice 2001). Mutersbaugh makes use of participatory mapping techniques and interviews in order to demonstrate how environmental and social factors, including certification processes, are shaping field-level spatiality of organic coffee production. Mutersbaugh's results show that collective actions among farmers that promote the viability of organic coffee production at the local level tend to widen the extent of the canopy coverage of shade trees, thus lessening the environmental drawbacks associated with the isolation and fragmentation of agroforests.

In chapter 3 ("Satellite Remote Sensing for Management and Monitoring of Certified Forestry: An Example from the Brazilian Amazon"), Jane M. Read examines the role of remotely sensed imagery for the certification of sustainable forestry. The certification of sustainable forest practices is a major approach in the global conservation and sustainability movements of recent decades. It was fueled through the emphasis on sustainable forestry in the wake of global meetings and agreements, such as the 1992 United Nations Conference on Environment and Development in Rio de Janeiro and the 1997 Kyoto Protocol on Climate Change, as well as the prominent activity of global environmental NGOs, particularly the Forest Stewardship Council (FSC). Read's chapter focuses on the significance for conservation management territories of two basic types of satellite imagery, widespread Landsat-7 data with 30 m pixel size and the new higher resolution (1 m and 4 m) IKONOS satellite data (Read 2003). Her chapter evaluates the application of these remotely sensed images to management and monitoring issues involving sustainable forestry operations in Brazil, which is among the top countries in the world in terms of the extent of the logging of humid tropical forests and also the high level of interest in sustainable tropical forestry. Read finds that the two types of satellite imagery are potentially complementary for use in certification activities. Moreover, the usefulness and compatibility of these techniques suggest guidelines for the spatial parameters of productive conservation territories for sustainable forestry.

In chapter 4 ("Productive Conservation and Its Representation: The Case of

Beekeeping in the Brazilian Amazon"), J. Christopher Brown examines two of the main spatial underpinnings of beekeeping, which has been widely promoted as a key to sustainability throughout tropical and subtropical regions of the world. Indeed, honey production through beekeeping is currently one of the most popular "local" ecodevelopment activities for environmental management areas, such as the buffer zones of biosphere reserves, and for major global biodiversity hot spots, such as the state of Rondônia in the western Brazilian Amazon. First, Brown historicizes the spatial diffusion of beekeeping, promoted initially under British rulers in Africa and subsequently by global conservation and sustainability organizations, throughout the tropics. Second, Brown evaluates the spatial-environmental attributes of beekeeping, particularly its suitability to different agricultural activities and forest types. In conducting these evaluations Brown applies the method of discourse analysis, through which extensive scientific and policy literature on each topic is subjected to close textual interpretation. Brown combines this textual analysis with socioecological analysis from his own extensive field studies on beekeeping in Rondônia (Brown 2001). Brown finds two disparate yet defining and seldom acknowledged spatialities of tropical beekeeping—the network of international experts in development organizations and the preference of bees for abundant flower sources in disturbed landscapes, even cutover forests—that are important for future application of this activity. Brown's chapter recommends that these spatialities become explicitly incorporated into the promotion of beekeeping in conservation and globalization policy circles.

Linking Scales in Livelihood Analysis and Global Environmental Science
The majority of conservation management areas are increasingly found in environments that support the low-impact use of resources through agriculture, livestock raising, and other livelihood activities. As a defining feature of contemporary conservation, the prevalence of this resource use highlights the growing importance of understanding the livelihoods of these resource users. *Livelihood analysis* refers to the broadly social evaluation (including economic, political, and cultural factors) of environment-related allocations of labor, land, and other personal, institutional, and collective resources (Blaikie and Brookfield 1987; Friedmann and Rangan 1993). In developing countries, the livelihoods of resource users typically depend on a diversified mixture of work activities. Even at the level of the single household, these resource users typically undertake activities that range from gardening and nonwage work to extensive wage work in the informal sector and salary work in some cases (Coomes and Barham 1997). Their livelihoods also depend often on migration, frequently leading to the creation of social ties that span rural-urban and transnational spaces.

The livelihoods of inhabitants residing in and near conservation areas are crucial to the scaling of the environmental resource systems that are tied to agriculture, livestock, and other activities. Activities such as procuring seeds of the food plant complexes that represent agrobiodiversity and herding livestock in or near protected areas result in multiple scales of resource use. The scales of such livelihood activities do not necessarily correspond, however, to the scaling of environmental management (Brunckhorst and Rollings 1999; Hobbs 1998). For example, zonation-based management, which is common to plans for sustainability in many conservation areas, may differ from the flows and movements of local agropastoral resources (M. Turner 1999; Zimmerer 2003a). One major challenge, then, is linking the scales of environmental change issues and resource use practices to those of the expanding plans for management in conservation areas. Central to the analysis of environmental scale issues is the role of experts in environmental science (Forsyth 2003; Stott and Sullivan 2000). Experts in the biogeophysical and social sciences make assumptions and adopt frameworks about scale that guide the elaboration of the multiplying number of scientific models and publicized reasoning about human-induced environmental change. The scale features of scientific models are reflective of the cause-effect relations that are contained at the core of these models. Similarly, the assumptions and frameworks regarding scale in these models are typically used to guide or target the increasing number of policy prescriptions and projects that are informed through environmental science. As a result, the assumptions and reasoning about scale are often replete with a variety of political implications—and thus a politics of science—that are increasingly influential and important to bring into the open for discussion and analysis.

In chapter 5 ("Urban House-Lot Gardens and Agrodiversity in Santarém, Pará, Brazil: Spaces of Conservation That Link Urban with Rural"), Antoinette M. G. A. WinklerPrins investigates the nature of seed sourcing for the tropical trees, shrubs, and herbs of urban house-lot gardens that are highly diverse in terms of both agrobiodiversity per se (the ecological and taxonomic diversity of agricultural species and varieties; Brush 2004; Zimmerer 2003b) and agrodiversity (the diversity of agricultural management techniques and environments; Brookfield 2000). Urban house-lot gardens, which are common in Brazil and other developing countries, are valued highly for the domestic production of food, fiber, beverages, and medicine (WinklerPrins 2002). These gardens, also referred to as dooryard gardens and kitchen gardens, are especially common in medium-sized cities (100,000–500,000 inhabitants) of the Brazilian Amazon, such as Santarém, which is among the world's fastest-growing urban areas. WinklerPrins finds that the high diversity of house-lot gardens is the result of the exchange, gift-giving, and knowledge-sharing networks of women members of urban households. Especially important to the supply of seed and planting material for the gardens, and thus the scaling of this resource-use system, are the social networks that connect these urban-ites to family members and acquaintances in surrounding rural areas.

In chapter 6 ("Multilevel Geographies of Seed Networks and Seed Use in Re-lation to Agrobiodiversity Conservation in the Andean Countries"), I exam-ine the spatial organization and scaling of seed flows for the diverse Andean potatoes and ulluco, regionally important food staples that are grown through-out the Andean countries from Venezuela to Chile, with a special importance in Peru, Bolivia, and Ecuador. The biological complexes of the Andean potato and ulluco crops have become a focus of expanding new conservation efforts, since both food plants are important to local livelihoods, including those of many farm families residing in or near recently designated protected areas, and since the diverse Andean potatoes are the primary source of biological diversity for the cultivated potato of central importance to global food sup-ply. To date most conservation efforts and management plants, based to an extent on models of international environmental science, have adopted a framework of community-based conservation in targeting these complexes of food plant agrobiodiversity in the Andean countries. I find that the scale of the individual rural or peasant community is present in these seed flows but that it is only one of several scales of the seed transfers that actively sup-port the agrobiodiversity of Andean potatoes and ulluco. My findings show that this agrobiodiversity is dependent upon a series of interlinked scales of seed flows—including the notable importance of multicommunity seed flows—that farmers create in response to broad-based environmental and so-cial rationales.

In chapter 7 ("Shifting Scales, Lines, and Lives: The Politics of Conservation Science and Development in the Sahel"), Matthew D. Turner interrogates the selection of spatial scales and management boundaries that are chosen by environmental scientists and conservation practitioners who specialize in the environmental change issues of the African Sahel, most particularly the concerns over desertification and overgrazing that have helped define this world region as an iconic case for global environmental management and conservation during recent decades. Turner focuses on the scientific ideas and practices in this arena that are associated with the disciplines or ap-proaches of range science, ecology, conservation biology, conservation plan-ning, soil science, remote sensing, and geography. These scientific undertak-ings are increasingly grouped under terms such as *global environmental change science* and *global science*. Turner finds that the choice of spatial scale among these scientists is far more than a product of good and bad science. These sci-entists, who are typically well intentioned and genuinely committed to im-proving sustainability, undertake activities that incorporate and reinforce a "subtle and complex politics of scale and categorization that works to reorder the political and material lives [affecting such livelihood practices as herd mobility] of those that find themselves within a conservation area." Turner also analyzes how the politics of science in Sahelian environmental change is also created *within scientific practice* due to the influences of technology support, research funding, predominant theoretical models, and conservation prerogatives.

Transnational and Border Issues in Global Conservation Management
Border issues involve the activities and concerns that arise in conjunction with the existence or establishment of boundaries for the purpose of environmental management including conservation. A basic tension exists in the border issues that multiply with globalization's increasing influence on conservation. On the one hand, the number and type of management units is proliferating in both developed and developing countries worldwide. Officially designated conservation areas alone saw an increase from an estimated 1200 units in 1975 to more than 22,000 units by the late 1990s (IUCN and WCMC 1998). The actual number of areas is much higher for environmental management designations in general, since these estimates do not include many places where land use and environmental management projects have been implemented without an official conservation designation. It is worth noting that the existence of boundary issues is also proliferating within management units, since conservation and sustainability models increasingly rely on multiple, within-unit spatial designations (e.g., the three-zone schema of transition, buffer, and core areas of the mainstay model of international biosphere reserves; Batisse 1997; Hadley 2002; Stevens 1997). On the other hand, and existing in tension with this proliferation of units and boundaries, is the growing recognition of the spatial processes of environments that are often characterized by multiple levels or are more complex than may have been realized in earlier planning for management and conservation (Allen 1998; Clark, Tuxill, and Ashton 2003).

In chapter 8 ("Conservation Initiatives and 'Transnationalization' in the Mekong River Basin"), Chris Sneddon addresses the undertaking of transboundary environmental management and conservation initiatives in the Mekong Basin countries of Vietnam, Cambodia, Thailand, and Laos PDR. These transboundary initiatives are being organized at the multi- or transnational level by various global conservation organizations that have contributed to the establishment of a significant network of conservation territories in the region. These initiatives are concurrent with plans fostered by the Asian Development Bank (ADB), referred to as the Greater Mekong Subregion (GMS) initiative, to promote the Mekong as a "natural" economic and governance territory that crosses national borders in order to take account of the defining riparian realities of the Mekong watershed and the ecoregions that are contained within its range (Sneddon and Nguyen 2001). The transnational characteristics of the project are emblematic of the global trend toward multicountry designs, with peninsular Southeast Asia and the Mekong Basin serving as leading examples due to the small to medium size of the countries and the high environmental value and potential vulnerability of places there. Sneddon finds that certain basin-scale governance initiatives in the Mekong region, especially river basin development, are likely to be at odds with the associated initiatives for biodiversity conservation. Sneddon's chapter demonstrates how these conflicts within large transnational or transboundary conservation areas involve not only internal border issues but also interactions with still larger political and economic forces.

In chapter 9 ("A Transnational Perspective on National Protected Areas and Ecoregions in the Tropical Andean Countries"), Rodrigo Sierra focuses on the tensions between the growing recognition of global-scale priorities of planning for biodiversity conservation, on the one hand, and the national and local scales of organization that continue to exert strong influences, on the other hand. Sierra's study is based on the case of the tropical Andean countries, namely Venezuela, Colombia, Ecuador, Peru, and Bolivia. The landscapes of this transnational region are increasingly recognized as containing a highly concentrated share of the world's most renowned biodiversity hot spots that are both uniquely valuable and subject to certain changes of land use that pose a serious threat (Mittermeier et al. 1998; Sierra, Campos, and Chamberlin 2002). Sierra designs and conducts an analysis of data on the coverage of protected areas in terms of ecoregion representation in order to examine the transnational dimensions and effectiveness of these areal designations. Since the ecoregions are biogeophysical units, the coverage of these categories often implies a transnational dimension that involves one or more international boundaries. Sierra finds that national borders continue to represent significant divides in coordinating the coverage of ecoregions through protected areas. This finding is extremely important for new global conservation initiatives in the tropical Andean region, including a world-renowned triad of conservation corridor projects—the Vilcabamba-Amboró corridor of Bolivia and Peru, the Condor-Cutucú corridor of Ecuador and Peru, and the Chocó-Manabí corridor of Colombia and Ecuador—that are international designs being planned and implement by global conservation organizations.

In chapter 10 ("Development of Peru's Protected-Area System: Historical Continuity of Conservation Goals"), Kenneth R. Young and Lily O. Rodríguez examine the development of protected areas in Peru. Their analysis is focused on the interactions of global conservation organizations and the roles of diverse Peruvian supporters of conservation, who are the nationally based advocates of protected areas in Peru and have included many citizen scientists, local conservation advocates, and foreigners based over the long term in Peru. Peru stands out as one of the most environmentally diverse countries in the world (Rodríguez and Young 2000). For this reason, a number of modern-day environmental conservationists have expressed a paramount interest in protecting key areas of Peru, both in more strictly regulated landscapes and in ones that are oriented more toward sustainable resource utilization. Following a historical analysis of Peruvian conservation, in which they describe the contributions of both Peruvian and non-Peruvian scientists since the 1800s, Young and Rodríguez provide an assessment of recent and ongoing trends in protected-area conservation in the different environmental regions of Peru. Young and Rodríguez find that the ecoregions of coastal desert and Andean mountain environments have been historically valued among local and national conservation interests in Peru. By contrast, these coastal and Andean places have figured less importantly in conservation agendas during recent decades due to the predominance of attention, effort, and resources accorded to the conservation of humid tropical forests. The influence of globalization on Peruvian protected-area conservation accounts for much of this shift. Increasingly, a premium is placed on the protection of rain forest habitats because of their high biodiversity value by global conservation organizations and through global networks of environmental science and conservation knowledge.

Decentralization and Environmental Governance in Globalization
Decentralization, broadly defined, refers to the dispersal or deconcentration of government administrative and financial functions from the national to more local levels, with a correspondingly greater role for civil society institutions and international actors. Globalization institutions have used decentralization as a chief tool for creating the local framework of conservation and development. Devolution, which refers to the strategic restructuring of political power and governance functions to more local units, has been a handmaiden of much environmental globalization. Reasons for the ties of decentralization (and devolution) to globalization stem from the general commitment of global institutions to the diminution or dismantling of central government administrative and governance capacities that are regarded as overly bureaucratic, costly, autocratic or inadequately democratic, and, in some cases, systemically corrupt. Global financial institutions, such as the World Bank and International Monetary Fund, have widely enforced decentralization and devolution policies as requirements for debt refinancing and loan restructuring among poorer nations. Perhaps not surprisingly, neoliberal-led decentralization has become one of the spatially distinguishing features of the global expansion of conservation and environmental management during the past couple of decades (Dupar and Badenoch 2002; Lutz and Caldecott 1996; Ribot and Larson 2005; Wily 2003).

Community-based resource management (CBRM), municipality-scale programs for forestry and other resource management, and "village management" (*gestion des terroirs*) are just a few of the many environmental governance initiatives that are associated with decentralization. Global conservation organizations have often supported these sorts of initiatives for environmental governance in many areas worldwide. The location of such initiatives is especially common in and near protected areas, since global conservation organizations are prone to regard the decentralized style of programs as a means of achieving sustainability in certain locales (Brandon, Redford, and Sanderson 1998; Sundberg 1998; Wells and Brandon 1992; Western and Wright 1994). Although the decentralization initiatives may seem to suggest a kind of compatibility with the local knowledge and cultural milieu of residents, they also typically raise thorny issues about access to resources and political representation that can worsen the social inequalities of a local community, ethnic group, or region (Escobar 1998; Neumann 1995, 1998; Nietschmann 1995; Stevens 1997; Stonich 2001; Zerner 2000). The privatization and legal titling of land are frequently upheld as a necessary precondition of decentralized resource management. Privatization and land titling are integral parts of globalization throughout most of the world, particularly under neoliberal policies, as government and financial institutions have pushed for the development of global markets. Environmental governance changes associated with private property management range from the parcelization of community or *ejido* lands in Mexico to decollectivization and the establishment or formalization of private property in China, Africa, Southeast Asia, and the postsocialist transitions of central Europe (Goldman 1998; Lavigne 1999; M. Turner 1999).

In chapter 11 ("Conservation, Globalization, and Democratization: Exploring the Contradictions in the Maya Biosphere Reserve, Guatemala"), Juanita Sundberg examines several of the decentralization and globalization processes that are central to the ongoing implementation of the Maya Biosphere Reserve (MBR) in the tropical lowlands of Guatemala. The MBR is one of the exemplars of the global United Nations Educational, Scientific, and Cultural Organization (UNESCO)/Man and the Biosphere Program (Sundberg 1998). It is also a cornerstone of the Mesoamerican Biological Corridor, which is described later in this chapter as a prime example of transnational conservation and demonstrates the major interface of agriculture and resource use. Sundberg describes the background to the establishment of the MBR during the 1980s, which featured global conservation organizations that acted in concert with national conservation agencies and a new legal code for a national System of Protected Areas. Sundberg traces how the administration of the MBR was decentralized, as the Guatemalan government devolved the implementation of the MBR to a bilateral aid agency (the United States

Agency for International Development [USAID]), which subsequently subcontracted the operation to three United States–based NGOs. Sundberg's chapter analyzes how the decentralization of the MBR created environmental governance issues involving different groups of people living in and near the protected area. Issues of governance in conservation management are inextricably tied to the broad political challenges of democratization nationwide in Guatemala. Moreover, the impact on decision making tends to be felt most acutely among less powerful groups within the local society, such as farm migrants to the MBR region and indigenous women.

In chapter 12 ("Decentralization, Land Policy, and the Politics of Scale in Burkina Faso"), Leslie C. Gray offers an analysis of the decentralization of land management policy for development and sustainability purposes during recent years in the Sahelian country of Burkina Faso in West Africa (see also Gray 2002). Gray uses the techniques of field interviews and ethnographic participant observation with villagers, local officials, and policymakers. Responding to the pressures of international financial institutions, particularly the World Bank, this decentralization of rural resource use is based on land use zoning and the clarification of land rights, mostly through land privatization. The approach to decentralization in Burkina Faso depends on the promotion of village-level institutions of resource management, a model commonly referred to as *gestion des terroirs*. A majority of countries in West Africa have recently implemented the *gestion des terroirs* approach, which imbues the study of Burkina Faso with regionwide implications. Gray finds that local politics often thwarts attempts to establish the zonification of land use, due in part to the multiple scales at which village residents make use of a wide variety of resource areas including managed forests, cultivated areas, and fallow sites. Gray's study also demonstrates how local decentralization must be coordinated with a greater effectiveness of national governance in order to adequately address the zonation of land use planning and the socially just adjudication of land rights.

In chapter 13 ("Fences, Ecologies, and Changes in Pastoral Life: Sandy Land Reclamation in Uxin Ju, Inner Mongolia, China"), Hong Jiang addresses the role of anti-desertification initiatives that are closely associated with decollectivization and the legal, political, and economic transition to private property in the Inner Mongolia region of northern China. This decollectivization entails the shift from community or collective resource management to a private property regime. Decollectivization, privatization, and the breakup of rural community properties are primary expressions of a general sort of decentralization and globalization throughout Asia, Africa, Central and Eastern Europe, Russia, and Latin America. This shift in property rights regimes has been guided by powerful global organizations, most notably the World Bank and the International Monetary Fund, as well as national governments themselves, which has been the case in China. Indeed, China is the largest example of decollectivization worldwide (Jia and Lin 1994; McMillan and Naughton 1996). In Jiang's case study, decollectivization has entwined with the combined undertakings of fence building and tree and shrub planting projects for anti-desertification purposes. In her research, which is based on extensive field interviews (and remote sensing image analysis; see Jiang 2004), Jiang finds that decollectivization, along with the building of fences and vegetation planting as part of anti-desertification campaigns, has led to contradictory outcomes for local livelihoods and environments. The impact of these new borders is associated with successful tree and shrub establishment; yet, at the same time, these changes have reduced groundwater levels and created additional difficulties for those local residents that rely on nomadic or transhumance pastoralism.

Overview of the Main Premises on Conservation and Globalization

Global Conservation Expansions

Conservation areas worldwide are reported by global organizations to have expanded more than ten times in areal coverage during the past three decades (IUCN and WCMC 1998; Zimmerer, Galt, and Buck 2004). The area of publicly designated protected areas grew to more than 12.2 million km^2 by 1997, and it has expanded since then to the estimated total of 15.1 million km^2 in 2003. This section is designed to provide an overview of the global conservation changes at the levels of world regions and individual countries. It adopts a perspective that is distinct from yet complementary to the previous quantitative and cartographic portraits of the global-scale distributions of wilderness and human disturbances (Hannah et al. 1994; Sanderson et al. 2002; Wackernagel and Rees 1996). These existing portraits, which remain crucial to informing worldwide conservation priorities, are typically based on the identification of types of ecosystems or ecoregions and areal categories of human-induced habitat change that reflect the extent of impact. By contrast, the approach here is focused on globalization. The blend of the environmental and social sciences in this volume is based on a view that sees globalization as working its way through national governments, with the likelihood that world region–level trends or organizations are also evident. Worth reiterating is that quantitative estimates used in this section are coarse approximations that are nonetheless useful at the global scale; moreover, the interest in these quantitative estimates includes their often important role in the rhetoric and narratives of presenting global conservation.[4]

---

4. Methodologically this subsection and the following one are dependent on a series of research techniques that are needed for assessing the sources, creation, use, and limitations of global data for

A diverse array of international institutions has been central to the expanding designation and management of protected areas since the 1970s. Global institutions that include the United Nations and the IUCN (the World Conservation Union) as well as international NGOs with global influence (Conservation International [CI], The Nature Conservancy [TNC], Worldwide Fund for Nature, World Resources Institute [WRI]) have been leaders in the designation and implementation of conservation management across the world. The period of the past couple of decades has also been marked by the growth of global conservation institutions specifically for the creation and coordination of protected areas. Specialized conservation agencies for global protected areas have included the World Conservation Monitoring Centre (WCMC), part of the IUCN, and the World Commission on Protected Areas (WCPA). These global conservation organizations have often coordinated their efforts through various agreements ("partnering") and comanagement arrangements with multilateral as well as national and local institutions throughout both developed and developing countries (Bryant and Bailey 1997). It is notable, although perhaps unsurprising, that certain key institutions of national governments have remained influential in protected-area expansion, and that enhanced influence is exerted through numerous local organizations, including a severalfold increase in the number of NGOs.

Global conservation expansion, as reflected in the evidence presented here, is apparent in changes at the worldwide scale, at the level of world regions and individual countries, and, finally, in the case of specific new areal designs, such as the case of transboundary protected areas. Our analysis of conservation changes at these levels relies in part on the country-level estimates of conservation areas worldwide that were compiled and published through global conservation organizations for 1985 (IUCN 1985) and 1997 (IUCN and WCMC 1998). These data sources are the closest available match to the period leading to the present, and they are commonly regarded as the best available estimates of global conservation areas (Lightfoot 1994; Seager

------

environmental management and protected-area conservation in particular. This analysis is referred to as source-critical use of global data for both qualitative and quantitative analysis, which is considered a requisite for considering the collection and compilation of information from many countries (Miller and Edwards 2001). Global institutions are essential information managers in the creation, design, monitoring, and presentation of published data, as well as unpublished information, which exists in an impressive array and staggering quantity. Useful too in this analysis is the use of computerized cartography, including geographic information systems (GIS), for the display of conservation and globalization data that are spatially complex and prone to change over time (Zimmerer, Galt, and Buck 2004). Cartographic visualization techniques are used to make spatial and temporal assessments and also to conduct exploratory analysis of map design and data. The cross-temporal visualization of spatial patterns is important to this chapter since the relations of conservation and globalization are dynamic, and hence these issues involve changes in both locational ("where") and attribute ("what") components (Seager 1995).

1995). These data sources are considered generally adequate with regard to the global estimates of areal coverage of conservation units, although caution is advisable in basing conclusions entirely on the quantitative analysis of these data.[5]

The worldwide coverage of conservation areas was greatly expanded in the twelve-year period between 1985 and 1997, as seen in figure 1.2. At the global level, the estimated coverage showed a significant increase from an estimated 3.5 percent (approximately 5,290,000 km^2) to 8.8 percent (approximately 12,240,000 km^2). Several sizable differences that emerged in the estimated areas and percent-protected measures of the late 1990s were continuations of existing contrasts at the level of world regions and individual countries that had existed previously. By 1997, designated protected-area coverage had been much expanded, and it continues to expand today, although it is still sharply differentiated. In terms of aggregate area, approximately one-third of the global conservation expansion of the recent past has taken place in the Western Hemisphere (North America, Central America and the Caribbean, and South America). Sizable increases occurred also in the estimated conservation coverage of South Asia, Sub-Saharan Africa, and Southeast Asia. Differences evident in the areal expansion of global conservation lend support to the so-called growth-and-differentiation view of globalization as constituting spatial dynamics that depend upon, and often widen, the differences within and among world regions, countries, and economic sectors (Chase 2002; Dicken 1992; Harvey 2000; Tomlinson 1999).

A considerable share of the recent global expansion of protected areas has corresponded primarily to humid tropical regions, although this concentration is not as marked as may be thought (Zimmerer, Galt, and Buck 2004). Nonetheless, regions of the humid tropics included several "big gainers," perhaps most notably a number of countries in Central America, the Caribbean, and South America (fig. 1.2; Dinerstein 1995; Lightfoot 1994; Rodríguez and Young 2000; Zimmerer, Galt, and Buck 2004). Global conservation organizations were notably effective in helping raise the protected-area coverage of these big gainers. The analysis also reflects how this aspect of the influence of these organizations is more pronounced in certain countries (and world regions by extension). As a result, the global conservation expansion is not tending toward either global-scale evenness or a level of selectiveness that suggests the targeting of specific habitats. Indeed, the overall change seems to indicate that national-level influences over conservation, which coexist and interact with ones that are global, have remained equally important in

5. The global data on conservation areas definitely do not contain the accuracy or reliability needed to address the much-discussed uncertainty that surrounds such issues as the management and enforcement efficiency or the "paper parks" debate (Bruner et al. 2001).

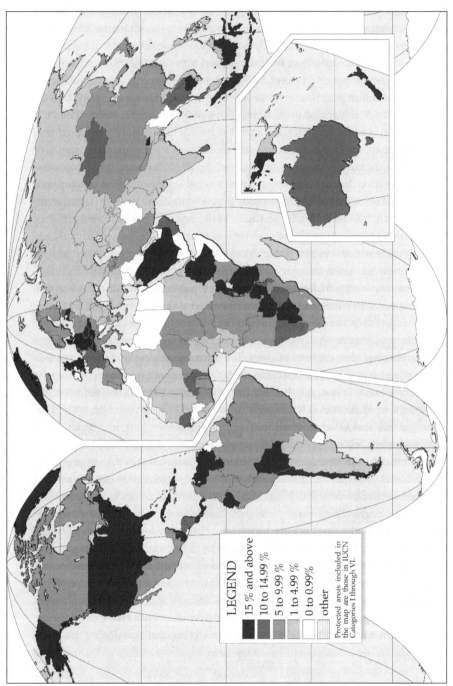

LEGEND

■ 15 % and above
■ 10 to 14.99 %
■ 5 to 9.99 %
■ 1 to 4.99 %
□ 0 to 0.99 %
▨ other

Protected areas included in
the map are those in IUCN
Categories I through VI.

FIGURE 1.2. Global-scale map of the estimated coverage of protected areas, 1997

determining the worldwide expansion of protected areas (Zimmerer, Galt, and Buck 2004).

Transboundary or transfrontier protected areas (TPAs) are a specific example of the growing link of conservation to globalization. These are protected-area complexes that result where management areas for environmental protection in different countries are adjoined across international boundaries (IUCN and WCPA 2001; Zbicz 1999; Zbicz and Green 1997). The design of these areas may comprise separate units that are connected or joined into a single TPA. One example of a TPA is the Peace Park, which combines management for conservation and sustainability with the promotion of peace and cooperation (IUCN and WCPA 2001). By 1999, a total of 136 TPAs had been created (fig. 1.3). These complexes involved ninety-eight different countries (Zbicz 1999). The TPAs are important not only for conservation coverage of cross-boundary ecoregions that were often ignored previously but also for creating powerful symbols of peace and cooperation.

Globalization is readily apparent in the coordination of these areas, since many TPAs were formed through the coordinating role of global institutions such as the European Union (EU) and United Nations as well as global conservation organizations such as the IUCN, CI, and TNC (Agrawal 2000; Duffy 2001; WRI 1992). In some world regions, such as Sub-Saharan Africa, institutional mixes active in TPAs were attentive also to serving and incorporating the business interests of transnational or global companies in such sectors as ecotourism (Hughes 2001; Söderbaum and Taylor 2003). In these places new sites of business opportunities have expanded in and around the TPAs. Business expansion has sometimes precipitated conflicts with existing practices of resource use and arrangements for resource access, for the inhabitants of TPAs, often small-scale land users, have sometimes been overlooked in TPA planning and management (Stevens 1997; Stonich 2001; Zerner 2000).

Globalization is evident also in that many TPAs have been established in geopolitical hot spots of relatively unpopulated and zealously guarded border regions that are prioritized as sites in need of systems of transnational management. Since many of these TPA areas are labeled "no man's lands" (WRI 1992), the marginal social and political status of the residents may have eased the establishment of TPAs and added new conflicts or possible conundrums for the local inhabitants (Agrawal 2000; Stevens 1997; Zerner 2000). The establishment of TPAs raised the prospect of local people's gains from TPA management through possible income-earning opportunities and territorial protection, while, at the same time, this establishment just as often threatened local livelihoods and resource access. Another sign of globalization is the timing of the expansion of TPAs. It is clear that the number and extent of TPAs has grown substantially since the 1980s. The total number of

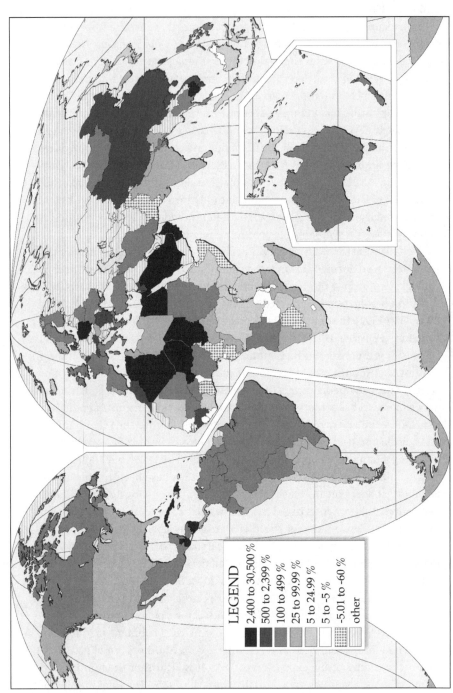

FIGURE 1.3. Global-scale map of the percentage change of estimated protected-area coverage, 1985–1997

LEGEND
2,400 to 30,500 %
500 to 2,399 %
100 to 499 %
25 to 99.99 %
5 to 24.99 %
5 to -5 %
-5.01 to -60 %
other

complexes rose from 41 to 136 during the 1985–99 period (fig. 1.4). As discussed above, TPAs became well established in a number of world regions where previously they were absent entirely (e.g., Central America, Southeast Asia, Russia/Central Asia).

The Interface of Conservation with Resource Use and Agriculture
The so-called third wave of conservation interest, which crystallized in the late 1980s and 1990s, has pushed for the expanded incorporation of sustainably utilized environments into global systems of protected areas. These utilized environments included such designations as buffer zones, managed forests, extractive reserves, and national wildlife refuges located in periurban areas. Globally the prioritization of sustainably utilized environments has reflected the awareness of a need for coordinating conservation efforts with local communities as well as the inherent value of these environments and the capacity they contribute to responding to environmental challenges such as climate change (Brandon, Redford, and Sanderson 1998; Ramankutty et al. 2002; Sarmiento 2000; Sneddon 2000; Western and Wright 1994; Young 1999; Zimmerer and Young 1998). The global policy emphasis on sustainability has been strong enough to trigger an energetic backlash, mainly since the late 1990s, both among anti-environment development interests and among proponents of primarily strict protection or so-called fortress conservation (Kramer, van Schaik, and Johnson 1997; Redford and Sanderson 2000; Terborgh 2000). To our minds, the apparent backlash against the combination of conservation and sustainability is still the more reason to examine the genesis of global interests in this conservation-sustainability nexus and the potential commitments to the future of conservation-minded resource use and agriculture.[6]

The Fourth World Congress on National Parks and Protected Areas, which was convened in Caracas, Venezuela, in 1992, was a major signpost for the growing convergence of conservation and sustainability in the period of the 1980s and 1990s. The Caracas Congress, as it is known, launched a conservation agenda that prioritized the protection and sustainable development of utilized environments (Barzetti 1993; IUCN 1992). Likewise, the Global Biodiversity Strategy and other policy frameworks that emerged from the 1992 United Nations Conference on Environment and Development in Rio de Janeiro (UNCED, or the "Earth Summit") placed an increased emphasis on

6. Important debate and policy repercussions surround the desirable balance of these two general types of management in global conservation strategy (Bruner et al. 2001; Pimm et al. 2001; Schwartzman, Moreira, and Nepstad 2000; Terborgh 2000). While both management types are widely agreed upon as necessary, one key area of debate and investigation is the proper balance between protected areas that tend toward strict preservation and those geared toward sustainable utilization.

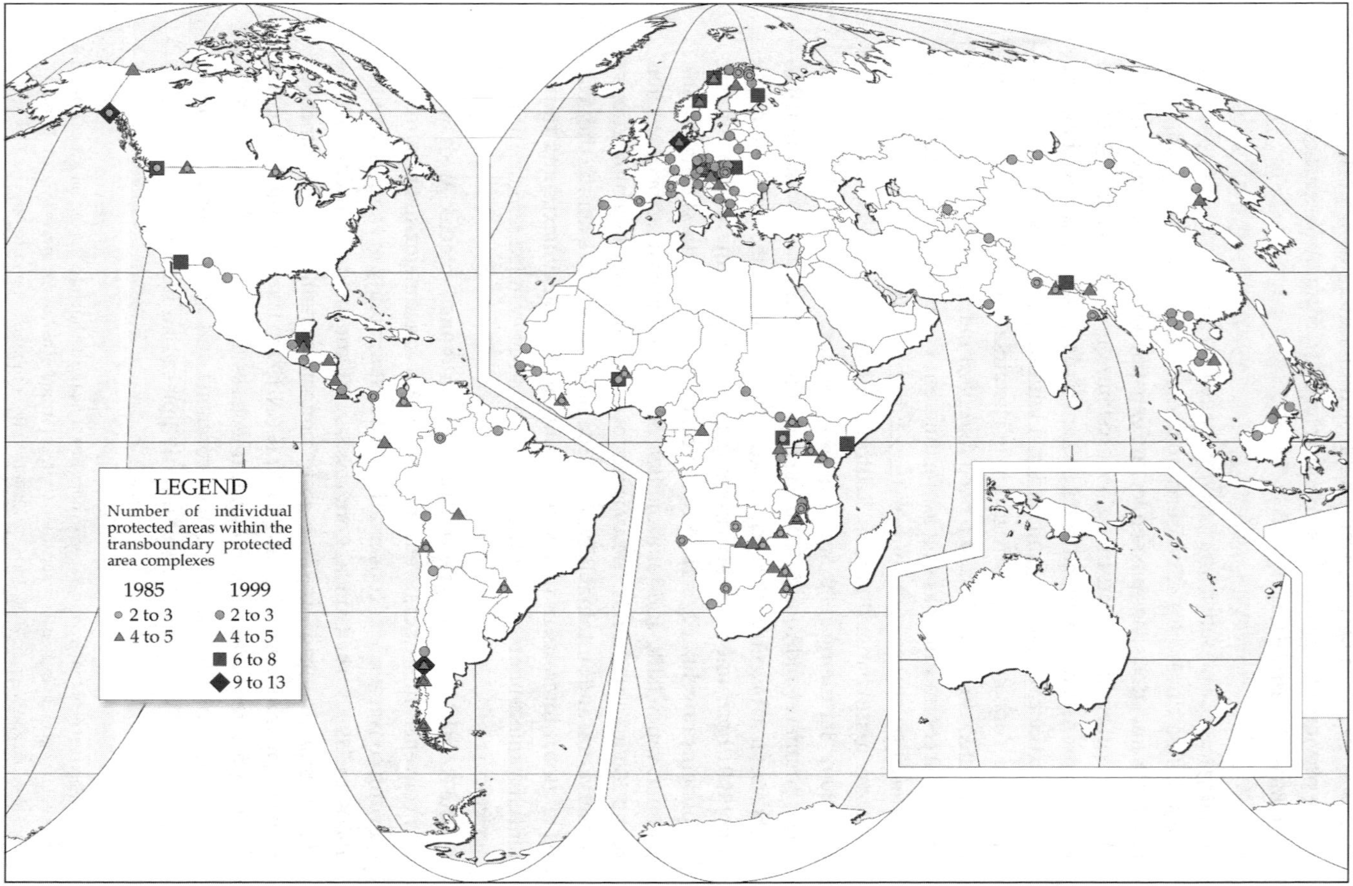

FIGURE 1.4. Global-scale map of transboundary protected areas (TPAs) established by 1999

LEGEND

Number of individual protected areas within the transboundary protected area complexes

1985
- 2 to 3
- 4 to 5

1999
- 2 to 3
- 4 to 5
- 6 to 8
- 9 to 13

sustainable development initiatives in such protected-area units as buffer zones, community-based conservation, and biosphere reserves and bioregional management (WRI 1992, 133–38).

The growing interest of global conservation organizations in the sustainability of utilized environments led to a selective increase in the core number of management categories for protected areas that is useful for our data analysis in this section. The 1992 Caracas Congress adjusted and enlarged its system of categories in order to distinguish among habitat/species management areas, protected landscapes/seascapes, and managed-resource protected areas. Its emphasis on utilized environments and sustainability clearly distinguished this congress from earlier global-level meetings on protected areas. Under these varied types of conservation management, it became possible to distinguish between protected areas that have been designated as either more strictly protected (primarily wilderness-style management) or sustainable utilization (designed for sustainable land use). As a result, the globalization of conservation interests represented in the third wave has led to a somewhat greater concentration on the incorporation of utilized environments since the 1980s (Phillips 2002).

The analysis of global-scale conservation data does suggest a shift, albeit minor in degree, away from more strictly protected areas and toward utilized protected areas at the global scale during the 1980s and 1990s (Zimmerer, Galt, and Buck 2004).[7] In 1985, the estimated coverage of utilized protected areas amounted to 55.75 percent of the global total of protected areas, while the more strictly protected areas covered 44.25 percent. By 1997, the estimated coverage of utilized protected areas had risen to 58.11 percent of the global total, while the coverage of more strictly protected areas represented 41.89 percent. Comparing these figures shows that a modest shift occurred

7. The methods used in this analysis were chosen in order to characterize and to map the relative global importance of more and less strictly protected areas during the worldwide expansion of conservation coverage in the 1985–97 period, which is reported in detail by Zimmerer, Galt, and Buck (2004). A brief summary of these methods is that quantitative techniques were used for the evaluation of data from the world conservation estimates of that time period (IUCN 1985; IUCN and WCMC 1998). The study defined more strictly protected areas as categories I–III of the IUCN Protected Area Management Categories (IUCN and WCMC 1998, xviii). These designations include strictly preserved Nature Reserves or Wilderness Areas (category I), National Parks (category II), and Natural Monuments (category III). Less strictly protected areas are defined as categories IV–VI of the IUCN system. The latter designations apply to Habitat/Species Management Areas (category IV), Protected Landscape/Seascapes (category V), and Managed Resource Protected Areas (category VI). The treatment of data prior to analysis relied on checking for consistency, standardization, and geographic unit. Since the specification of category VI was added to the IUCN data sets in the early 1990s the dates of establishment of individual units contained in the 1997 report were examined in order to determine which ones belonging to this category had already existed in 1985. In this way it was possible to distinguish the units that were created between 1985 and 1997.

toward the direction of utilized protected areas during the 1985–97 period. Notably, the minor degree of this shift did not incur the reduction of the areal coverage of the more strictly protected areas. Indeed, the estimated coverage of more strictly protected areas also increased significantly through the 1985–97 period (by approximately 3.0 million km²). Accordingly, the changing ratio is suggestive of the still larger increase of utilized protected areas during this period (an estimated addition of 4.1 million km²).

The global shift in the predominant management characteristics of conservation areas is evident in the disaggregated estimates that suggest certain differences of both direction and degree. Percent-wise, the extent of the estimated shift toward utilized protected areas is found to range from a high in Western/Mediterranean Europe to a low in Australia and the Pacific (Zimmerer, Galt, and Buck 2004). On the other hand, a total of five world regions are represented by estimates that suggest a shift in the opposite direction, toward more strictly protected areas. This direction of shift has been most pronounced in East Asian countries, followed by weak degrees of shift in North America, Sub-Saharan Africa, South Asia, and Southeast Asia. The point of this analysis is to suggest the growing tension in the trend toward incorporating resource and land use, such as agriculture, into global conservation; on the one hand, it is emphasized in the rhetoric of sustainability and the third wave of global conservation, while, on the other hand, the admittedly coarse quantitative approximations suggest a modest or minimal shift as well as considerable variability among countries and world regions.

The growing interface of conservation with agriculture and resource use, together with the expanding role of globalization, is evident in a variety of particular management designs. Conservation corridors are one of the best-known and most popular of the designs that incorporate extensive areas for agriculture and resource use. Conservation corridor projects typically comprise three types of environmental planning units: existing protected areas, proposed new protected areas, and new and existing corridors that connect these areas (Bennett 2004; Kaiser 2001; Soulé and Terborgh 1999). The current launching of conservation corridor projects was born in proposals of the 1980s that subsequently gained momentum through the implementation phases. A growing number of international corridor projects are designed as part of the increased interest in ecological networks among conservation organizations and sustainability proponents. Major corridor and closely related network projects that are currently in existence include the Baltic ecological networks, Tri-DOM (in central Africa), the Yellowstone to Yukon Conservation Initiative (Y2Y), the Terai Arc Landscape in Nepal, the Green Wood (the Netherlands), the Mesoamerican Biological Corridor in southern Mexico and Central America, and the Vilcabamba-Amboró Conservation Corridor in Peru and Bolivia.

Many conservation corridor projects at the international level, particularly those in developing countries, have involved the support of global political organizations and global treaties (e.g., the Global Environmental Facility [GEF], the United Nations Framework Convention on Climate Change [UNFCCC], Kyoto Climate Change Protocol, global financial institutions such as the World Bank, and global conservation organizations such as CI, TNC, and the Wildlife Conservation Society [WCS]). These global conservation organizations have been especially active in those corridor projects and ecological networks that have taken shape in developing countries, such as the Mesoamerican Biological Corridor and the Vilcabamba-Amboró Conservation Corridor of Latin American countries, the Terai project of South Asia (Nepal), and the Tri-DOM project of central Africa.

Of those, the Mesoamerican Biological Corridor (MBC) is currently one of the most advanced of the international conservation corridor projects in developing countries, which leads to its brief description here as basic background. The MBC is designed to connect the protected areas of eight countries from southern Mexico to southern Panama. Globalization and strict conservation are clearly central to the MBC, since it was first envisioned in the 1980s and early 1990s by international conservation biologists (including many in the United States) who recognized the crucial importance of connecting the fragmented protected areas of Central America and southern Mexico. A report of the 1992 Caracas Congress had summed up this conservation priority: "one of the region's characteristics is that 68% of its protected areas are small (under 10,000 ha) and, taken together, scarcely cover 350,144 has of the total land user protection . . . just five large areas cover a total of 2.7 million hectares, or 50% of the regional system" (Barzetti 1993, 102). (One of these five protected-area cornerstones of the MBC project is the MBR, which is discussed in chap. 11.)

Global funding agencies and initiative partners that are central to the MBC are the World Bank, the United Nations, the GEF, TNC, the WRI, the WCS, and CI. While global conservation organizations have continued to laud the MBC project and the participation of local groups as key to its success (Miller, Chang, and Johnson 2001), this appraisal is unlike that of groups of indigenous and human rights activists, including the global networks that support them. These groups have expressed concern and alarm that the MBC is "captive of the Plan Puebla Panama," which calls for economic integration through the growth of markets, including the presumed expansion of resource markets, in the countries from southern Mexico through Panama (Martinez 2001; World Rainforest Movement 2001). One lens into their concerns is the high degree of spatial overlap between the MBC conservation corridor and the business corridor that is represented in Plan Puebla Panama; this overlap of transnational spatial arrangements for business and conservation is also a notable feature of the Maputo Development Corridor of southern Africa (Hughes 2001; Söderbaum and Taylor 2003; see Sneddon, chap. 8 in this volume, for a Southeast Asian case study).

The emphasis on sustainable utilization is also evident in the evolution of the MBC, as demonstrated in the gradual unfolding of the spatial organization and design planning, since the interests representing agriculture and resource use exert a growing influence. Initially it was planned that the primary trend of the areal arrangements of the MBC would tend toward the increased cohesion of protected areas. Spatial descriptors of this planned evolution of the MBC have placed emphasis on the goals of connectivity and network consolidation (Miller, Chang, and Johnson 2001). Interestingly, these modes of planned integration tend to be associated with both globalization processes and the corridor principles of conservation biology and ecological science (Dicken 1992; Soulé and Terborgh 1999; Tomlinson 1999). Yet the most noticeable change thus far in the overall configuration of MBC areas is contrary to these plans and predictions.

By 1996 the corridor had evolved to resemble a "braided network" (Archie Carr III, quoted in Kaiser 2001). Establishment of the separate braided segments has come about as a result of several factors, prominent among them the decision making associated with agriculture and resource use in the countries of the MBC (Edelman 1995; Zimmerer, Galt, and Buck 2004).[8] These adjustments toward the braided network have become a fundamental aspect of the conservation corridor project, rather than mere fine-tuning. Evolution of the MBC is most likely a foreshadowing of one of the major future directions of conservation initiatives and globalization. The expanding design and implementation of corridors for nature protection in areas of developing countries seem destined to bring conservation—including a component that is conspicuously international and global in scope—into ever-closer contact with agriculture, resource use, and livelihoods. Corridor projects thus promise to multiply the areal coverage of this expanding interface and magnify the importance of these issues (Zimmerer 2005).[9]

8. Since the coordination of the MBC involves the national governments of each country, as well as the Central American Commission for Environment and Development, the evolution of the MBC project has been strongly shaped by the needs of national governments and, to a degree, the within-country constituencies of agricultural and resource use groups (Kaiser 2001).

9. Another sort of dual purpose is also evident in maps of conservation corridor projects such as shown here (fig. 1.4). Many such maps are intended both to show reality and to shape reality, a goal that may verge on the contradictory. While both maps in our example were designed to represent the reality of the Mesoamerican Biological Corridor as accurately as possible at the time, these maps were also prepared as part of funding proposals. The importance of these maps as attempts to shape future reality needs to be seen as part of the general financial appeal of such corridor projects as MBC (Kaiser 2001, 2197).

Summary

Each chapter of our book offers specific insights and contributions to sustain our main arguments on globalization and conservation. These insights and arguments are centered on four spatioenvironmental themes, namely the spatiality of global conservation and sustainability projects, scale linkages in livelihood analysis and global environmental science, border issues in global conservation management, and decentralization and environmental governance in globalization. Our book can be summarized as enabling a view of global conservation in a way that considers the coupling of seemingly incongruous tendencies that are at the core of environmental globalization. On the one hand, environmental globalization is marked by increasingly salient global interconnections, allowing ideas, information, financing, and people to cover greater distance in shorter time, a phenomenon known as time-space compression. On the other hand, environmental globalization demonstrates the persistence or enhancement of various differences at the world-region and country levels, and a heightened importance of the "local."

From this vantage point we see these dynamic trends in conservation and the linkages to development as together an integral part of environment-related globalization, rather than as artifacts of the incomplete or imperfect application of a hypothetical or hoped-for paradigm of global resource management. The array of globalization-related differences and similarities that are being "produced" in conservation worldwide, when taken collectively, show the need for conceptual focal points, such as environmental networks, conservation territories, and linked scales of human-environment interaction, that illuminate effects that can be thought of as intermediate-level. Planning that is based on the functioning of these intermediate-level effects is a necessary complement to the mainly global and local poles that have tended to define the predominant geographical foci of environmental globalization. Developing focused examinations of these intermediate-level effects involving human-environment interaction and spatial dynamics, as we do in this volume, is of growing importance to conservation planning and management that take place as an increasingly integral yet complex part of globalization.

Acknowledgments

The research for this chapter and the concluding one was made possible as part of my activities supported through a fellowship of the John Simon Guggenheim Foundation during the academic year 2002–3. This research has benefited from numerous collaborations, discussions, and presentations. I am grateful to my collaborators, Ryan Galt and Margaret Buck, in this regard. In addition, Ryan designed and created the maps that I used. My earliest presentation of the research occurred in the context of the Environmental Breakfast Seminar, where Bill Cronon and Nancy Langston, my colleagues at the University of Wisconsin–Madison, offered generous insights and criticisms. My knowledge of environmental conservation is much richer through the interactions of more than fifteen years with my departmental colleague Thomas Vale. I have presented the research in this chapter on numerous occasions, most recently as the Taafe Lecture in the Department of Geography at Ohio State University and at the School of Forestry and Environmental Studies of Yale University. In these places I received many helpful comments and suggestions. Notwithstanding my incomplete rendering I would like to offer thanks for the inputs and insights of Paul Robbins, Kendra McSweeney, Becky Mansfield, Larry Brown, and Nancy Ettlinger at OSU and Michael Dove, Gus Speth, Enrique Mayer, Carol Carpenter, Rob Mendelsohn, Dan Esty, Avery Cohen, Alder Keleman, and John Tuxill at Yale. The chapter also owes to feedback from Kathy McAfee and Ivette Perfecto on my keynote address to the Yale conference "Sustainable Agriculture in the Americas."

References

Agrawal, A. 2000. Adaptive management in transboundary protected areas: The Bialowieza National Park and Biosphere Reserve as a case study. *Environmental Conservation* 27:326–33.

——. 2005. *Environmentality: Technologies of government and the making of subjects.* Durham, NC: Duke University Press.

Agrawal, A., and C. C. Gibson. 1999. Enchantment and disenchantment: The role of community in natural resource conservation. *World Development* 27 (4): 629–49.

Allen, T. F. H. 1998. The landscape "level" is dead: Persuading the family to take it off the respirator. In *Ecological scale: Theory and applications,* ed. D. L. Peterson and V. T. Parker, 35–54. New York: Columbia University Press.

Amin, A. 2002. Spatialities of globalisation. *Environment and Planning A* 34:385–99.

Barzetti, V., ed. 1993. *Parks and progress: Protected areas and economic development in Latin America and the Caribbean.* Cambridge, UK, and Washington, DC: IUCN (World Conservation Union) and IADB (Inter-American Development Bank).

Bassett, T., and K. B. Zuéli. 2000. Environmental discourses and the Ivorian savanna. *Annals of the Association of American Geographers* 89 (3): 377–401.

Batisse, M. 1997. Biosphere reserves: A challenge for biodiversity conservation and regional development. *Environment* 39 (5): 7–33.

Batterbury, S., T. Forsyth, and K. Thomson. 1997. Environmental transformations in developing countries: Hybrid research and democratic policy. *Geographical Journal* 163 (2): 126–32.

Bennett, G. 2004. *Integrating biodiversity conservation and sustainable land use: Lessons learned from ecological networks.* Gland, Switzerland, and Cambridge, UK: IUCN.

Berkes, F. 2004. Rethinking community-based conservation. *Conservation Biology* 18:621–30.

Blaikie, P., and H. Brookfield, eds. 1987. *Land degradation and society.* New York: Methuen.

Brah, A., M. J. Hickman, and M. Mac an Ghaill, eds. 1999. *Global futures: Migration, environment, and globalization.* London: Macmillan.

Brandon, K., K. H. Redford, and S. E. Sanderson, eds. 1998. *Parks in peril: People, politics, and protected areas.* Washington, DC: The Nature Conservancy.

Braun, B., and N. Castree, eds. 1998. *Remaking reality: Nature at the millennium.* London: Routledge.

Bray, D. W., and M. W. Bray. 2002. Beyond neoliberal globalization: Another world. *Latin American Perspectives* 29:117–26.

Brechin, S. R., ed. 2003. *Contested nature: Promoting international biodiversity conservation with social justice in the twenty-first century.* Albany: State University of New York Press.

Brechin, S. R., P. R. Wilhusen, C. L. Fortwangler, and P. C. West. 2002. Beyond the square wheel: Toward a more comprehensive understanding of biodiversity conservation as social and political process. *Society and Natural Resources* 15:41–64.

Brookfield, H. 2000. *Exploring agrodiversity.* New York: Columbia University Press.

Brosius, J. P., A. L. Tsing, and C. Zerner. 1998. Representing communities: Histories and politics of community-based natural resource management. *Society and Natural Resources* 11 (2): 157–69.

Brown, J. C. 2001. Responding to deforestation: Productive conservation, the World Bank, and beekeeping in Rondonia, Brazil. *Professional Geographer* 53 (1): 106–18.

Brunckhorst, D. J., and N. M. Rollings. 1999. Linking ecological and social functions of landscapes: I. Influencing resource governance. *Natural Areas Journal* 19 (1): 57–64.

Bruner, A. G., R. E. Gullison, R. E. Rice, and G. A. B. da Fonseca. 2001. Effectiveness of parks in protecting tropical biodiversity. *Science* 291:125–26.

Brush, S. B. 2004. *Farmer's bounty: Locating crop diversity in the contemporary world.* New Haven, CT: Yale University Press.

Bryant, R., and S. Bailey. 1997. *Third world political ecology.* London: Routledge.

Buttel, F. H., A. P. Hawkins, and A. G. Power. 1990. From limits to growth to global change: Constraints and contradictions in the evolution of environmental science and ideology. *Global Environmental Change* 1 (1): 57–66.

Buttel, F. H., and P. J. Taylor. 1992. Environmental sociology and global environmental change: A critical assessment. *Society & Natural Resources* 5 (3): 211–30.

Castro, G., and I. Locker. 2000. *Mapping conservation investments: An assessment of biodiversity funding in Latin America and the Caribbean.* Washington, DC: Biodiversity Support Program.

Cavallaro, V., and Dansero, E. 1998. Sustainable development: Global or local? *GeoJournal* 45 (1–2): 33–40.

Chase, J., ed. 2002. *The spaces of neoliberalism: Land, place, and family in Latin America.* Bloomfield, CT: Kumarian.

Christen, C., S. Herculano, K. Hochstetler, R. Prell, M. Price, and J. T. Roberts. 1998. Latin American environmentalism: Comparative views. *Studies in Comparative International Development* 33 (2): 58–87.

Clark, T. W., J. Tuxill, and M. S. Ashton, eds. 2003. Appraising AMISCONDE at La Amistad Biosphere Reserve, Costa Rica: Finding effective conservation and development. *Journal of Sustainable Forestry* 16 (1/2).

Coomes, O., and B. Barham. 1997. Rain forest extraction and conservation in Amazonia. *Geographical Journal* 163 (2): 180–88.

Cosgrove, D. 2001. *Apollo's eye: A cartographic genealogy of the earth in the Western imagination.* Baltimore: Johns Hopkins University Press.

Daniels, R., and T. J. Bassett. 2002. The spaces of conservation and development around Lake Nakuru National Park, Kenya. *Professional Geographer* 54 (4): 481–90.

Dicken, P. 1992. *Global shift: The internationalization of economic activity.* New York: Guilford.

Dinerstein, E. 1995. *A conservation assessment of the terrestrial ecoregions of Latin America and the Caribbean.* Washington, DC: World Bank.

Duffy, R. 2001. Peace parks: The paradox of globalization. *Geopolitics* 6 (2): 1–26.

Dupar, M., and N. Badenoch. 2002. *Environment, livelihoods, and local institutions: Decentralization in mainland Southeast Asia.* Washington, DC: World Resources Institute.

Edelman, M. 1995. Rethinking the hamburger thesis: Deforestation and the crisis of Central America's beef exports. In *The social causes of environmental destruction in Latin America,* ed. M. Painter and W. Durham, 25–62. Ann Arbor: University of Michigan Press.

Ellwood, W. 2001. *The no-nonsense guide to globalization.* London: Verso.

Escobar, A. 1995. *Encountering development: The making and unmaking of the third world.* Princeton, NJ: Princeton University Press.

———. 1998. Whose knowledge, whose nature? *Journal of Political Ecology* 5:53–82.

———. 1999. After nature: Steps to an antiessentialist political ecology. *Current Anthropology* 40:1–16.

Forsyth, T. 2001. Environmental social movements in Thailand: How important is class? *Asian Journal of Social Science* 29 (1): 35–51.

———. 2003. *Critical political ecology: The politics of environmental science.* London: Routledge.

Frank, D. J. 1997. Science, nature, and the globalization of the environment, 1870–1990. *Social Forces* 76 (2): 409–36.

Friedmann, J., and H. Rangan, eds. 1993. *In defense of livelihood: Comparative studies on environmental action.* West Hartford, CT: Kumarian.

Goldman, M., ed. 1998. *Privatizing nature: Political struggles for the global commons.* New Brunswick, NJ: Rutgers University Press.

Gray, L. C. 2002. Environmental policy, land rights and conflict: Rethinking community natural resource management programs in Burkina Faso. *Environment and Planning D: Society and Space* 20 (2): 167–82.

Grove, R. H. 1996. *Green imperialism: Colonial expansion, tropical island Edens, and the origins of environmentalism, 1600–1860.* Cambridge: Cambridge University Press.

Guha, R. 2000. *Environmentalism: A global history.* New York: Longman.

Guthman, J. 2004. *Agrarian dreams: The paradox of organic farming in California.* Berkeley: University of California Press.

Gwynne, R., and C. Kay. 2004. *Latin America transformed: Globalization and modernity.* 2nd ed. London: Arnold.

Hadley, M., ed. 2002. *Biosphere reserves: Special places for people and nature.* Paris: United Nations Educational, Scientific, and Cultural Organization (UNESCO).

Haenn, N. 2002. Nature regimes in southern Mexico: A history of power and environment. *Ethnology* 41:1–26.

Hall, A., ed. 2000. *Amazonia at the crossroads: The challenges of sustainable development.* London: Institute of Latin American Studies.

Hannah, L., D. Lohse, C. Hutchinson, J. L. Carr, and A. Lankerani. 1994. A preliminary inventory of human disturbance of world ecosystems. *Ambio* 23:246–50.

Harvey, D. 1996. *Justice, nature and the geography of difference.* Cambridge, MA: Blackwell.

———. 2000. *Spaces of hope.* Berkeley and Los Angeles: University of California Press.

Hecht, S. B. 2004. Invisible forests: The political ecology of forest resurgence in El Salvador. In *Liberation ecologies: Environment, development, social movements,* 2nd ed., ed. R. Peet and M. Watts, 64–104. London: Routledge.

Hobbs, R. J. 1998. Managing ecological systems and processes. In *Ecological scale: Theory and applications,* ed. D. L. Peterson and V. T. Parker, 459–84. New York: Columbia University Press.

Holling, C. S., and G. K. Meffe. 1996. Command and control and the pathology of natural resource management. *Conservation Biology* 10 (2): 328–37.

Huber, E., and F. Solt. 2004. Successes and failures of neoliberalism. *Latin American Research Review* 39 (3): 150–64.

Hughes, D. 2001. Rezoned for business: How ecotourism unlocked Black farmland in Eastern Zimbabwe. *Journal of Agrarian Change* 1:575–99.

IUCN (World Conservation Union). 1985. *1985 United Nations list of national parks and protected areas.* Cambridge, UK, and Gland, Switzerland: IUCN.

——. 1992. Parks for life: A new beginning. *IUCN Bulletin* 23:10.

IUCN (World Conservation Union) and WCMC (World Conservation Monitoring Centre). 1998. *1997 United Nations list of protected areas.* Cambridge, UK: IUCN Publications.

IUCN (World Conservation Union) and WCPA (World Commission on Protected Areas). 2001. *Transboundary protected areas for peace and cooperation.* Gland, Switzerland: IUCN.

Jasanoff, S., ed. 2004. *States of knowledge: The co-production of science and social order.* London: Routledge.

Jia, H., and Z. Lin. 1994. *Changing central-local relations in China.* Boulder, CO: Westview.

Jiang, H. 2004. Cooperation, land use, and the environment in Uxin Ju: A changing landscape of a Mongolian-Chinese borderland in China. *Annals of the Association of American Geographers* 94 (1): 117–39.

Johnston, R. J., ed. 2000. *The dictionary of human geography.* 4th ed. Oxford: Blackwell.

Johnston, R. J., P. J. Taylor, and M. J. Watts. 2002. *Geographies of global changes: Remapping the world.* 2nd ed. Oxford: Blackwell.

Kaiser, J. 2001. Bold corridor project confronts political reality. *Science* 293:2196–99.

Kellert, S. R., J. N. Mehta, S. A. Ebbin, and L. L. Lichtenfeld. 2000. Community natural resource management: Promise, rhetoric, and reality. *Society and Natural Resources* 13:705–15.

Kramer, R., C. van Schaik, and J. Johnson, eds. 1997. *Last stand: Protected areas and the defense of tropical biodiversity.* Oxford: Oxford University Press.

Lavigne, M. 1999. *The economics of transition: From socialist economy to market economy.* New York: St. Martin's.

Lightfoot, D. R. 1994. An assessment of the relationship between development and institutionally preserved lands. *Area* 26 (2): 112–22.

Lutz, E., and J. Caldecott, eds. 1996. *Decentralization and biodiversity conservation.* Washington, DC: World Bank.

Marston, S. A. 2000. The social construction of scale. *Progress in Human Geography* 24:219–41.

Martinez, R. 2001. Mesoamerican Biological Corridor: Captive of the Plan Puebla Panama. *Proceso Sur.* Global Exchange newsletter, 27 October. Available at http://www.globalexchange.org.

McCarthy, J., and S. Prudham. 2004. Neoliberal nature and the nature of neoliberalism. *Geoforum* 35:275–83.

McMillan, J., and B. Naughton. 1996. *Reforming Asian socialism: The growth of market institutions.* Ann Arbor: University of Michigan Press.

Miller, C. A., and P. N. Edwards, eds. 2001. *Changing the atmosphere: Expert knowledge and environmental governance.* Cambridge, MA: MIT Press.

Miller, K., E. Chang, and N. Johnson. 2001. *Defining common ground for the Mesoamerican Biological Corridor.* Washington, DC: World Resources Institute.

Mittermeier, R., N. Myers, J. Thomsen, G. da Fonseca, and S. Oliveieri. 1998. Biodiversity hotspots and major tropical wilderness areas: Approaches to setting conservation priorities. *Conservation Biology* 12:516–20.

Mol, A. P. J. 2001. *Globalization and environmental reform: The ecological modernization of the global economy.* Cambridge, MA: MIT Press.

Mooney, H. A. 1998. *The globalization of ecological thought.* Oldendorf, Germany: Ecology Institute.

Morehouse, B. J. 1996. *A place called the Grand Canyon: Contested geographies.* Tucson: University of Arizona Press.

Murdoch, J., and T. Marsden. 1995. The spatialisation of politics: Local and national actor-spaces in environmental conflict. *Transactions of the Institute of British Geographers,* n.s., 20:368–80.

Mutersbaugh, T. 2002. Migration, common property, and communal labor: Cultural politics and agency in a Mexican village. *Political Geography* 21:473–94.

Naughton-Treves, L. 1997. Farming the forest edge: Vulnerable places and people around Kibale National Park, Uganda. *Geographical Review* 87 (1): 27–46.

Neumann, R. P. 1995. Local challenges to global agendas: Conservation, economic liberalization, and the pastoralists' rights movement in Tanzania. *Antipode* 27 (4): 363–82.

——. 1998. *Imposing wilderness: Struggles over livelihood and nature preservation in Africa.* Berkeley: University of California Press.

——. 2004. Nature-state-territory: Toward a critical theorization of conservation enclosures. In *Liberation ecologies: Environment, development, social movement,* 2nd ed., ed. R. Peet and M. Watts, 195–217. London: Routledge.

——. 2005. *Making political ecology.* London: Arnold.

Nietschmann, B. 1995. Conservación, autodeterminación y el area protegida Costa Miskita, Nicaragua. *Mesoamérica* 29:1–55.

Paulson, S., and L. L. Gezon, eds. 2004. *Political ecology across spaces, scales, and social groups.* New Brunswick, NJ: Rutgers University Press.

Peck, J. 2004. Geography and public policy: Constructions of neoliberalism. *Progress in Human Geography* 28:392–405.

Peet, R., and M. Watts, eds. 2004. *Liberation ecologies: Environment, development, social movement.* 2nd ed. London: Routledge.

Peluso, N. L., and M. Watts, eds. 2001. *Violent environments.* Ithaca, NY: Cornell University Press.

Perreault, T. 2005. State restructuring and the scale politics of rural water governance in Bolivia. *Environment and Planning A* 37:263–84.

Perreault, T., and P. Martin. 2005. Geographies of neoliberalism in Latin America: Introduction. *Environment and Planning A* 37:191–201.

Phillips, A. 2002. *Management guidelines for IUCN category V protected areas (protected landscapes/seascapes).* Cambridge, UK: IUCN.

Pimm, S., M. Ayres, A. Balmford, G. Branch, K. Brandon, T. Brooks, R. Bustamante, et al. 2001. Can we defy nature's end? *Science* 293:2207–8.

Ramankutty, N., J. A. Foley, J. Norman, and K. McSweeney. 2002. The global distribution of cultivable lands: Current patterns and sensitivity to possible climate change. *Global Ecology & Biogeography* 11:377–92.

Ravenga, C., S. Murray, J. Abramovitz, and A. Hammond. 1998. *Watersheds of the world: Ecological value and vulnerability.* Washington, DC: World Resources Institute and Worldwatch Institute.

Read, J. M. 2003. Spatial analyses of logging impacts in Amazonia using remotely sensed data. *Photogrammetric Engineering and Remote Sensing* 69 (3): 275–82.

Redford, K. H., and C. Padoch, eds. 1992. *Conservation of neotropical forests: Working from traditional resource use.* New York: Columbia University Press.

Redford, K. H., and S. E. Sanderson. 2000. Extracting humans from nature. *Conservation Biology* 14 (5): 1362–64.

Ribot, J. C. 1999. Decentralisation, participation, and accountability in Sahelian forestry: Legal instruments of political-administrative control. *Africa* 69 (1): 23–65.

Ribot, J. C., and A. M. Larson, eds. 2005. *Democratic decentralisation through a natural resource lens.* London: Routledge.

Rice, R. A. 2001. Noble goals and challenging terrain: Organic and fair trade coffee movements in the global marketplace. *Journal of Agricultural and Environmental Ethics* 14 (1): 39–66.

Robbins, P. 2004. *Political ecology: A critical introduction.* Oxford: Blackwell.

Roberts, J. T. 1998. Emerging global environmental standards: Prospects and perils. *Journal of Developing Societies* 14:144–63.

Roberts, J. T., and N. D. Thanos. 2003. *Trouble in paradise: Globalization and environmental crises in Latin America.* London: Routledge.

Rodríguez, L. O., and K. R. Young. 2000. Biological diversity of Peru: Determining priority areas for conservation. *Ambio* 29:329–37.

Rudel, T. K. 2002. Paths of destruction and regeneration: Globalization and forests in the tropics. *Rural Sociology* 67:622–36.

Sanderson, E. W., M. Jafteh, M. A. Levy, K. H. Redford, A. V. Wannebo, and G. Woolmer. 2002. The human footprint and the last of the wild. *BioScience* 52:891–904.

Sarmiento, F. O. 2000. Breaking mountain paradigms: Ecological effects on human impacts in managed Tropandean landscapes. *Ambio* 29:423–31.

Schroeder, R. A. 1999. Geographies of environmental intervention in Africa. *Progress in Human Geography* 23:359–78.

Schwartzman, S., A. Moreira, and D. Nepstad. 2000. Rethinking tropical forest conservation: Perils in parks. *Conservation Biology* 14 (4): 1351–57.

Scott, J. C. 1998. *Seeing like a state: How certain schemes to improve the human condition have failed.* New Haven, CT: Yale University Press.

Seager, J. 1995. *The new state of the earth atlas.* 2nd ed. New York: Simon and Schuster.

Sierra, R., F. Campos, and J. Chamberlin. 2002. Conservation priorities in continental Ecuador: A study based on landscape and species level biodiversity patterns. *Landscape and Urban Planning* 59: 95–110.

Sneddon, C. S. 2000. "Sustainability" in ecological economics, ecology, and livelihoods: A review. *Progress in Human Geography* 24 (4): 521–49.

Sneddon, C. S., and B. T. Nguyen. 2001. Politics, ecology and water: The Mekong Delta and development of the Lower Mekong Basin. In *Living with environmental change: Social vulnerability, adaptation and resilience in Vietnam,* ed. W. N. Adger, P. M. Kelly, and Nguyen Huu Ninh, 234–62. London: Routledge.

Söderbaum, F., and I. Taylor, eds. 2003. *Regionalism and uneven development in Southern Africa.* Aldershot, UK: Ashgate.

Solbrig, O. T., R. Paarlberg, and F. de Castri, eds. 2001. *Globalization and the rural environment.* Cambridge, MA: Harvard University Press.

Soulé, M. E., and J. Terborgh, eds. 1999. *Continental conservation: Scientific foundations of regional reserve networks.* Washington, DC: Island.

Speth, J. G. 2003. *Worlds apart: Globalization and the environment.* Washington, DC: Island.

Stevens, S., ed. 1997. *Conservation through survival: Indigenous peoples and protected areas.* Washington, DC: Island.

Stiglitz, J. E. 2002. *Globalization and its discontents.* New York: W. W. Norton.

St. Martin, K. 2001. Making space for community resource management in fisheries. *Annals of the Association of American Geographers* 91 (1): 122–42.

Stonich, S. C. 2001. *Endangered peoples of Latin America: Struggles to survive and thrive.* Westport, CT: Greenwood.

Stott, P., and S. Sullivan, eds. 2000. *Political ecology: Science, myth, and power.* London: Arnold.

Sundberg, J. 1998. Strategies for authenticity, space, and place in the Maya Biosphere Reserve, Petén, Guatemala. *Yearbook, Conference of Latin Americanist Geographers* 24:85–96.

Swyngedouw, E. 1997. Neither global nor local: "Glocalization" and the politics of scale. In *Spaces of globalization: Reasserting the power of the local,* ed. K. R. Cox, 137–66. New York: Guilford.

Terborgh, J. 2000. The fate of tropical forests: A matter of stewardship. *Conservation Biology* 14 (5): 1358–61.

———, ed. 2002. *Making parks work: Strategies for preserving tropical nature.* Washington, DC: Island.

Tomlinson, J. 1999. *Globalization and culture.* Chicago: University of Chicago Press.

Turner, B. L. II, J. Geoghegan, E. Keys, P. Klepeis, D. Lawrence, P. M. Mendoza, S. Manson, et al. 2001. Deforestation in the southern Yucatán peninsula: An integrative approach. *Forest Ecology and Management* 154:355–70.

Turner, M., ed. 1999. *Central-local relations in Asia-Pacific: Convergence or divergence.* Hampshire, UK: Macmillan.

Turner, M. D. 1999a. Merging local and regional analyses of land-use change: The case of livestock in the Sahel. *Annals of the Association of American Geographers* 89 (2): 191–219.

———. 1999b. No space for participation: Pastoralist narratives and the etiology of park-herder conflict in Southeastern Niger. *Land Degradation and Development* 10:345–63.

———. 2003. Methodological reflections on the use of remote sensing and geographic information science in human ecological research. *Human Ecology* 31 (2): 255–79.

Wackernagel, M., and W. Rees. 1996. *Our ecological footprint: Reducing human impact on earth.* Bariola Island, BC: New Society.

Watts, M. J., and D. Goodman. 1997. Agrarian questions: Nature, culture, and industry in fin-de-siècle agro-food systems. In *Globalising food: Agrarian questions and global restructuring,* ed. D. Goodman and M. J. Watts, 1–34. London: Routledge.

Wells, M. P., and K. E. Brandon. 1992. *People and parks: Linking protected area management with local communities.* Washington, DC: World Bank, World Wildlife Fund, and U.S. Agency for International Development.

———. 1993. The principles and practice of buffer zones and local participation in biodiversity conservation. *Ambio* 22 (2–3): 157–62.

Western, D., and R. M. Wright, eds. 1994. *Natural connections: Perspectives in community-based conservation.* Washington, DC: Island.

Wiens, J. A. 1989. Spatial scaling in ecology. *Functional Ecology* 3:385–97.

Wily, L. A. 2003. *Governance and land relations: A review of decentralisation of land administration and management in Africa.* London: International Institute for Environment and Development.

WinklerPrins, A. M. G. A. 2002. House-lot gardens in Santarém, Pará, Brazil: Linking rural with urban. *Urban Ecosystems* 6:43–65.

Wiseman, R., and L. Hopkins, eds. 2001. *Sowing the seeds for sustainability: Agriculture, biodiversity, economics and society; Proceedings of the eight interactive Session held at the Second IUCN World Conservation Congress, Amman, Jordan, 7 October 2000.* Gland, Switzerland: IUCN.

World Rainforest Movement. 2001. The indigenous and biological corridor in Central America. *World Rainforest Movement Bulletin* 44.

Worster, D. 1990. The ecology of order and chaos. *Environmental History Review* 14:1–18.

WRI (World Resources Institute). 1990. *World resources, 1990–1991.* Oxford: Oxford University Press.

———. 1992. *World resources, 1992–1993.* Oxford: Oxford University Press.

———. 1994. *World resources, 1994–1995.* Oxford: Oxford University Press.

WRI (World Resources Institute) and IIED (International Institute for Environment and Development). 1986. *World resources, 1986.* New York: Basic Books.

Young, E. 1999. Balancing conservation with development in small-scale fisheries: Is ecotourism an empty promise? *Human Ecology* 27 (4): 581–620.

Zbicz, D. C. 1999. Transfrontier ecosystems and internationally adjoining protected areas.

Zbicz, D. C., and M. Green. 1997. Status of the world's transfrontier protected areas. *Parks* 7 (3): 5–10.

Zerner, C., ed. 2000. *People, plants, and justice: The politics of nature conservation.* New York: Columbia University Press.

Zimmerer, K. S. 2000. The reworking of conservation geographies: Nonequilibrium landscapes and nature-society hybrids. *Annals of the Association of American Geographers* 90 (2): 356–69.

———. 2003a. Environmental zonation and mountain agriculture in Peru and Bolivia: Toward a model of overlapping patchworks and agrobiodiversity conservation. In *Political ecology: An integrative approach to geography and environment-development studies,* ed. K. S. Zimmerer and T. J. Bassett, 141–69. New York: Guilford.

———. 2003b. Geographies of seed networks for food plants (potato, ulluco) and approaches to agrobiodiversity conservation in the Andean countries. *Society and Natural Resources* 16 (7): 583–601.

———. 2005. An expanding interface with agriculture will change global conservation. In *Our lands, our food: Farmers' movements, trade, and the environment in the Americas,* ed. J. Cohen, 25–33. New Haven, CT: Yale School of Forestry and Environmental Studies.

Zimmerer, K. S., and T. J. Bassett. 2003. Approaching political ecology: Society, nature, and scale in human-environment studies. In *Political ecology: An integrative approach to geography and environment-development studies,* ed. K. S. Zimmerer and T. J. Bassett, 1–28. New York: Guilford.

Zimmerer, K. S., and E. D. Carter. 2002. Conservation and sustainability in Latin America. *Yearbook of the Conference of Latin Americanist Geographers* 27:22–43.

Zimmerer, K. S., R. E. Galt, and M. V. Buck. 2004. Globalization and multi-spatial trends in the coverage of protected-area conservation (1980–2000). *Ambio* 33 (8): 520–29.

Zimmerer, K. S., and K. R. Young, eds. 1998. *Nature's geography: New lessons for conservation in developing countries.* Madison: University of Wisconsin Press.

PART I

Spatialities in Global Conservation and Sustainability Projects

As protected-area coverage expands worldwide, so do the number and variety of spatial arrangements for conservation. Of these spatial configurations, the most familiar to the general "geographical imagination" is that of conservation territories: contiguous, well-demarcated units that are labeled here as "territories" in order to emphasize the importance of fixed location and boundaries to these units. Examples include parks, nature reserves, and buffer zones. Globally, conservation territories are estimated to have grown in number from about 15,000 in the late 1960s to more than 200,000 in 2000. Conservation networks represent a contrasting type of spatial arrangement that is typically designated for the dual purpose of conservation and sustainable use. The latter goal is emphasized in this section. Examples of such networks include conservation corridors (discussed in chap. 1), agricultural landscapes and greenways, the institutions and individuals that are associated with alternate "green" markets and the land use that is influenced through them, and the social networks of conservation experts, advocates, and planners.

Conservation networks are spatially extensive, comprise multiple sites, and often are discontinuous, but may contain discrete conservation territories (such as parks) within them. Sustainable use is commonly a more important goal in ecological networks than it is in bounded conservation units. Our volume addresses several of the sociospatial issues that arise in the network-type arrangements of conservation and sustainability projects that range from organic coffee (chap. 2) to sustainable forestry (chap. 3) and eco-development projects based on beekeeping (chap. 4).

The introduction of this section requires taking stock of a few basic tensions, or possible contradictions, that characterize the design of spatial arrangements for conservation and sustainable use in globalization-led changes. The first tension is between the goal of fixing and demarcating territories and networks of conservation, on the one hand, and the reality of international flows, dynamic social and ecological change, and nearly constant rearranging of spatial units that are a product of globalization, on the other hand. A distinguishing feature of globalization in realms other than environmental conservation—for example, those of economics, politics, and culture—is the overriding tendency toward the ongoing spatial reassortment of territories. In these other realms the location and characteristics of territories and networks is typically fluid, dynamic, or stable for only a limited time ("fixed") before undergoing spatial rearrangements within the context of global change. By contrast, global conservation and sustainable use are based on the design of territories and networks that aim for stability and fixity for relatively long periods. This tension in spatial dynamics is also apparent in the design templates of units for global conservation and sustainable use. A reliance on natural local templates for these purposes, such as unfragmented agroecosystems of organic coffee (chap. 2), is significantly increased under globalization. It suggests a tendency that appears antithetical, at least at first glance, to what might be expected from global systems of integration, including the certification programs that are increasingly common for organic coffee and various other global commodities that represent sustainable use.

The second tension exists between the global networks of high-quality spatial information created and distributed through new spatial technologies, on the one hand, and the rapidly growing use of these technologies for local management for conservation and sustainable use, on the other hand. New improvements in the quality, cost, and accessibility of spatial technologies, such as remote sensing, geographic information systems (GIS), and the global positioning system (GPS), are integral to the global expansion of conservation and its interface with agriculture and resource use. These technologies are helpful in identifying, demarcating, and monitoring conservation areas. Increasingly they are being used, via global agreements and coordinating organizations, to manage conservation and sustainable use in more dynamic and integrated, but more remote, ways. The networks for this coordination often occur in conjunction with global certification programs (exemplified in the case of sustainable forestry in chap. 3). Whereas use of spatial technologies in the past usually focused on the design or analysis of benchmark conditions, existing situations, and the spatial delimitation of management plans, now these technologies are considered for use in fine-scale and real-time activities such as sustainable forestry at the stand level (areas as small as 100 m^2).

A definite tension emerges in bringing the large data capacity and powerful analytical potential of spatial technologies to local resource management, perhaps especially in developing countries. This tension arises in part from an obvious issue of accessibility: how local land users and forest managers gain access to and employ new data and technologies. It also brings to the fore the multifaceted politics of the new technological innovations. The politics of this particular tension, embodied in the expression "conservation and sustainability for whom?" are still little known in the case of these rapidly expanding geospatial technology applications to sustainable forestry. Still-existing on-the-ground experience with conventional mapping that has been long used in demarcating forest and wildlife areas has led to abundant, heated political tensions in various social and environmental settings throughout the world. This type of political tension foreshadows the issues that are arising with regard to access to and impact of the geospatial technologies.

A third tension is evident in the profusion of ecodevelopment or productive conservation projects worldwide. This trend has been strongly shaped by support networks involving global conservation and development organizations, but at the same time most of these projects are defined, presented, and justified as "local." Indeed, the claim that ecodevelopment projects are local is fairly indicative of the interests and rationales of global organizations, who use the term *local* in a dual sense, to suggest both low environmental impact and adequate social equity for development, often with valid reasons. Yet the active growth of global ecodevelopment is found to exert pressures on local projects that are at odds with the commonplace assumptions of what is local.

Beekeeping, the topic of chapter 5, is one of the clearest cases of this inherent tension between the global and the local in the ecodevelopment projects. For the past decade, global conservation organizations have widely supported beekeeping as a favored type of project for sustainable development with low environmental impact and relatively affordable costs of operation, which make it accessible to rural inhabitants across a range of income levels. The global popularity of beekeeping during this time did not arise de novo as a local sustainability project for tropical forest areas. Beekeeping, like various other favored local models of ecodevelopment involving cottage industries and low-impact extractive activities (e.g., basketry, tropical tree fruit gathering), is a long-established development strategy with a narrative produced and reshaped by international networks of experience and influence. Current interest in these popular ideas of local ecodevelopment is built on existing international or global networks of institutions, experience, finance, and influence. Chapter 5 analyzes one of the most important examples of the defining tension that exists in the mutual dependence of the local and the global in ecodevelopment.

2 Certifying Biodiversity: Conservation Networks, Landscape Connectivity, and Certified Agriculture in Southern Mexico

TAD MUTERSBAUGH

Many of the world's remaining areas of high conservation value are frag-mented, as they are broken into small, dispersed patches surrounded by and incorporated into human activity zones such as agricultural fields, forests, and firewood reserves (Laurance and Bierregaard 1997). Dispersed patches provide environmental services—particularly when incorporated into conservation networks—as reservoirs for endemic species and as "islands" that form "stepping stones" that facilitate both international migration of bird and other animal species and regional movement between conservation territories such as wildlife reserves. Networks of discontinuous or human-connected landscapes (in contrast to the large contiguous blocklike character-istics of protected areas) provide an important complement to conservation territories, particularly in areas where a reserve of sufficient size to encom-pass fragmented, dispersed spaces is not economically or politically feasible. For these cases, conservation networks provide an alternative model to con-servation territories. Linking together the efforts of thousands of small farm-ers, local authorities, and environment and development nongovernmental organizations (NGOs), these networks can provide environmental protection in places and at spatial scales that would be impractical for conservation ter-ritories.

This chapter will examine one such model of network conservation, namely, certified organic coffee production in Mexico. This conservation model unites the sustainability efforts of thousands of socially networked family coffee producers, each with only a few hectares of coffee, into a spatial conservation network that is participatory and democratic while providing an effective means to provide environmental services ranging from species protection to carbon sequestration, clean water supplies, and erosion con-trol. At a "meso-scale" of regions within nations, this participatory, produc-tive conservation framework is made possible by grassroots organizations. In the case of the Oaxacan Statewide Peasant Union presented here, coffee farmers are actively integrated at regional and village scales into assessment, design, and implementation of ecologically sound agricultural practices (CEPCO [Coordinadora Estatal de Productores de Cafe Oaxaqueños] 1996; for other examples of Mexican peasant unions and environmental issues see Hernandez Castillo and Nigh 1998; Whatmore and Thorne 1997; Bray, Sanchez, and Murphy 2002). The certified organic producer villages in this study are members of this Oaxacan coffee farmer organization, which includes 12,500 coffee producer families (comprising approximately 100,000 persons including dependents) residing in 200 villages, integrated into forty-four regional organizations. At the higher organizational (and spatial) scales, Mexican national certifiers Certimex and the Organic Crop Improvement Association–Mexico (OCIA-Mexico), contractually linked to international certifiers (IMO-CONTROL and OCIA-International) and organic labelers (Naturland, OCIA), support regional efforts with regulatory oversight.

However, even as a dispersed or conservation network strategy overcomes constraints associated with conventional conservation territories, this network approach, which is termed here "soft path" in contrast to the "hard boundary" distinctions of conventional conservation territories, confronts unique ecological, political, and economic challenges. These tensions form a second key theme of this chapter. In the broadest terms, conservation networks confront profoundly disruptive and unpredictable logics or processes that are characteristic of economic, social, and cultural globalization. These illogics, in the sense that they undermine sound environmental management and sustainability, include highly variable product markets, governmental policy changes, and, particularly in the Mexican case, widespread labor migration. Thus, even though networks promote and indeed require active, popular participation in place of an exclusion of local people often associated with limited-access "fortress"-style parks (see Sundberg, chap. 11 in this volume; Orlove and Brush 1996), they nevertheless retain the shortcomings of "command and control" conservation management (Holling and Meffe 1996). These pitfalls are embedded in certification schemas, externally mandated social and ecological norms, and external technologies used to assess the efficacy of conservation efforts (see Read, chap. 3 in this volume; Mutersbaugh 2004). And as Brown (chap. 4 in this volume) notes, the very global environment and development initiatives that certification schemas mean to evaluate are themselves productive of local illogics, particularly where local conservation efforts are weakly developed. Social and economic tensions have ecological repercussions, introducing tensions between desirable environmental outcomes and sustainable social outcomes within networks. Key goals of a conservation network may include incorporation of the most important local biodiversity hot spots into conserved (organic) parcels and identification of a spatiality that is a purposefully supported style of landscape that promotes landscape connectivity—that is to say, a spatiality that promotes exchange between conserved parcels. Some sociospatial factors produce ecological tensions, which are examined in the final section of this paper. For example, socially produced farm spatial distributions may enhance diversity by aiding species with characteristics such as wide geographic range (e.g., pines), an ability to exploit dispersed habitats (e.g., birds or butterflies), or preference for agricultural habitats. In another example, conservation norms may also be viewed as a social product devised and disseminated within a transnational conservation network: to what degree, then, are conservation norms science based? How are they adapted to local ecological and agronomic circumstances, and how is farmer participation rewarded?

Network Conservation via Economic Incentives

Researchers of many stripes from university research ecologists to World Bank officials propose the use of market-based incentives to achieve ecological diversity and species preservation goals via, for instance, support for productive conservation (environment and development) projects or protection of biological corridors and reserve buffers (Rice 2001; Rice and Ward 1996; Smithsonian Migratory Bird Center 2001; World Bank 2001; Inter-Agency Technical Committee 1999). Certified agriculture, in which farming methods are based upon ecologically informed production norms, is one specific proposal within this arena that would extend biodiversity conservation beyond traditional biological reserves to cover areas that are ecologically sensitive and yet spatially dispersed. In general, certified agriculture has the advantage of (a) constraining protected resources to specific, geographically encoded sites, (b) utilizing a currently existing network of inspection staff created by organic certification agencies, (c) working with peasant groups with a significant, documented history of successful compliance with production norms, and (d) having a system for education of service personnel at levels including national and transnational certifier agencies, regional and statewide producer organizations, and village support personnel.

Despite a measure of success in extending certified agriculture to producer groups (Hernandez Castillo and Nigh 1998; Rice 2001; Bray, Sanchez, and Murphy 2002), key problems remain. First, many critical ecological zones are not involved in certified production. Limitations include conversion costs, lack of producer administrative capacity, and agronomic aspects of farming (Mutersbaugh 2002; Rice 2001). Second, even where organic production is initiated, general increases in biodiversity may be difficult to achieve. Among coffee producers who participate in certification programs, there exists a diversity of production practices with varying impacts on biodiversity.

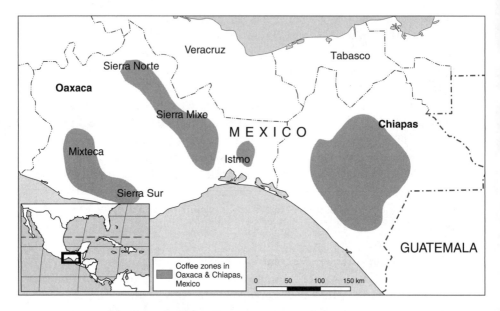

FIGURE 2.1. Map of study area: coffee-growing areas of Oaxaca and Chiapas

Organic certifiers, however, have recognized the importance of biodiversity conservation and are increasingly incorporating biodiversity into production norms, requiring a greater diversity of plant species used as shade for coffee.

The coffee producing areas in the southern Mexican state of Oaxaca are the focus of the present study (fig. 2.1). Located in endangered and fragile mesophytic tropical ("cloud") forest and upper-elevation tropical woodland, coffee lands vital to ecological preservation efforts in the Americas are under threat. A long-term decline in real coffee prices is resulting in migration, coffee plot abandonment, and changes of land use to labor-saving activities such as goat herding. Since a high percentage of Oaxacan coffee producers are indigenous producers (Bray, Sanchez, and Murphy 2002), abandonment of coffee cultivation results in cultural loss as well. Thus, improving farmer welfare through linkages to coffee certification programs may help meet the aims of cultural and ecological conservation.

Conservation and Spatiality in Certified Organic Agriculture

This section has two objectives. The first is to present a sense of the spatialities of organic production, whose spatial properties and the social-environmental processes that form them serve to express both the promise and the tensions of certified agriculture as a conservation strategy. As we will see, conservation network spatialities are—as in the case for conservation territories—formed quite apart from the goals of environmental preservation, as a result of cultural and social histories expressed in the organization of space. A second objective is to introduce the network of labelers, certifiers, regulatory agencies, and producer organizations that organizes the movement of organic products from farm to consumer and guarantees rigorous organic product standards. This second, transnational organizational tier supports conservation networks by providing certification procedures that ensure that consumer products originate in certified plots and are not blended with noncertified goods, conservation norms spelling out environmentally friendly agronomic and processing methods, and economic payments that compensate producers for the extra work required by organic production. Organic coffee is one instance of certified "green labeling" initiatives that comprise a US$24 billion global organic trade and encompass 24 million hectares of land (Willer and Yussefi 2004). At least 200 certifiers undertake global certification activities for organic coffee; eight operate in Mexico (Gómez Tovar, Gómez Cruz, and Schwentesius Rindermann 1999).

Conservation Geography of Certified Agricultural Plots
If certified sustainable agriculture is to realize its full potential as a conservation strategy, it must not only be economically and ecologically sound, but also spread its benefits such that areas of high conservation value—those areas deemed most important to biodiversity conservation—are incorporated to the greatest degree possible. Research undertaken in three communities in the Mixteca and Sierra Sur regions of Oaxaca, Mexico, will be used to illustrate microgeographic aspects of certified plots. First, however, it is useful to consider the macro distribution of certified organic agricultural plots in relation to areas important to the conservation of biodiversity. Although in global terms there is not a high degree of overlap between areas of high conservation value and certified agriculture, for some areas and crops, such as Mexican coffee, production zones include recognized centers of high species diversity and endemism, particularly those located at the borders of tropical and temperate elevational zones (Moguel and Toledo 1999).

Village "communal" land areas (75–300 square kilometers) demonstrate a similar distribution pattern: while many organic plots are located in areas of high conservation value, coverage is uneven, the result of a social and political history of organizational development. In village studies of certified organic coffee organizations, field surveys found a wide diversity in cultivation and processing techniques within villages, yet a relative uniformity across individual producer fields (Mutersbaugh 2002). Certified organic producer organization statistics for villages discussed below are presented in table 2.1. Villages differ by property type (communal/private), language (Spanish/ Mixtec/Zapotec), wealth (kilos per producer), and well-being (education levels, infant mortality).

TABLE 2.1.   Certified organic producers in surveyed villages

	Number of producers	Property type	Kilos/producer (average/high/low)	Language spoken
Village A	58	Private	792/4,900/100	Zapotec/Spanish
Village B	65	Common property	256/1,500/85	Mixtec
Village C	32	Common property	1,400/9,868/92	Zapotec/Spanish

Note: Communal property = Propiedad comunitaria; kilo/producer from 2003–4 cycle.

The maps in figures 2.2, 2.3, and 2.4, produced by certified organic coffee producer organizations in the Mixteca and Sierra Sur regions of Oaxaca, provide a sense of the social context of conservation networks. These maps are a product of both village and transnational social contexts. At the village level, local organic technical officers produce the maps, carefully delineating trail networks and locating plots with respect to landmarks such as rivers, mountains, and neighboring producers. At the transnational level, the maps are shaped by the purpose for which they are made, namely, to facilitate organic certification: each producer is demarcated by her or his organic number (a number which must appear on each piece of paperwork corresponding to the producer). In a sense, then, these maps rest at the intersection between transnational certification and local production, and to prefigure subsequent discussion, it may be noted that these maps also embed key tensions between social and conservation values.

Two characteristics, uneven distribution and clustering, may be identified in the maps presented, and are a result of an interaction of social and biophysical dynamics. Interviews with coffee producers indicate that uneven distribution of plots results from differences in suitability of coffee producing terrain, and from political and economic factors. In figure 2.2 (Village A in table 2.1), plots are widely dispersed as villagers have sought the best available lands for coffee production. Many areas are unsuitable due to south-facing, coastal slopes, and so coffee is located in the higher elevations to the north of village lands and in cooler, moister *jollas* (valleys) that provide slopes with a northerly aspect that protects coffee from sun and the hot, dry winds that rise from the tropical lowlands. Political factors also affect plot distribution: a history of social conflict between government-aligned and non-aligned villagers has resulted in social divisions that prevent many producers from joining organic producer groups. Economic factors limit participation but do not exert local-level distributional effects.

Figure 2.3 (table 2.1, Village B) demonstrates a somewhat greater density of organic plots and a more even distribution than in figure 2.2. This is in part biophysical, the result of a more temperate location in the interior Mixtec region lacking strong microclimatic differentiation on north- versus

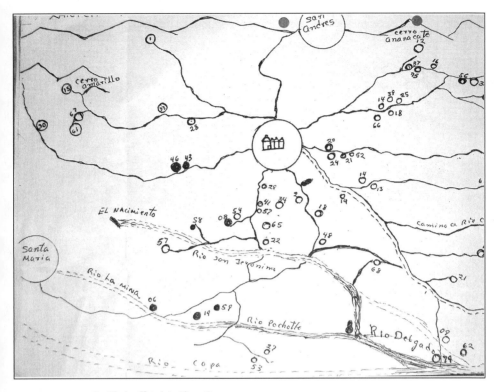

FIGURE 2.2. Certified coffee plots: Sierra Sur

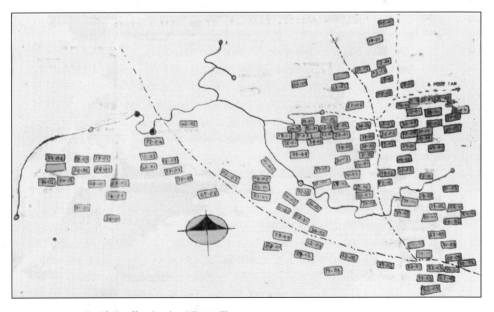

FIGURE 2.3. Certified coffee plots in a Mixtec village

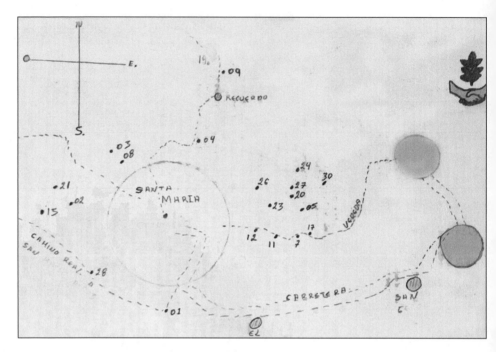

FIGURE 2.4. Certified coffee plots in a Zapotec village

south-facing slopes. It also, however, reflects a lack of political divisions in this particular Mixtec community such that all village families are able to participate in certified organic coffee production.

An accentuated "clustering" effect is seen most clearly in figure 2.4 (Village C in table 2.1) due to the smaller scale and the mapmaker's visual method. An examination of plot ownership documents in this case finds that this clustering is reflective of kinship relations, that is, extended kin networks tend to join certified organic producer organizations as a group, and since their plots are often adjacent, this contributes to clustering. Certification practices also affect clustering. During interviews, producers noted that obtaining organic certification is easier if other, neighboring plots are also certified organic since producers do not have to worry about loss of organic certification due to chemical applications on neighboring plots.

Biodiversity and Certified Organic Agriculture
In a transnational social context, the maps provide plot locations required by transnational certification activities. Figure 2.5 shows "audit trail" monitoring connections and documentary instruments (to the right) of which these maps form an essential component. Producer plots form the first, and most rigorously inspected, point along the audit trail; the maps ensure that each

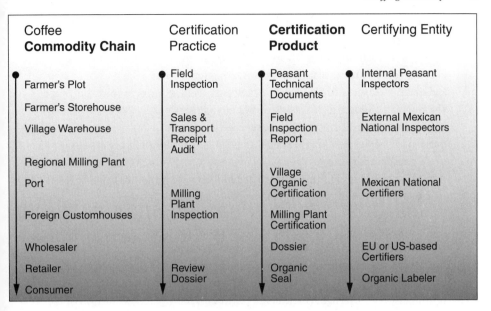

Coffee **Commodity Chain**	Certification Practice	**Certification Product**	Certifying Entity
Farmer's Plot	Field Inspection	Peasant Technical Documents	Internal Peasant Inspectors
Farmer's Storehouse			
Village Warehouse	Sales & Transport Receipt Audit	Field Inspection Report	External Mexican National Inspectors
Regional Milling Plant		Village Organic Certification	Mexican National Certifiers
Port	Milling Plant Inspection		
Foreign Customhouses		Milling Plant Certification	
Wholesaler		Dossier	EU or US-based Certifiers
Retailer	Review Dossier	Organic Seal	Organic Labeler
Consumer			

FIGURE 2.5. Processes of certification in organic coffee production

plot has a separate identity and is inspected on an annual basis by certifier-approved inspectors. Once certified organic, the coffee's "chain of custody" is traced by a document trail through certified organic spaces such as trucks, warehouses, and shipping containers to protect imparted qualities (e.g., biodiversity-enhancing production methods) from contamination by banned materials or mixing with noncertified goods. This certification trail serves as the basis for a payment system in which certified coffee receives a price well above market rates, thereby encouraging producers to convert to and continue biodiverse organic coffee production, which is more labor-intensive than nonorganic production: just to convert one hectare of traditional coffee to certified organic requires about 268 laborer-days (Bray, Sanchez, and Murphy 2002). Perhaps most important, higher prices make organic producers less likely to convert coffee plots to pasture, a worst-case scenario in terms of habitat conservation.

In return for higher prices, certified (or "active") organic coffee producers undertake horticultural and processing activities designed to protect the environment. These activities are based on norms established by international organic organizations such as the International Federation of Organic Agriculture Movements (IFOAM) and subsequently adapted to local social, agronomic, and ecological contexts by locally based certifying agencies. Organic production requirements of ecological import may be divided into conservation and biodiversity norms. (These norms, it may be added, reach beyond

coffee to cover all crops grown by a farm family, thereby adding to the beneficial effect.) Conservation norms eschew the use of environmentally destructive agrochemicals such as pesticides, herbicides, and fertilizers, ex-
cepting some natural compounds and biological controls such as beneficial
(predacious) insects or fungi that attack pest species. Agrochemicals disrupt
the normal biological processes of healthy ecosystems by poisoning (pesticides), disrupting hormonal processes (herbicides), or favoring the growth of
particular species (fertilizers; see Tilman 1982). Conservation norms also promote good practices to protect air, water, and soil quality. For coffee production these practices include converting coffee waste pulp into organic fertilizer, using catchment basins for coffee-processing wastewater, terracing, and
hand clearing of brush. These norms eliminate degrading activities such as
dumping coffee pulp into waterways—a common practice that suffocates
aquatic life as rotting pulp depletes oxygen levels.

Biodiversity-oriented norms constitute a recent expansion in the notion
of what should be considered organic. Within the transnational organic community, North American certifiers in particular have placed attention on developing biodiversity standards—for example, preservation and encourage-
ment of trademarked "bird-friendly" shade coffee (Rice and Ward 1996).
Traditionally cultivated Mexican coffee plots have been relatively biodiverse,
particularly in Oaxaca. Plots are farmed in the manner of "ranchos," in which
coffee forms one component of a complex gardenlike production structure
that integrates dozens of plants with economic, medicinal, or biological
value (Beaucage 1997; Sosa Maldonado, Escamilla Prado, and Díaz Cárdenas
1999). This cultivation system is complex in terms of diversity and structure,
with multilevel canopy and terrain modifications creating an abundance of
niches with varied microclimates (Nestel 1995). "Coffee gardens" are considered traditional cultivation systems, but it is important to realize, following
Padoch, Harwell, and Susanto (1998), that indigenous agroforestry systems
constantly evolve in response to changing market conditions, cultural values,
household subsistence needs, and labor availability, all the while conserving
and often enhancing biological diversity.

Strong evidence that Mexican coffee agroforest gardens conserve biodiversity has been forthcoming in recent studies. Surveys have documented
the role of coffee in supporting the overwintering of many bird species (Donald 2004), but perhaps more important, research shows that coffee plots,
playing the role of a high-quality matrix linking forest habitat patches, increase landscape connectivity thereby enhancing biodiversity at the landscape scale (see the appendix). Biological censuses in the El Triunfo Biosphere
Reserve show that coffee plots increase bird diversity, particularly in the case
of generalist species. However, the authors caution that replacement of habitat patches by shade coffee would reduce the diversity of forest specialist spe-

cies (Tejeda-Cruz and Sutherland 2004). Frog studies show a similar result, with frog biodiversity increased when coffee plots form a matrix between forest habitat patches (Pineda and Halffter 2004). And in a bee study, increasing coffee land-use intensity resulted in an increase in the diversity of solitary bee species, although overall bee diversity remained static due to a decline in social species (Klein et al. 2002).

Biodiversity norms respect these traditional systems and use them as models for enhancing the biodiversity of "conventional" agrochemical-reliant coffee plots and/or plots in which the original shade canopy has been significantly modified or eliminated. For example, Moguel and Toledo (1999) note that for some areas of Southern Mexico, notably Chiapas, 1980s-era Mexican government programs encouraged the introduction of "sun-tolerant" or so-called improved, modern coffee varieties. This led to the removal of fauna-friendly shade trees and the widespread replacement of native shade trees by Cuajiniquil (*Inga* spp.), a leguminous tree species that casts an even, thin shade and fertilizes coffee through the capture of nitrogen via root nodules. Sun coffee systems are even more widespread in other Latin American countries such as Colombia and Costa Rica, where, like Chiapas, coffee production has been dominated by well-capitalized, larger farms of over twenty hectares and a central focus of environmentally destructive agricultural modernization policies. Sun conversion has had less purchase in Oaxaca, although as in Chiapas, 1980s programs encouraged Cuajiniquil introductions.

Organic coffee requires a biodiverse shade canopy of local tree species. Where absent, this is accomplished by forming seedling nurseries of native species and progressively introducing them into coffee plots according to a production schedule agreed upon by producer and certifying agency. The identification of native species is accomplished in two ways. First, researchers have conducted field surveys of biodiversity within coffee plots and ecological studies of intraplot relations (Rice 2003): these studies serve to inform certifier practice with respect to diversity through the elaboration of specific biodiversity norms such as the Smithsonian Migratory Bird Center's Bird-Friendly norms and certification. Second, village inspectors conduct local vegetation studies in collaboration with NGOs such as Servicios Ambientales Oaxaqueños (SAO). Plot surveys are augmented by discussions with community elders to learn of tree species, whether previously common or not, that have been excluded and are in need of rescue from local extinction to establish a baseline. These studies then form the basis for determining which trees are to be reincorporated into coffee plots.

Via conservation practices that bridge local and transnational scales, social networks may form a basis for environmental conservation networks. However, as noted, tensions between conservation and social networks lead to unintended consequences. These "conservation conundrums" may be assessed through an examination of the certification practices—inspections, producer training, document production, and village organizational management—required by certified organic agriculture. Several points of tension are evident. First, a costly and regimented certification structure makes it difficult for the poorest farmers to participate in organic production (Michelsen 2001; Rice 2001). Second, certification does not easily incorporate the biodiversity knowledge of indigenous producers (Beaucage 1997; Moguel and Toledo 1999; Perfecto et al. 1996; Perez-Grovas Garza 1998). In response, sustainable agriculture organizations strive to devise a certified organic agriculture that meets goals of consumer legibility and producer manageability (Certimex 2000; Rice 2001; Gómez Tovar, Gómez Cruz, and Schwentesius Rindermann 1999; Naturland 1999; Rice and McLean 1999; Waridel and Teitelbaum 1999) and yet also makes a significant contribution to biodiversity conservation.

Limits to Certification as Productive Conservation?

As the above discussion indicates, a promising trend toward expanded, producer-driven biodiversity conservation engenders profound tensions between social and ecological aspects of networks. Subsequent sections detail several key tensions. First, social networks are quite expensive to manage, and current organic or fair trade price premiums are not attractive enough to substantially extend an organic agriculture-based conservation network. Second, agronomic imperatives associated with the production of high-quality coffee limit zones where coffee farming may be profitably pursued. Third, the transnational, multisited context of biodiversity norms production lends itself to, at best, confusion in the application of norms, and at worst, charges of "ecological neocolonialism" in negotiations over the local application of norms.

Economics of Network Conservation

Two economic factors limit the returns to organic/fair trade production and hence its ability to incorporate new producers within a conservation network. The first factor is price: at current levels, which at this writing have not seen a significant increase in a decade and hence have declined in real value, the prices are not high enough to entice new producers to join organic production. Although organic/fair trade coffee receives approximately US$1.20/lb upon export, producers pay shipping, certification, and administrative costs that reduce their on-farm payment to US$0.80/lb. This price falls in the middle of a recent Mexican "cost band" analysis (Technoserve 2004), which means that coffee producers earn only enough to cover production costs.

A second economic issue revolves around high local administrative labor outlays. For villages in this study, leaders of producer organizations spent up to sixty days per year on activities such as inspections and document management, while peasant inspectors and community technical officers, often the same persons acting in a double capacity, spent forty days per year (Mutersbaugh 2002). Since this labor is often performed as unpaid *cargo* service labor, a form of communal service work common in Southern Mexican Indian villages (Mutersbaugh 2002; Cohen 2001), the effect of organic farming is to compensate farmers but not regional officers who coordinate certified production. At a village level, the combination of high costs and high administrative overhead means that only relatively wealthy villages, or those possessed of strong indigenous governance mechanisms, are able to easily implement certified organic agriculture (Mutersbaugh 2002).

At a household level, recalling table 2.1 above, we note that significant inequality exists in producer villages. Poorer producers find the significant outlays to pay for certifications burdensome while they likewise receive relatively less benefit. Migration also plays an important role. Families that are labor-poor for demographic reasons such as age structure (Chayanov [1925] 1986) or migration also find it difficult undertake the time-consuming cultivation and terrain modification labor (e.g., terracing) required by certified organic agriculture. Migration forms a principal economic alternative for many households and has increased rapidly in the wake of the decade-long coffee crisis, producing effects ranging from increased labor costs (in the Mixteca Baja region of Southwest Oaxaca) to a striking increase in the number of women organic coffee producers. As a consequence of certification, owners must be present in the plots to be certified, and because of this many migrating men have transferred land titles to women. However, insofar as certification favors literate producers—often those individuals most likely to migrate—families in which literate household members migrate may be at a disadvantage (Mutersbaugh 2002; Cohen 2001).

Together, these economic dynamics have the effect of limiting potential contributions of certified organic farmers to network conservation. Low prices, high (unremunerated) administrative workloads, and migration all contribute to constrain the expansion of organic-coffee-based conservation networks. Without a strong economic basis, it becomes difficult to undertake intensified conservation activities, although it must be said that many producers engage in these practices and, indeed, continue with heavy administrative workloads for year after year. This uncompensated contribution is explained by the strong commitment that organic producers have toward conservation, a sentiment I have heard expressed in innumerable interviews and participant observation activities.

Agronomic Imperatives: Coffee versus Conservation
In assessing spatial aspects of organic producer maps (figs. 2.2, 2.3, and 2.4), we noted that coffee is part of an agricultural landscape. Plots are arranged

around a village center, with foot trails leading to north-facing slopes and moist valleys most suitable for coffee production. Households utilize a relatively small portion of the available habitat for coffee production: plots too far from the household require too much transit time; plots in suboptimal agricultural zones do not yield a marketable product. Constraints due to distance—coffee plots are often located up to an hour and a half's walk from the home—are readily appreciated. Marketability, however, merits additional discussion. Given the low prices paid to organic/fair trade producers, it is no longer sufficient to produce organic and fair trade certified coffees: they must also be of premium, gourmet quality to receive higher prices and ensure market access. This quality imperative restricts coffee to between 1,000 and 1,700 meters (for the best natural sweetness and acidity) and humid, fertile, and well-shaded zones (to increase bean weight and yield), and has the effect of further restricting lands available for integration into conservation networks.

Another factor to be considered is the tension between yield-increasing agronomic norms and biodiversity norms. Many practices that increase yields, even when organic, are not necessarily optimal from a biodiversity perspective. For example, a relatively thin forest canopy is better for coffee production than heavy shade, hence the recommendation for planting of Cuajiniquil mentioned above, yet does not provide adequate nesting sites for birds. Other practices include removal of epiphytes from coffee trees and fertilization with organic compost, neither of which promotes biodiversity. In effect, many agronomic norms reduce coffee plot biodiversity (and hence matrix quality), yet a strict adherence to biodiversity enhancement would reduce organic coffee incomes and limit the economic purchase of market-based network conservation.

Certifying Biodiversity? Transnational Norms versus Local Knowledge
In the event that the economic and social issues constraining certified agriculture are resolved, there remains the problem of how knowledge about appropriate environmental practices is created, codified, and communicated. Organic agriculture technicians often lack sufficient information of either "scientific" or "indigenous" character about coffee agroforest systems (Moguel and Toledo 1999), forest conservation (Laurance and Bierregaard 1997), and variation in local environments. In addition, in ways examined in greater detail below, the practice of certification also places certain constraints on knowledge transfer. Together these dynamics contribute to conservation failures (Padoch, Harwell, and Susanto 1998; Bassett and Zuéli 2000). In principle, scientific and indigenous knowledge studies of coffee plot biodiversity and traditional production methods (Perfecto et al. 1996; Moguel and Toledo 1999; Beaucage 1997; Rice 2001; Greenberg, Bichier, and Sterling 1997)

should permit the emergence of a consensus view as to the best organic methods for biodiversity conservation. In practice, however, many factors complicate these efforts.

To begin, one must consider the problem of adapting conservation geography to organic coffee plots situated in varied environmental contexts. The organic agricultural movement has long favored a "process-oriented" approach that focuses on a set of practices defined as "organic," such as composting, nonuse of synthetic agrochemicals (pesticides, herbicides), and soil management. This approach rests on the assumption that if environmentally destructive practices are eliminated, the farm environment will return to a more "natural" state. However, as organic agriculture has reached increasingly into the global South, Northern organic norms have required fundamental modification, as much for social as for ecological reasons (Mutersbaugh 2002). And, at the same time, the process-based approach has been undercut by transnational product certification norms that prohibit producers from participating in organic norm evaluation (Michelsen 2001; Mutersbaugh 2004).

A "typological" approach in which biodiversity treatments are based upon landscape categorizations may provide an alternative approach. Moguel and Toledo's (1999) study, for example, categorizes Mexican organic coffee production into five different types: traditional rustic, traditional polyculture, commercial polyculture, shaded monoculture, and unshaded monoculture. The implication is that it may be possible to create management schemes that would differentially apply certified organic processes by coffee type, nudging each toward a more diverse and environmentally friendly form. This approach accepts that organic production is shaped by historical, cultural, and agrotechnical contingencies. A difficulty, however, is that actual diversity of coffee production types in Mexico is far greater than any categorization is able to represent; hence, categorization does not fully resolve the difficulties created by the regulatory separation of farmers from norms producers by transnational standards such as the European Union (EU) 2092/91 (Barrett, Browne, and Harris 2001), USDA National Organic Program, and International Organization for Standardization (ISO) guide 65 (ISO 2000). These norms require an organizational separation, and limited communication, between certifying agencies and farmers such that erstwhile allies in conservation find themselves divided. With respect to biodiversity, the knowledge-based practices of indigenous producers are assessed by organizations divorced from the meso-level context: complexity is missed and the legibility of indigenous practices is reduced (Padoch, Harwell, and Susanto 1998; Scott 1998, 11; Mutersbaugh 2004). In short, international standards regiment interorganizational links. In practice, organizations find it necessary, and possible, to enter into negotiations (Mutersbaugh 2004), yet the "everyday practice" of interorganizational communication is constrained. This sociospatial separation can lead to adverse consequences, as related by one organic inspector I interviewed:

> If one goes to [regional organization X] it appears strange, denuded. There aren't any trees in the coffee plots. They used to have trees, but then the [producer] organization received a letter from a [European] labeler asking that they build terraces to protect the slopes from erosion. Well, the producers felled their trees across the slope and staked them out. That's the easiest way to build terraces, but it gets rid of a lot of the trees. When the certifiers came to inspect the coffee plots again [the following year] . . . it was too late. (Organic inspector, personal communication, July 2000)

In light of this, I want to argue, following Padoch, Harwell, and Susanto (1998), that any best practices approach to biodiversity conservation through certification, particularly at a meso level, must involve a far more participatory and communicative structure. Organic agriculture arose as a social movement with farm-level practitioners as the major producers and protectors of environmental integrity (see Michelsen 2001; Rice 2001). When this realization is joined to an understanding that biodiversity is cultivated and created not just by individual households but on a much wider scale as a part of local governance (Bassett and Zuéli 2000; Neumann 1998; Freidberg 2001; Schroeder 1999) and cultural practice (Zimmerer 1996; Carney and Watts 1991), it becomes evident that grassroots organizations can and must play a key role in biodiversity conservation.

Conclusion

Certified organic agriculture has the potential to make a significant contribution to, and provide a general model for, dispersed conservation. As a component of a comprehensive conservation strategy, it is particularly applicable to places where the institutions of parks, buffer zones, and corridors may be problematic, and it would permit resources to be channeled to persons or groups who protect and conserve biodiversity. By implementing a management scheme that protects the ecological complexity of the agrarian landscape, certified agriculture may promote biodiversity—even where the plots themselves do not contain endangered endemics—by increasing landscape connectivity. In order to achieve this goal, however, certified agriculture must overcome a number of hurdles. First, at present, organizational costs are supported by charging affiliated producers a share of their "organic premium" and relying on local governance to support information- and knowledge-intensive certification activities. This funding model, however, is inadequate. A twofold solution would be to (a) provide training grants and (b) improve the economy of certification by using the administrative labor resulting from certified organic/fair trade activities to manage additional paid conservation activities. Training grants would support the very significant costs of training village peasant inspectors and the activities of producer organization leaders. The use of organic managerial talent to manage additional conservation services would result in an increased economy of scale. Additional activities would include (a) other agricultural and animal husbandry activities and (b) conservation activities such as carbon sequestration, biodiversity conservation, and water capture in a more robust conservation package. These services could be paid from a number of sources, including governments, international conservation organizations, and NGOs, in order to complement fair trade and organic market income. In this context, organic coffee can form only one component of a broader strategy.

Second, conservation in general, and certified agriculture in particular, needs to hold to the participatory stance long championed by sustainable agriculture. Neither international organizations nor national and regional governments possess the financial and managerial resources necessary to save fragmented and dispersed biotic resources. By drawing on the strengths of organic product producer social movements—and recognizing that biodiversity conservation is a goal shared by peoples the world over—it may be possible to build and sustain the managerial capacity that is large enough, with sufficient geographic distribution, to implement effective conservation.

Third, the problem of identifying key zones of biodiversity with communities, and working toward a conservation network that maximizes the landscape connectivity of certified organic parcels, must be addressed. The geography of certified plots raises questions about how, given its social and economic character, certified organic agriculture may best be adapted to the project of biodiversity conservation. Given recent research on landscape connectivity, and on the important role played by coffee plots as a high-quality matrix that both separates and connects forest habitat patches, the role of farmers in maintaining appropriate matrix quality that maximizes biodiversity at the landscape level is crucial.

The possibility of producer-organized conservation, expressed in the mapmaking and map-distribution process detailed above, speaks to the ways that groups produce and codify local knowledge about place, and also to the difficulties of translating local knowledge into a form intelligible to transnational certifiers (Mutersbaugh 2004). Can map making also be used to identify and address issues of biodiversity conservation? The aim would be to encourage producers to survey village lands for biological hot spots and threatened zones, and more generally to identify processes of environmental degradation such as forest-to-pasture conversion. This done, it may by possible to codify biodiversity information on maps so as to enable a transnational "conservation conversation" based upon both a profound knowledge of local geography and increasingly sophisticated satellite imagery. This exchange could facilitate the extension of sustainable agriculture to areas that are threatened or that have exceptionally high ecological diversity, identify plots to serve as matrix templates (rather than depending overmuch on plot diversity norms generated by certifier agencies), and bring questions of landscape connectivity to the fore.

Pulling these threads together, I would argue that certified agriculture, as a model of network conservation, makes a unique and essential contribution to biodiversity conservation at a "meta level," above the individual peasant plot, yet connected into a broader conservation geography including conservation territories. Conservation is a complex project that requires the skilled labor of indigenous peoples and conservation scientists: it is more than households or villages are capable of, yet also beyond the managerial capacity of national governments debilitated by structural adjustment. The experience of CEPCO and other Mexican coffee producer unions demonstrates the feasibility of an approach based upon certified agriculture, yet this meta-level organization is not effectively financed under current organic market schemes. This research, then, would suggest that conservation initiatives expand incentives to support community-driven biodiversity conservation.

Appendix: Conservation Implications of Landscape Connectivity and Matrix Quality

Conservation geography's concern with animal and plant survival has long been shaped by an appreciation of landscape patterns. Following Levins (1969), animal and plant species may be viewed as metapopulations, or "a 'population of populations' occupying some percentage of suitable habitat patches" (Vandermeer and Carvajal 2001). In other words, any particular plant or animal experiences a landscape, such as a Oaxacan coffee-producing zone, as divided into habitat patches (places that provide sufficient resources for it to live and reproduce) and matrix (areas that do not provide sufficient resources). Each habitat patch has its own "local" population, and the metapopulation constitutes the union of local populations. Following upon the findings of island biogeography, it is understood that local "extinctions" may, and often do, occur at any given patch, but that the metapopulation is sustained because individuals from other, nearby patches repopulate vacant patches. Species conservation, then, depends—in a straightforward formulation—upon large, high-quality habitat patches, high patch density, and a high-quality intervening matrix, each of which enhances the ability of the plant or animal to reach and repopulate vacant patches.

The contemporary problem confronted by conservation geography is that, driven principally by expansions in human settlement, agriculture, and industry, landscapes across the globe are characterized by a simultaneous decline in habitat patch size and a growing distance between patches. This declining landscape connectivity—that is to say, the relatively greater distances between and size of remaining habitat patches—adversely effects the reproduction of metapopulations.

The goal of network conservation, then, is to remedy declining landscape connectivity and thus contribute to species conservation. This may be accomplished in two ways. First, habitat patches may be increased in size and number, thereby increasing both the reproductive space and the likelihood that vacant fragments will be recolonized. In particular, species that are intrinsically rare due to specialization have been found to become scarcer as fragmentation increases and connectivity decreases (Bonte et al. 2004). Second, the matrix between habitat patches may be increased in quality to facilitate movement between patches.

However, to bring in a bit more complexity, both theoretical and empirical studies indicate that too great an increase in matrix quality can lead to unstable population dynamics and diversity reduction. Theoretical models indicate that a very high-quality matrix facilitates interpatch movement to the point where population dynamics may become unstable (Vandermeer and Carvajal 2001). This interesting result finds support in empirical studies of frog populations—undertaken in Mexican coffee plots—indicating that the (coffee plot) matrix between forest habitat patches *increases* frog species diversity because landscape-level "gamma" diversity is determined to a greater extent by interpatch spatial turnover ("beta" diversity related to movement between patches) than by "alpha" intrapatch diversity (Pineda and Halffter 2004). Plants in particular seem to eschew a model in which the matrix and habitat components are represented as a simple binary (Murphy and Lovett-Doust 2004).

As an initiative within the context of network conservation, certified organic coffee production attempts to improve the ecological quality of coffee plots. The research cited above indicates that both theoretical and empirical work supports the idea that properly managed coffee agroforest systems can increase biodiversity by providing a matrix of appropriate quality to support habitat patches and increase landscape-level diversity. Indeed, a wide array of scientific studies has begun to highlight the importance of landscape connectivity and fragmentation in tropical forests (Castro and Fernandez 2004; Vasconcelos and Luizao 2004; Estrada et al. 1998). Although I review recent work on coffee agroforest systems, this article focuses principally upon ways in which conservation geography is as much social as ecological, because its success requires coffee farmers to engage in conservation activities. Additional theoretical and empirical bases for farmer-based conservation are found in Zimmerer (1996, 1999) and Padoch, Harwell, and Susanto (1998).

References

Barrett, H. R., A. W. Browne, and P. J. C. Harris. 2001. Smallholder farmers and organic certification: Accessing the EU market from the developing world. *Biological Agriculture and Horticulture* 19 (2): 183–99.

Bassett, T. J., and K. B. Zuéli. 2000. Environmental discourses and the Ivorian Savanna. *Annals of the Association of American Geographers* 90 (1): 67–95.

Beaucage, P. 1997. Integrating innovation: The traditional Nahua coffee-orchard (Sierra Norte de Puebla, Mexico). *Journal of Ethnobiology* 17 (1): 45–67.

Bonte, D., L. Baert, L. Lens, and J. P. Maelfait. 2004. Effects of aerial dispersal, habitat specialisation, and landscape structure on spider distribution across fragmented grey dunes. *Ecography* 27 (3): 343–49.

Bray, D. B., J. L. P. Sanchez, and E. C. Murphy. 2002. Social dimensions of organic coffee production in Mexico: Lessons for eco-labeling initiatives. *Society and Natural Resource* 15 (5): 429–46.

Carney, J., and M. Watts. 1991. Disciplining women? Rice, mechanization, and the evolution of Mandinka gender relations in Senegambia. *Signs: Journal of Women in Culture and Society* 16 (4): 651–81.

Castro, E. B. V. de, and F. A. S. Fernandez. 2004. Determinants of differential extinction vulnerabilities of small mammals in Atlantic forest fragments in Brazil. *Biological Conservation* 119 (1): 73–80.

CEPCO (Coordinadora Estatal de Productores de Cafe Oaxaqueños). 1996. *Annual report 1996.* Oaxaca: CEPCO.

Certimex. 2000. *Normas para la produccion y procesamientos de productos ecologicos.* Oaxaca, Mexico: Certimex.

Chayanov, A. V. [1925] 1986. *The theory of peasant economy.* Madison: University of Wisconsin Press.

Cohen, J. 2001. Transnational migration in rural Oaxaca, Mexico: Dependency, development and the household. *American Anthropologist* 103 (4): 954–67.

Donald, P. F. 2004. Biodiversity impacts of some agricultural commodity production systems. *Conservation Biology* 18 (1): 17–37.

Estrada, A., R. Coates-Estrada, A. A. Dadda, and P. Cammarano. 1998. Dung and carrion beetles in tropical rain forest fragments and agricultural habitats at Los Tuxtlas, Mexico. *Journal of Tropical Ecology* 14:577–93.

Freidberg, S. 2001. Gardening on the edge: The social conditions of unsustainability on an African urban periphery. *Annals of the Association of American Geographers* 91 (2): 349–69.

Gómez Tovar, L., M. A. Gómez Cruz, and R. Schwentesius Rindermann. 1999. *Desafíos de la agricultura organica: Comercializacion y certificacion.* Mexico City: Grupo Mundi-Prensa.

Greenberg, R., P. Bichier, and J. Sterling. 1997. Bird populations in rustic and planted shade coffee plantations of eastern Chiapas, Mexico. *Biotropica* 29:501–14.

Hernandez Castillo, R. A., and R. Nigh. 1998. Global processes and local identity among Mayan coffee growers in Chiapas, Mexico. *American Anthropologist* 100:136–47.

Holling, C. S., and G. K. Meffe. 1996. Command and control and the pathology of natural resource management. *Conservation Biology* 10 (2): 328–37.

Inter-Agency Technical Committee (Economic Commission for Latin America and the Caribbean [ECLAC], United Nations Environment Programme [UNEP], United Nations Development Programme [UNDP], Inter-American Development Bank [IADB], and World Bank [WB]). 1999. Environmental strategies for sustainable development in Latin America and the Caribbean: 1999 regional action plan for the period 2000–2001. In *Proceedings of the Eleventh meeting of the Forum of Ministers of the Environment of Latin America and the Caribbean.* 10–13 March, Lima, Peru.

ISO (International Organization for Standardization). 2000. WTO, ISO, and world trade: The agreement on technical barriers to trade (TBT). Available at http://www.iso.ch/wtotbt/wtotbt.htm.

Klein, A. M., I. Steffan-Dewenter, D. Buchori, and T. Tscharntke. 2002. Effects of land-use intensity in tropical agroforestry systems on coffee flower-visiting and trap-nesting bees and wasps. *Conservation Biology* 16 (4): 1003–14.

Laurance, W. F., and R. O. Bierregaard. 1997. A crisis in the making. In *Tropical forest remnants,* ed. W. F. Laurance and R. O. Bierregaard, Jr., xi–xv. Chicago: University of Chicago Press.

Levins, R. 1969. Some demographic and genetic consequences of environmental heterogeneity for biological control. *Bulletin of the Entomological Society of America* 15:237–40.

Michelsen, J. 2001. Organic farming in a regulatory perspective. *Sociologia Ruralis* 41 (1): 65–84.

Moguel, P., and V. M. Toledo. 1999. Biodiversity conservation in traditional coffee systems of Mexico. *Conservation Biology* 13:11–21.

Murphy, H. T., and J. Lovett-Doust. 2004. Context and connectivity in plant metapopulations and landscape mosaics: Does the matrix matter? *Oikos* 105 (1): 3–14.

Mutersbaugh, T. 2002. "The number is the beast": Certified organic coffee and certification labor in a Mexican village. *Environment and Planning A* 34:1165–84.

———. 2004. Serve and certify: Paradoxes of service work in organic-coffee certification. *Environment and Planning D* 22 (4): 533.

Naturland. 1999. Desarrollo de sistemas de control internos en organizaciones de productores (organizaciones de pequeños agricultores). UC/12/95 Naturland-Verband e.V., Kleinhaderner Weg. 1. D-82166. Gräfelfing, Germany: Naturland.

Nestel, D. 1995. Coffee in Mexico: International market, agricultural landscape and ecology. *Ecological Economics* 15:165–79.

Neumann, R. 1998. Imposing wilderness: Struggles over livelihood and nature preservation in Africa. Berkeley: University of California Press.

Orlove, B. S., and S. B. Brush. 1996. Anthropology and the conservation of biodiversity. *Annual Review of Anthropology* 25:329–52.

Padoch, C., E. Harwell, and A. Susanto. 1998. Swidden, sawah, and in-between: Agricultural transformation in Borneo. *Human Ecology* 26 (1): 3–20.

Perez-Grovas Garza, V. A. 1998. Evaluacion de al sustentabilidad del sistema de produccion de café organico en la Union de Ejidos Majomut en la Region de Los Altos de Chiapas. Master's thesis, Universidad Autonoma Chapingo, Mexico.

Perfecto, I., R. A. Rice, R. Greenberg, and M. E. VanderVoort. 1996. Shade coffee: A disappearing refuge for biodiversity. *BioScience* 46:598–608.

Pineda, E., and G. Halffter. 2004. Species diversity and habitat fragmentation: Frogs in a tropical montane landscape in Mexico. *Biological Conservation* 117 (5): 499–508.

Rice, P. D., and J. McLean. 1999. *Sustainable coffee at the crossroads.* Washington, DC: Consumer's Choice Council.

Rice, R. A. 2001. Noble goals and challenging terrain: Organic and fair trade coffee movements in the global marketplace. *Journal of Agricultural and Environmental Ethics* 14:39–66.

———. 2003. *Manual de café bajo sombra* [Manual for shade-grown coffee]. Washington, DC: Smithsonian Migratory Bird Center, National Zoological Park.

Rice, R. A., and J. R. Ward. 1996. *Coffee, conservation, and commerce in the Western Hemisphere.* Washington, DC: Smithsonian Migratory Bird Center.

Schroeder, R. 1999. *Shady practices: Agroforestry and gender politics in the Gambia.* Berkeley: University of California Press.

Scott, J. C. 1998. *Seeing like a state: How certain schemes to improve the human condition have failed.* New Haven, CT: Yale University Press.

Smithsonian Migratory Bird Center (SMBC). 2001. http://nationalzoo.si.edu/ConservationAndScience/MigratoryBirds/Coffee/Bird_Friendly/.

Sosa Maldonado, L., E. Escamilla Prado, and S. Díaz Cárdenas. 1999. Café orgánico: Producción y certificación en México. *El Jarocho Verde* 11:13–25.

Technoserve. 2004. *Business solutions to the coffee crisis.* Norwalk, CT: Technoserve. Available at http://www.technoserve.org/TNSCoffeeReport_Master.pdf.

Tejeda-Cruz, C., and W. J. Sutherland. 2004. Bird responses to shade coffee production. *Animal Conservation* 7:169–79.

Tilman, D. 1982. *Resource competition and community structure.* Princeton, NJ: Princeton University Press.

Vandermeer, J., and R. Carvajal. 2001. Metapopulation dynamics and the quality of the matrix. *American Naturalist* 158 (3): 211–20.

Vasconcelos, H. L., and F. J. Luizao. 2004. Litter production and litter nutrient concentrations in a fragmented Amazonian landscape. *Ecological Applications* 14 (3): 884–92.

Waridel, L., and S. Teitelbaum. 1999. *Fair trade.* Montreal: EquiTerre.

Whatmore, S., and L. Thorne. 1997. Nourishing networks. In *Globalising food: Agrarian questions and global restructuring,* ed. D. Goodman and M. Watts, 287–304. New York: Routledge.

Willer, H., and M. Yussefi. 2004. *Organic agriculture worldwide.* Bonn, Germany: Stiftung Ökologie & Landbau.

World Bank. 2001. Mesoamerican Biological Corridors project. Available at http://www.worldbank.org/html/extdr/newprojects/lc2001133.htm.

Zimmerer, K. 1996. *Changing fortunes: Biodiversity and peasant livelihood in the Peruvian Andes.* Berkeley: University of California Press.

———. 1999. Overlapping patchworks of mountain agriculture in Peru and Bolivia: Toward a regional-global landscape model. *Human Ecology* 27 (1): 135–65.

3 Satellite Remote Sensing for Management and Monitoring of Certified Forestry: An Example from the Brazilian Amazon

JANE M. READ

Rapid and extensive deforestation, particularly in the tropics, has led to widespread concerns about sustainability of the world's forests (Myers 1993; Whitmore 1997).[1] Selective logging, while less overtly destructive than clearcutting, results in the degradation of large areas of forest by making forests vulnerable to fire and wind damage (Nepstad et al. 1999; Gerwing 2002). Conventional selective logging in the Brazilian Amazon alone has been estimated to degrade up to 15,000 km^2 per year of standing forests, an area almost equivalent (to up to 90 percent) of the area that was clear-cut in 1996 (Nepstad et al. 1999). While many tropical forests are currently protected in traditional conservation units, such as parks and reserves, extensive areas are also being protected through alternative conservation territories or networks (Zimmerer 2000). Among others, these include sustainable management strategies, such as extractive reserves, as well those with less obvious spatial manifestations, such as certification programs and conservation networks (e.g., see Mutersbaugh, chap. 2 in this volume, and Brown, chap. 4 in this volume). This chapter addresses one such alternative conservation strategy, sustainable forestry, and examines the use of remote sensing as a potential tool to aid in its implementation and certification in the Brazilian Amazon.

International environmental organizations and the forestry community have recognized the need for improved methods of forest management, resulting in a shift away from sustainable yield to sustainable forest management models (Gale 1998; Bergen, Colwell, and Sapio 2000). Forest certification is promoted as a tool to help reduce the amount of damage from logging operations and promote sustainable logging. Certified forestry can be thought of as creating a type of productive conservation territory that is based on

1. *Deforestation* here refers both to clear-cutting and cryptic deforestation (forest degradation without clear-cutting).

sound environmental management, and that has been made viable as a result of international, national, and local environmental pressures. While forest certification is often promoted at international and national levels, and policies and markets are global and national in nature, it is at the local level that implementation of certification strategies is most challenging.

With the emphasis on sustainable forest management practices, the needs of the forestry community at the local level have changed. Tools for rapid site assessments and preharvesting inventories are generally lacking. An integral part of forest certification systems are reduced-impact logging (RIL) methods, which aim to minimize logging impacts to soils and standing vegetation (Sist 2000). Such methods have been found to result in substantial reductions in damage to standing trees (Bertault and Sist 1997; Pereira et al. 2001). Reduced-impact logging strategies highlight the need for increased understandings of the type, extent, and condition of forests, as well as improved knowledge of fine-grained ecological characteristics and processes. This is particularly true for forestry activities in humid tropical forests with high species diversity and complex forest structure. A better understanding of how humid tropical forests and tree species function under selective logging conditions is crucial to the development of improved management practices, such as RIL strategies. In addition, logging certification programs need tools for assessing and monitoring certified operations that can confer the necessary credibility in international and national marketplaces (Scholz 1999).

Spatial information technologies, including geographic information systems (GIS), global positioning systems (GPS), and remote sensing, can play an important role in information gathering and implementation of certified systems. In light of recent sensor developments, satellite remote sensing, which up until now has not been employed routinely in forestry applications, has potential for addressing site-level and global issues of tropical forest management and conservation. Aerial photography, traditionally used in temperate-zone forestry, is in many ways inferior to satellite imagery: interpretation is subjective and labor intensive (Franklin 2001), and it is not possible to acquire data simultaneously over large areas. This chapter explores and discusses the application of multispectral satellite remote sensing to certified forestry in the Brazilian Amazon. A brief introduction to forest certification is followed by a discussion of remote sensing and its potential as a tool to aid in the implementation of forest certification.

Forest Certification

Forest certification has been gaining ground since the late 1980s (Vogt et al. 1999). Environmental nongovernmental organizations (NGOs) and concerned forest product producers and retailers were important drivers of certification (Bass et al. 2001). In addition, two environmental events—the 1992 United Nations Conference on Environment and Development, held in Rio de Janeiro, and the 1997 Kyoto Protocol on Climate Change—fueled international recognition of the need for information about the world's forests and improved methods of forest management, and led to increased interest in certification as a tool to address public concerns about deforestation (Fanzeres and Vogt 1999; Franklin 2001). The purpose of certification is to publicly recognize products derived from forests that have been managed according to a set of principles and criteria that are considered environmentally, socially, and economically appropriate (Fanzeres and Vogt 1999; FSC 2004).

Two key players in international forestry certification are the International Organization for Standardization (ISO) and the Forest Stewardship Council (FSC). The ISO approaches certification from an environmental management systems perspective, developing standards for such systems, whereas the FSC takes a performance-based approach by defining performance standards to be attained by forestry operations (Schlaepfer and Elliott 2000). Of the several certification schemes that exist worldwide (see Bass et al. 2001), the FSC has been the most widely recognized and adopted internationally (Fanzeres and Vogt 1999).

The FSC, founded in 1993, is an international NGO based in Oaxaca, Mexico. Using ten principles and criteria of forest management that focus on the local context of environmental, social, and economic values (FSC 2004), the FSC accredits certification bodies to carry out certifications. There are over 29 million hectares of FSC-certified forests worldwide (UNEP-WCMC et al. 2004). Forest Stewardship Council certification has been most widely embraced in developed nations, with Europe and North America accounting for 81 percent of the total global area (fig. 3.1; FSC 2004). Certification of developing nations' forests has not received as much attention (only 12 percent lie in Latin America), despite the fact that much of the impetus for certification is grounded in concerns about slowing tropical deforestation (Fanzeres and Vogt 1999).

While the potential advantages of certification to producers and retailers are clear (i.e., sustainable production, expanding markets, improved reputation and credibility), there are major barriers to certification that include the up-front costs of certification, lack of access to markets, lack of policies to promote certified forestry, and other social, cultural, and technological barriers (Bass et al. 2001). At the site level, the primary areas where forest management enterprises have difficulties in meeting certification criteria vary between developed and developing nations (Thornber 1999). In developing countries these include management, monitoring, and social issues, whereas enterprises in developed nations tend to have difficulties meeting requirements relating to environmental performance (Thornber 1999).

For moist tropical forests the lack of quantitative data available on

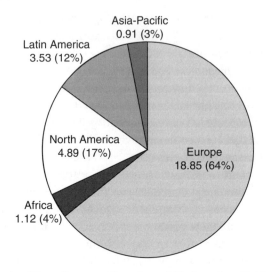

FIGURE 3.1. Area in millions of hectares of Forest Stewardship Council certified forests by region
Source: UNEP-WCMC et al. 2004; figures from 10 July 2002.

impacts of logging operations on forest fauna, flora and the forest as a whole (Pro-Natura, IIED, and GTZ 2000) contributes to difficulties with certification, which are further exacerbated by the lack of understanding of individual species and forest dynamics due to the diversity and complexity of tropical forest systems (but see, e.g., Bertault and Sist 1997; Barreto et al. 1998; Gerwing 2002; Slik, Verburg, and Kebler 2002; and Wilson and Sader 2002 for recent studies on effects of logging on tropical forests). Certification practices, including RIL techniques, are based on our "best" or current understandings of tropical forest structure and function, and further basic research is needed to improve on these practices if certification is to be widely accepted as a conservation strategy. For example, a basic requirement for determining appropriate species-specific extraction volumes and harvest rotation intervals, regeneration time, is unknown for many potentially harvestable species (Asbjornsen and Alatorre 1999).

Forest Certification in Brazil

Development of certification standards in the forest sector in Brazil began in the early 1990s with two initiatives, both of which are aimed at national and international markets. The first, a national ecolabeling scheme, CERFLOR, developed under the auspices of the national standards body (ABNT) in association with the Brazilian Society for Silviculture (Sociedad Brasileira de Silvicultura, or SBS), was planned to be finalized and operational in 2002 (SBS 2004). The second initiative, which began in 1993, is the FSC Brazilian National Initiative, which was officially founded and acknowledged by the FSC in September 2001 (FSC 2004). The FSC National Initiative is based on the FSC criteria and principles for good forest management (Ghazali and Simula 1998) and to date has developed criteria and principles for plantation forests and natural Amazonian upland forests (FSC 2004). Institutions accredited by the FSC that carry out certifications in Brazil are IMAFLORA, a Brazilian NGO that works in conjunction with the Rainforest Alliance's SmartWood program; Scientific Certification Systems (SCS); and Société Générale de Surveillance (SGS) Forestry QUALIFOR program (SGS 2004). The IMAFLORA/Rainforest Alliance certification accounts for the majority of certifications.

Although Brazil, with 1.05 million ha of FSC-certified forests, is the fourth-ranked country in the world by area of FSC-certified forests, and is one of the world's major producers and consumers of tropical logs and sawnwood, only a fraction (1 million m³ yr⁻¹) of Brazil's total timber production is produced as certified timber (UNEP-WCMC et al. 2004; FSC 2004).[2] Only seven of the twenty-three FSC-certified sites in Brazil are in natural (not plantation) upland Amazonian forests (table 3.1; FSC 2004). Companies with the largest areas of certified forests, such as Precious Woods Amazon, Gethal, and Cikel, were among the first to become certified; high costs, especially those needed up front for certification, may have restricted a number of small and medium-sized private and community enterprises from becoming certified (Macqueen et al. 2003; May, da Vinha, and Macqueen 2003). In a study of Brazilian timber extraction companies, it was found that medium-sized and larger enterprises (i.e., those annually producing >10,000 m³) are more concerned about sustainability than smaller companies (Sustainable Tree Crops Program 2001, cited in May, da Vinha, and Macqueen 2003, 19). In addition to high costs of certification and implementation, a number of factors work to the disadvantage of smaller companies: land tenure issues, competition from unmanaged extraction as well as nontropical timber, lack of training and technical assistance, lack of forest zoning, and difficulties in obtaining an approved management plan (May, da Vinha, and Macqueen 2003). At the same time, pressure from foreign investors concerned with reputation and maintenance of future export markets (e.g., Precious Woods Amazon from Switzerland and Gethal from Germany; Macqueen et al. 2003) is important in driving the larger companies to become certified (Viana et al. 2002).

While the larger certified enterprises export the majority of their timber abroad, most timber extracted from Amazonian forests (86 percent) goes to the internal Brazilian timber market (Viana et al. 2002). The formation of a Brazilian buyers' group for certified forest products in 2000 (May, da Vinha, and Macqueen 2003) has been instrumental in getting certified timber into

2. Brazil produced 126 million m³ and 37.1 million m³ of tropical logs and sawnwood, respectively (year 2000 figures: International Tropical Timber Organization 2001).

TABLE 3.1.   Certified natural forests in the Brazilian Amazon, June 2002

Owner	Location	Area (ha)
Precious Woods Amazon	Itacoatiara, AM	80,571
Gethal Amazonas	Manicoré, AM	40,862
Juruá Florestal Ltda.	Moju, PA	12,000
Cikel Brasil Verde S/A	Paragominas, PA	140,658
Maracaí Florestal & Industrial Ltda.	Sinop, MT	8,172
Lisboa Madeireira Ltda.	Icoaraci, PA	45,738
Associação dos Moradores e Produtores do Projeto Agroextrativista Chico Mendes	Xapuri, AC	900

Source: FSC 2004.

the national market (Viana et al. 2002). In fact, internal demand for certified natural timber from the Amazon (>3 million m³ per year) far exceeds current output (Macqueen et al. 2003), pointing to the need for increased production of certified timber in the natural Amazonian forests (Viana et al. 2002).

Mil Madeireira Itacoatiara Ltda., Amazonas, Brazil: An Example

Mil Madeireira Itacoatiara Ltda. is an FSC-certified RIL operation located approximately 140 km east of Manaus, Brazil (fig. 3.2). The vegetation at Mil Madeireira is classified as tropical moist forest (Holdridge 1947) and extends over terrain characterized by flat plateaus interrupted by steep slopes. Elevation ranges from 60 to 120 m above sea level. The area experiences a dry season from July to October and the wettest months from February through April, with mean annual rainfall of 2100 mm and mean temperature of 26 °C (RADAMBRASIL 1976). Eighty thousand hectares of moist tropical forest at the site owned and operated by Precious Woods Amazon (PWA) were certified by IMAFLORA/SmartWood in 1997, and represent the first FSC-certified forests in the Amazon region (Precious Woods 2004). In 2001 PWA purchased an additional 47,000 ha of adjacent forest to meet self-sufficiency requirements based on a revised cutting cycle of thirty to thirty-five years, instead of the twenty years previously thought to be adequate (J. Rogério, personal communication). Mil Madeireira makes up almost 45 percent by area of the total natural upland certified forests in Brazil (FSC 2004). Approximately 75 percent of the timber (mostly sawnwood) from Mil Madeireira is exported to Europe (Precious Woods 2004), where prices for certified timber can be 10–15 percent higher than normal (Scholz 1999).

Precious Woods Amazon employs various strategies to reduce the biophysical impacts of forestry impacts on soil and standing vegetation. These include the following: controlled cutting cycles of thirty to thirty-five years (i.e., 3,000–5,000 ha yr⁻¹); controlled timber volumes (15–20 m³ ha⁻¹); careful siting of extraction roads to reduce the total area in roads while at the same

FIGURE 3.2. Location of Mil Madeireira Itacoatiara Ltda.

time enabling maximum access to timber areas; avoidance of harvesting near streams and on steep terrain; directional felling; and ensuring that a proportion of seed trees is left standing (J. Rogério, personal communication; Precious Woods 2004). In addition, PWA encourages research at the site, which is vital for improving understanding of the impacts of low-impact logging over time on forest dynamics and sustainability, and for refining future management and policy practices.

Key to the management of the operation is the inventorying and mapping of physical site characteristics, and pre- and postlogging inventories of the fifty-five managed timber species (Precious Woods 2004). Precious Woods Amazon makes extensive use of GIS for operational and planning purposes. Field surveys locate all trees greater than 50 cm diameter at breast height, and the resulting inventories are combined in a GIS with other data layers of site characteristics. These include management sectors and blocks, buffer zones around streams where logging is prohibited, road types, and streams. Foresters make use of maps derived from the GIS to guide later field surveys. Landsat-7 ETM+ scenes of the area are used for remote assessment of the sites, as well as for display and mapping purposes (J. Rogério, personal communication).

While PWA already employs GIS and to some extent remote sensing in management of its forests, other potential uses of these spatial information technologies likely exist in areas of ongoing management of forests, pre- and postharvest inventories, and recertification requirements. Logging enterprises in developing nations tend to have difficulty in meeting certification criteria in two areas that could be addressed, to some extent, through remote sensing: environmental impact assessment (FSC Principle 6) and data collection for monitoring and assessment (FSC Principle 8, Thornber 1999). In addition to serving as tools to assist with forest inventories and postharvest monitoring, remote sensing and GIS are particularly suited to assist in areas related to the ecology of the forests and overall landscape management. These tools could help fill the need for more research on the ecological impacts of extraction activities (Viana et al. 2002) and criteria for sustainable management (Macqueen et al. 2003).

Remote Sensing in Forestry

Remotely sensed data have several advantages over ground-based data, including rapid data acquisition over large geographic areas, potential global coverage, unbiased data sets, and repeat coverage over short timescales. In forestry, remote sensing has proven to be a useful source of data for various site-mapping purposes, although it is unlikely to ever completely replace the need for ground-based data (Wynne and Carter 1997; Bergen, Colwell, and Sapio 2000). Despite the existence of alternative digital sources, aerial photography has traditionally been, and remains, the most common source of remotely sensed data in forestry (Bergen, Colwell, and Sapio 2000; Caylor 2000; Olson and Weber 2000; Franklin 2001). Franklin (2001) cites low cost and ease of use as the two main advantages of aerial photographs. The most affordable and available satellite data sources used in forestry are multispectral sensors that include the Advanced Very High Resolution Radiometer (AVHRR), Landsat Multispectral Scanner (MSS), Landsat Thematic Mapper (TM), and Système Pour l'Observation de la Terre (SPOT) sensors (Wynne and Carter 1997). Each of these sensors has different characteristics that affects its usefulness for specific tasks (table 3.2): the number of spectral bands and spatial resolution determine information content, while the orbit cycle determines the potential frequency of data collection (Lillesand, Kiefer, and Chipman 2004). A limitation in using these conventional satellite remotely sensed data types for operational site-specific mapping is their relatively coarse spatial resolution (Bergen, Colwell, and Sapio 2000; Roller 2000; Franklin 2001). Typical uses of aerial photography that are not routinely possible with intermediate-resolution satellite sensor data include determination of tree size classes and percent crown closure, species identification, and stereoscopic viewing (Caylor 2000).

Recent improvements in satellite remote sensing technologies, however,

TABLE 3.2.   Characteristics of conventional satellite remote sensors used in forestry

Sensor	Number of bands	Resolution	Orbit cycle
AVHRR	5 multispectral	1.1 km (at nadir)	4–9 days/833 or 870 km
Landsat-MSS	4 multispectral	80 m	16 days/705 km
Landsat-TM	6 multispectral	30 m	16 days/705 km
	1 thermal	120 m	
SPOT-5 (HRG)	3 multispectral	10 m	26 days/822 km
	1 mid-infrared	20 m	
	1 panchromatic	5 m	

Source: Lillesand, Kiefer, and Chipman 2004; SPOT 2004.
Note: AVHRR: advanced very high resolution radiometer; MSS: multispectral scanner; TM: thematic mapper; SPOT (HRG): Système Pour l'Observation de la Terre (high-resolution geometric).

offer increased capabilities and promise to provide forest managers with important information heretofore not readily available. These improvements include (a) sensors with improved "high" spatial resolution (Green 2000); (b) hyperspectral sensors, which measure many narrow wavelength bands and have been shown to be useful for determining species composition (Martin et al. 1998; Ustin and Trabucco 2000); (c) light detection and ranging (LIDAR) systems, which directly measure vertical forest structure (Dubayah and Drake 2000); (d) improved resolution synthetic aperture radar (SAR) systems, which can penetrate clouds and is useful for mapping a variety of forest characteristics, including forest structure and moisture (Dobson 2000); and (e) development of new, derived products from sensor data, such as vegetation productivity data computed every eight to sixteen days from the Moderate Resolution Imaging Spectroradiometer (MODIS) that was launched in 1999 (Running, Queen, and Thornton 2000).

Multispectral high-spatial-resolution satellite remote sensing is possibly the most promising of the new remote sensing technologies for forestry applications based on accessibility, global coverage, and its similarity to the more familiar Landsat-MSS and Landsat-TM sensors, which makes interpretation of these data more intuitive for scientists not specializing in remote sensing. Three new higher-resolution sensors were launched relatively recently, and more are planned (Wynne and Carter 1997; Lillesand, Kiefer, and Chipman 2004). Landsat-7, an intermediate-resolution sensor, was launched in 1999 with the Enhanced Thematic Mapper (ETM+) sensor on board. The data differ from Landsat-TM in that there is an added 15 m panchromatic band. The commercial IKONOS satellite was launched the same year and offers data with improved spatial resolution over Landsat-7, but with fewer spectral bands (Lillesand, Kiefer, and Chipman 2004). Digital Globe's commercial QuickBird sensor, launched in 2001, provides 2.44 m multispectral and 0.61 m panchromatic data (Digital Globe 2004).

Comparison of Landsat-ETM+ and IKONOS Data at Mil Madeireira

For use in certified forestry, there are advantages and drawbacks to remote sensors of different resolutions. To assess the merits of remote sensors in certified forestry, I compare visual observations of intermediate-resolution Landsat-7 ETM+ data (28.5 m multispectral and 15 m panchromatic) and higher-resolution IKONOS Carterra data (4-m multispectral and 1-m panchromatic) over an area of approximately 6,500 ha of recently logged and old-growth (unlogged) forest at Mil Madeireira. The Landsat-7 and IKONOS data were acquired one day apart in March 2001.

Landsat-7 ETM+ data consist of seven bands ranging from the visible to thermal wavelengths, and are recorded as 8-bit data (digital number range 0–255). IKONOS multispectral data consist of four bands (blue, green, red, and near-infrared) comparable to the four corresponding ETM+ bands, and are recorded as 11-bit data (digital number range 0–2047). The availability of finer-resolution panchromatic data in conjunction with the coarser multispectral bands with both these data sets allows for enhanced display and visual interpretation through the creation of pan-sharpened images (merging of the higher-resolution panchromatic with the multispectral data).

Using the ETM+ data (fig. 3.3b) one can distinguish major logging features, including the sawmill, major roads, and patios. Some topographic features, such as river courses and higher ground, can be detected visually. However, little difference between the appearance of the canopy matrix (excluding areas in patios and major roads) of old-growth and logged areas was found (fig. 3.3b, d). The IKONOS data show greater detail than the ETM+ data (fig. 3.3a, c). For example, individual tree crowns can be distinguished on the IKONOS data (fig. 3.4). Major log transportation roads and patios, as well as minor roads and treefall gaps, can also be identified in the IKONOS image. The IKONOS data also show much greater spatial and spectral detail of the canopy matrix than do the ETM+ data. Read et al. (2003), during fieldwork in summer 2001, found that single-skid roads, and smaller and older treefall gaps, were not visible on the IKONOS images, but were not able to determine at what size or age these features become undetectable.

It is apparent that the higher resolution IKONOS data are more sensitive to logging features than ETM+ data, and if spatial resolution were the only variable IKONOS would be an obvious choice. However, ETM+ data has some important advantages over IKONOS: (a) higher spectral resolution (seven ETM+ bands versus four in IKONOS); (b) coverage (34,000 km^2 in an ETM+ scene compared to 65 km^2 in IKONOS); and (c) cost (an ETM+ scene costs approximately ten times less than an IKONOS scene). Thus, using either ETM+ or IKONOS data involves important trade-offs, and forestry managers must weigh these factors when acquiring satellite remote sensing data.

Choosing between remote sensors also depends on the intended use of the

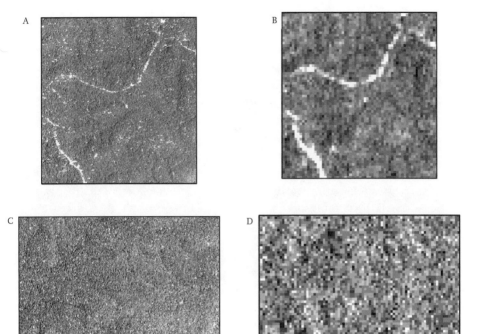

FIGURE 3.3. 4-m IKONOS data (A, logged forest; C, old-growth forest) and 30 m Landsat-ETM+ data (visible red band; B, logged forest; D, old-growth forest) (335 ha plots)

Note: Major logging roads and patios appear bright with both data types. Treefall gaps are only clearly visible as small bright patches on the IKONOS images.

data in certified forestry operations, whether for planning, operations, management, or monitoring at the site level. Planning tasks utilizing remote sensing data may include site context mapping—that is, mapping the extent, pattern, and changes of land cover and land use surrounding the forestry operation (Franklin 2001)—determining forest areas, and assessing landscape patterns, which include forest fragmentation. Intermediate-resolution data such as Landsat-TM and ETM+ are appropriate for these purposes, because of their relatively large scene sizes, the existence of historical databases of Landsat data for change detection, and their relatively low cost. The usefulness of Landsat and other intermediate resolution multispectral satellite remotely sensed datasets for these purposes has been demonstrated in the literature (Sader and Joyce 1988; Skole and Tucker 1993; Sader 1995; Brondizio et al. 1996; Dimyati et al. 1996; Steininger 1996; Tanner, Kapos, and Adams 1998; Peralta and Mather 2000; Read, Denslow, and Guzman 2001). Management and operational tasks involving decisions about where to site roads and patios and determining protection zones based on topographic features, such as steep topography, are likely to depend on a range of data types. Our obser-

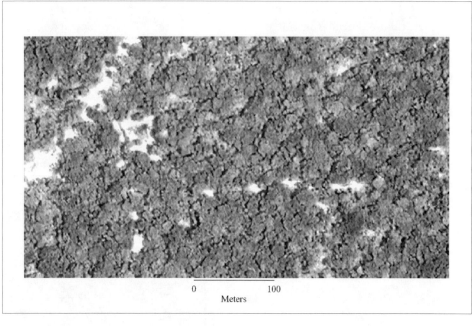

0   100
Meters

FIGURE 3.4. Pan-sharpened 1-m IKONOS image (visible red band) showing logged forest
Source: Adapted from Read et al. 2003.
Note: Logging patios, major and minor roads, and some tree gaps can be detected as bright pixels.

vations at Mil with Landsat ETM+ data demonstrate that these data could be helpful in these areas.

Other site-specific management and operational tasks relating to the method, intensity, timing, and specific locations of harvest require more detailed forest information, which ETM+ alone cannot resolve (Roller 2000). Determination of rotation cycles and harvest times is based on characteristics such as forest composition (age and species distributions) and structure (stand density, average tree height, standing biomass, and leaf area index), as well as changes in these characteristics over time (Franklin 2001). Monitoring tasks aimed at assessing biophysical criteria, such as the timing, extent, location, and intensity of logging, and the area coverage of roads and patios, incorporate many of the same data needs as management and operational tasks.

Observations at Mil Madeireira demonstrate that IKONOS data can better detect logging features (including treefall gaps) than ETM+ data, and hence give better estimates of canopy disturbances resulting from roads, patios, and canopy gap openings due to tree extractions. Bongers (2001) suggests that finer spatial resolutions would be required for detecting logging features, and this suggestion is upheld by the observations with IKONOS data at Mil Madeireira. ETM+ data, however, may produce indirect measures of logging extent and intensity. For example, Souza and Barreto (2000) were able to esti-mate the area affected by selective logging activities in an Amazonian upland forest based on the locations of logging patios, and Asner et al. (2002) also concluded that the location of patios could be used to map areas affected by logging activities.

For monitoring logging intensity, the ability to distinguish treefall gaps is important; patios only account for 1–2 percent of total harvest area, while treefall gaps account for a much higher proportion of canopy gap distur-bances resulting from selective logging in the eastern Amazon (Asner et al. 2002). Thus, better estimates of logging intensity can be derived by using more precise data on actual treefall gaps. Even with IKONOS data, however, not all minor roads and extraction gaps are readily visible; thus, direct mea-surement from a visual interpretation of the image alone will underestimate area in roads or intensity of logging. This does not preclude the possibility of indirect estimates based on such data. It should be possible to derive a rela-tionship between canopy damage visible in the imagery and intensity of log-ging, following the work of Pereira et al. (2001) in the eastern Amazon. My studies in Mil Madeireira indicate that spatial image processing methods are more sensitive to differing degrees of logging with IKONOS, whereas ETM+ could only distinguish extremes of disturbance (Read 2003).

The ability to identify individual tree crowns from IKONOS data has many potential uses in forestry. Clark et al. (2004) were able to measure crown area and crown growth for large trees in a lowland tropical wet forest site in Costa Rica using IKONOS data. Since it is generally the large, tall trees that are har-vested, this demonstrates good potential for informing both management and assessment of logging activities. The same study also reported significant correlations of IKONOS data with stand basal area, estimated aboveground biomass and percent canopy >15 m tall. Monitoring changes in these struc-tural parameters over time could provide important information on forest dynamics in response to logging.

Features associated with selective logging are difficult to detect with Landsat-TM data (Asner et al. 2002), in part because regrowth is rapid and gaps tend to close quickly (Stone and Lefebvre 1998; Souza and Barreto 2000). As gaps age, the spectral sensitivity of remotely sensed data to them decreases—especially in the tropics, where rapid regrowth will close gaps within a few years (Stone and Lefebvre 1998). The studies done at Mil Madeireira that are cited here (Read 2003; Read et al. 2003) focus on the impacts of recent (<1 yr) logging activities at Mil Madeireira, but it is not clear after what time IKONOS data can no longer detect logging features. Interestingly, Clark et al. (2004), working in Costa Rica, were able to detect forest that had been selectively logged more than twenty years ago by measuring the sizes of the largest crowns on an IKONOS image; selectively logged forest areas had fewer large crowns per area than old-growth forests. Further research in this area is needed.

Unbiased and credible monitoring tools are key to the success of forest certification programs (Scholz 1999). Although interpretations are always subjective (see, e.g., Robbins 2003b), theoretically there is potential for using remotely sensed data for certification criteria monitoring, provided that robust standardized and/or automated methods can be devised and that appropriate data sets are available.

Constraints to Using Remote Sensing in Certified Forestry
Several constraints hinder the adoption of new digital data and methods in the forestry sector. These are related to our existing knowledge of tropical moist forests and remote sensing and to issues relating to data and implementation. Probably the most fundamental issue affecting the adoption of satellite remote sensing in forestry is our limited understanding of how remotely sensed reflectances relate to forest structure and composition. Our understanding of the principles of radiation interactions with different biophysical properties at the scale of individual trees to a forest canopy (e.g., leaf chemistry, leaf angle, branches, trunks) derives from controlled laboratory settings and is confounded by the effects of the atmosphere and other extraneous factors that cannot be controlled for in the acquisition of "real-life" remotely sensed data (Franklin 2001). This means that remotely sensed data are often confined to analysis of inferential relationships between the data and parameter of interest (Bergen, Colwell, and Sapio 2000; Franklin 2001). At the same time, our limited understanding of complex tropical forest structure, composition, and dynamics means that establishing relationships in these forests between parameters of interest and remotely sensed data is difficult.

Further hindering the application of remote sensing to certified tropical forestry is our lack of understanding of what makes a tropical moist forest sustainable. Despite ongoing efforts (e.g., Phillips et al. 2004), more research is needed on tropical forest canopy structure, species demographics and distributions, gap formation and recovery rates, and the effects of selective logging on forest structure, composition, and dynamics. It has been demonstrated that high-spatial-resolution satellite remote sensing is appropriate for assisting in some of these research areas, and this potential for satellite remotely sensed data to inform research, management, and monitoring of tropical forests in the future will increase as new sensors are developed and made operational. Data from the QuickBird sensor, and 2 m multispectral and 0.5 meter panchromatic data soon to be available from the next IKONOS satellite (Space Imaging 2004), hold particular promise for application to selective logging.

Despite the many advances in acquisition of remotely sensed data, several constraints to their use exist. Some of these constraints relate to problems inherent in the data regardless of the environments they represent. First, there are problems related to availability of remotely sensed data. Some data do not have global coverage, and not all are commercially available. Data availability often dictates the type of data to be used, resulting in inappropriate spatial or spectral resolution data applied to a task. This can force a change in the design of a task, which otherwise might have been performed using different criteria (e.g., McCusker and Weiner 2003). Second, some data types are difficult to use. For example, hyperspectral data require complex image-processing methods, while high-resolution data, such as IKONOS, are voluminous, making interpretation very time consuming (McCusker and Weiner 2003; Lillesand, Kiefer, and Chipman 2004). Third, some data interpretation is not intuitive, particularly of lidar and radar. Finally, the high cost of acquiring and using remotely sensed data is a strong impediment to its widespread use in certified forestry (Franklin 2001). The data sets themselves are expensive, and technological infrastructure and expert personnel for analyzing remotely sensed data represent a significant financial investment for forestry operations in developing countries (Pro-Natura, IIED, and GTZ 2000). In addition to these general considerations, in the tropics the availability of cloud-free data is limited, and this is problematic, especially if multitemporal data sets at anniversary dates are required of the same area. In sum, there is no one single data type that can answer all the needs of the forestry sector, and an integrative approach combining the advantages and disadvantages of various remotely sensed data types, as well as field and other data types, is recommended (Bergen, Colwell, and Sapio 2000; Roller and Bergen 2000).

Discussion

Despite the constraints discussed in the previous section, earlier sections in this chapter made the case for remote sensing as having significant potential in assisting forestry enterprises in research, certification, and management of certified forests. In practice, the adoption of remote sensing and other spatial information technologies will likely depend on the scale and nature of the forestry enterprise and on the sociopolitical environment. For example, PWA is a subsidiary of a company with international investors and interests in the Brazilian states of Amazonas and Pará, as well as Central America. The majority of its product goes to the export market. The operation at Mil Madeireira, which operated at a loss prior to 2000, is now turning a profit, and the company continues to purchase additional lands at their Mil Madeireira site (Precious Woods 2004). The scale of their investments has enabled them to build their operations to the point at which they are now profitable.

The ability of small and medium-sized enterprises to compete in this arena, however, is not promising. Community-based organizations and small private companies are unlikely to have the financial resources to withstand the high up-front costs associated with certification, or access to market information and credit. In a recent report focusing on small and medium-sized Brazilian forest enterprises, May, da Vinha, and Macqueen (2003) conclude that strengthening cooperative associations among these groups would help overcome some of these obstacles. The success of the Brazilian buyers group has been important in enabling smaller producers to reach the Brazilian market and encouraging growth of that market (Viana et al. 2002). The fact that the majority of the larger companies' products are exported could be helping the smaller enterprises to survive at the national level. Based on the high up-front costs associated with the use of spatial information technologies, presumably the scale of the enterprise will determine whether remote sensing (and other technologies) is incorporated into individual management strategies, and will also determine the degree of reliance on these technologies. Increased efficiency and credibility of operations employing technologies would advantage the companies that can afford them.

Remotely sensed data are often cited as being unbiased, with the implication being that they are more credible than other sources of data. However, many studies have demonstrated that the interpretation and presentation of these data are subjective, being social constructions defined by the purpose and criteria used in analyses (Harris et al. 1995; St. Martin 2001; McCusker and Weiner 2003; Robbins 2003a, 2003b). The aura of legitimacy that inheres to digital technology itself, as well as the ease of producing clean output through digital cartography and GIS software, confers a sense of scientific authority to remotely sensed data that may be misleading and discourage methodological scrutiny (for a discussion of this issue see Turner, chap. 7 in this volume). Robbins (2003b), in a discussion of how remote sensing can be used to meet bureaucratic requirements, cautions that there is a danger that these technologies can be used to meet a specific but concealed agenda while appearing to be unbiased. Similarly, in the Brazilian forestry sector, various interest groups operating at different scales could use the apparent authority of remote sensing to substantiate political claims and further political agendas; on the other hand, the technology also could serve a useful validation purpose in disputes among these parties.

Acknowledging the political uses of remote sensing data, however, does not automatically delegitimize its role in tropical rain forest conservation. On the contrary, it is possible that the use of information technologies would confer greater credibility to certified logging operations, especially in the area of environmental impact assessments, which otherwise are difficult to implement and present.[3] Environmental impact assessments are crucial where credibility across scales—such as the forestry operation, the certification body, national and international timber markets, and, ultimately, international environmental concerns—is key to the success of the system. Remote sensing data could be used to develop standardized protocols for environmental impact assessments and address issues of error and uncertainty (Longley et al. 2001).

Finally, I should note that while the FSC principles and criteria for certified forestry include both environmental and socioeconomic conditions, this chapter has only addressed the application of remote sensing to environmental and forestry management aspects (i.e., within the physical space of the operation). The application of remote sensing to assist in understanding of socioeconomic issues, which often act beyond the physical spaces of the territory (Daniels and Bassett 2002), is not as straightforward (Liverman et al. 1998; Turner 2003).

Conclusion

Forest certification serves as an important tool for protecting tropical forests. In Brazil the development of certification processes is well advanced, and the area of forests already certified relative to other tropical forest countries is high. Certified forestry, however, requires management practices different from those of conventional forestry, which can present disincentives to certification. This chapter has explored how remote sensing could provide new data and methods for informing management and monitoring of certified forestry. Different remotely sensed data types have different roles to play in planning, management, and operational and monitoring tasks, and they may also have important applications within the broader context of formulation, assessment, implementation, and monitoring of international environmental policy (Sherbinin, Kline, and Raustiala 2002). However, many obstacles and tensions influence the use of remote sensing and other spatial information technologies, and the scale at which they are adopted. In the absence of policies to encourage sustainable forestry and the use of the technologies to make it viable in the Amazon, the larger forestry enterprises are better placed than small and community-based organizations to undertake sustainable management and adopt spatial information technologies. The conservation territories that result tend to be large tracts of land close to transportation routes that primarily export to Europe and Asia. Smaller

3. On its web site PWA refers to its computer databases and forest inventories as "highly sophisticated," presumably believing that this will lend them legitimacy in the eyes of readers (http://www.preciouswoods.ch/; accessed 14 July 2004).

enterprises are gaining ground, encouraged by the success of internal marketing strategies; however, in the absence of technical support and incentives they are less likely to adopt expensive information technologies. A national Brazilian network of carefully sited sustainable forestry conservation territories based on ecological and socioeconomic criteria has not yet been conceived. The development of such networks in Brazil and other tropical countries is a necessary condition for successful application of remote sensing strategies to improve the conservation of tropical forests.

References

Asbjornsen, H., and E. Alatorre. 1999. Challenges and opportunities for tropical timber certification: Mexico's experience. In *Forest certification: Roots, issues, challenges, and benefits*, ed. K. A. Vogt, B. C. Larson, J. C. Gordon, D. J. Vogt, and A. Fanzeres, 291–98. Boca Raton, FL: CRC Press.

Asner, G. P., M. Keller, R. Pereira, Jr., and J. C. Zweede. 2002. Remote sensing of selective logging in Amazonia: Assessing limitations based on detailed field observations, Landsat ETM+, and textural analysis. *Remote Sensing of Environment* 80 (3): 483–96.

Barreto, P., P. Amaral, E. Vidal, and C. Uhl. 1998. Costs and benefits of forest management for timber production in eastern Amazonia. *Forest Ecology and Management* 108:9–26.

Bass, S., K. Thornber, M. Markopoulos, S. Roberts, and M. Grieg-Gran. 2001. *Certification's impacts on forests, stakeholders and supply chains: Instruments for sustainable private sector forestry series.* London: International Institute for Environment and Development.

Bergen, K., J. Colwell, and F. Sapio. 2000. Remote sensing and forestry: Collaborative implementation for a new century of forest information solutions. *Journal of Forestry* 98 (6): 4–9.

Bertault, J. G., and P. Sist. 1997. An experimental comparison of different harvesting intensities with reduced-impact and conventional logging in East Kalimantan, Indonesia. *Forest Ecology and Management* 94:209–18.

Bongers, F. 2001. Methods to assess tropical rain forest canopy structure: An overview. *Plant Ecology* 153:263–77.

Brondizio, E., E. Moran, P. Mausel, and Y. Wu. 1996. Land cover in the Amazon estuary: Linking of the Thematic Mapper with botanical and historical data. *Photogrammetric Engineering and Remote Sensing* 62 (8): 921–29.

Caylor, J. 2000. Aerial photography in the next decade. *Journal of Forestry* 98 (6): 17–19.

Clark, D. B., J. M. Read, M. Clark, A. M. Cruz, M. F. Dotti, and D. A. Clark. 2004. Application of 1-m and 4-m resolution satellite data to studies of tree demography, stand structure and land-use classification in tropical rain forest landscapes. *Ecological Applications* 14 (1): 61–74.

Daniels, R., and T. J. Bassett. 2002. The spaces of conservation and development around Lake Nakuru National Park, Kenya. *The Professional Geographer* 54 (4): 481–90.

Digital Globe. 2004. http://www.digitalglobe.com. Accessed 14 July 2004.

Dimyati, M., K. Mizuno, S. Kobayashi, and T. Kitamura. 1996. An analysis of land-use/cover change using the combination of MSS Landsat and land use map: A case study in Yogyakarta, Indonesia. *International Journal of Remote Sensing* 17 (5): 931–44.

Dobson, M. C. 2000. Forest information from Synthetic Aperture Radar. *Journal of Forestry* 98 (6): 41–43.

Dubayah, R. O., and J. B. Drake. 2000. Lidar remote sensing for forestry. *Journal of Forestry* 98 (6): 44–46.

Fanzeres, A., and K. A. Vogt. 1999. Origins of the concept of forest certification. In *Forest certification: Roots, issues, challenges, and benefits,* ed. K. A. Vogt, B. C. Larson, J. C. Gordon, D. J. Vogt, and A. Fanzeres, 11–19. Boca Raton, FL: CRC Press.

Franklin, S. E. 2001. *Remote sensing for sustainable forest management.* Boca Raton, FL: Lewis Publishers.

FSC (Forest Stewardship Council). 2004. http://www.fsc.org. Accessed 14 July 2004.

Gale, F. P. 1998. *The tropical timber trade Regime.* New York: St. Martin's.

Gerwing, J. J. 2002. Degradation of forests through logging and fire in the eastern Brazilian Amazon. *Forest Ecology and Management* 157:131–41.

Ghazali, B. J., and M. Simula. 1998. Timber certification: progress and issues (Brazil). Kuala Lumpur and Helsinki: International Tropical Timber Organization.

Green, K. 2000. Selecting and interpreting high-resolution images. *Journal of Forestry* 98 (6): 37–39.

Harris, T. M., D. Weiner, T. Warner, and R. Levin. 1995. Pursuing social goals through participatory GIS: Redressing South Africa's historical political ecology. In *Ground truth: The social implications of geographic information systems,* ed. J. Pickles, 196–222. New York: Guilford.

Holdridge, L. 1947. Determination of world plant formations from simple climatic data. *Science* 105:367–68.

International Tropical Timber Organization (ITTO). 2001. Annual review and assessment of the world timber situation 2001. Yokohama, Japan: ITTO.

Lillesand, T. M., R. W. Kiefer, and J. W. Chipman. 2004. *Remote sensing and image interpretation.* 5th ed. New York: Wiley.

Liverman, D., E. F. Moran, R. R. Rindfuss, and P. C. Stern, eds. 1998. *People and pixels: Linking remote sensing and social science.* Washington, DC: National Academy Press.

Longley, P. A., M. F. Goodchild, D. J. Macguire, and D. W. Rhind. 2001. *Geographic information systems and science.* Chichester, UK: Wiley.

Macqueen, D. J., M. Grieg-Gran, E. Lima, J. MacGregor, F. Merry, V. Prochnik, N. Scotland, R. Smeraldi, and C. E. F. Young. 2003. *Growing exports: The Brazilian tropical timber industry and international markets.* IIED Small and Medium Enterprise Series no. 1. London: International Institute for Environment and Development.

Martin, M. E., S. D. Newman, J. D. Aber, and R. G. Congalton. 1998. Determining forest species composition using high spectral resolution remote sensing data. *Remote Sensing of Environment* 65: 249–54.

May, P., V. G. da Vinha, and D. J. Macqueen. 2003. Small and medium forest enterprise in Brazil. Rio de Janeiro and London: Grupo de Economia do Meio Ambiente e Desenvolvimento Sustentável and International Institute for Environment and Development.

McCusker, B., and D. Weiner. 2003. GIS representations of nature, political ecology, and the study of land use and land cover change in South Africa. In *Political ecology: An integrative approach to geography and environment-development studies,* ed. K. S. Zimmerer and T. J. Bassett, 201–18. New York: Guilford.

Myers, N. 1993. Tropical forests: The main deforestation fronts. *Environmental Conservation* 20:9–16.

Nepstad, D. C., A. Verissimo, A. Alencar, C. Nobre, E. Lima, P. Lefebvre, P. Schlesinger, C. Potter, P. Moutinho, E. Mendoza, M. Cochrane, and V. Brooks. 1999. Large-scale impoverishment of Amazonian forests by logging and fire. *Nature* 398:505–8.

Olson, C. J., and F. P. Weber. 2000. Foresters' roles in remote sensing. *Journal of Forestry* 98 (6): 11–12.

Peralta, P., and P. Mather. 2000. An analysis of deforestation patterns in the extractive reserves of Acre, Amazonia from satellite imagery: A landscape ecological approach. *International Journal of Remote Sensing* 21 (13–14): 2555–70.

Pereira, R. J., J. C. Zweede, G. P. Asner, and M. M. Keller. 2001. Forest canopy damage and recovery in reduced impact and conventional selective logging in Eastern Para, Brazil. *Forest Ecology and Management* 100:1–15.

Phillips, P. D., C. P. de Azevedo, B. Degen, I. S. Thompson, J. N. M. Silva, and P. R. van Gardingen. 2004. An individual-based spatially explicit simulation model for strategic forest management planning in the eastern Amazon. *Ecological Modelling* 173:335–54.

Precious Woods. 2004. http://www.preciouswoods.ch/. Accessed 14 July 2004.

90 JANE M. READ

Pro-Natura, IIED (International Institute for Environment and Development), and GTZ (Deutsche Gesellschaft für Technische Zusammenarbeit). 2000. Barriers to forest certification in the Brazilian Amazon: The importance of costs. Rio de Janeiro, Brazil: Pro-Natura/IIED/GTZ.

RADAMBRASIL. 1976. Levantamento de Recursos Naturais: Ministério de Minas e Energia. Rio de Janeiro: Departamento Nacional de Produção Mineral.

Read, J. M. 2003. Spatial analyses of logging impacts in Amazonia using remotely-sensed data. *Photogrammetric Engineering & Remote Sensing* 69 (3): 275–82.

Read, J. M., D. B. Clark, E. M. Venticinque, and M. P. Moreira. 2003. Application of merged 1-m and 4-m resolution satellite data to research and management in tropical forests. *Journal of Applied Ecology* 40 (3): 592–600.

Read, J. M., J. S. Denslow, and S. M. Guzman. 2001. Documenting land-cover history of a humid tropical environment in northeastern Costa Rica using time-series remotely sensed data. In *GIS and remote sensing applications in biogeography and ecology,* ed. A. C. Millington, S. D. Walsh, and P. E. Osborne, 69–89. Boston: Kluwer.

Robbins, P. 2003a. Beyond ground truth: GIS and the environmental knowledge of herders, professional foresters, and other traditional communities. *Human Ecology* 31 (2): 233–53.

———. 2003b. Fixed categories in a portable landscape: The causes and consequences of land cover categorization. In *Political ecology: An integrative approach to geography and environment-development studies,* ed. K. S. Zimmerer and T. J. Bassett, 181–200. New York: Guilford.

Roller, N. 2000. Intermediate multispectral satellite sensors. *Journal of Forestry* 98 (6): 32–35.

Roller, N., and K. Bergen. 2000. Integrating data and information for effective forest management. *Journal of Forestry* 98 (6): 61–63.

Running, S. W., L. Queen, and M. Thornton. 2000. The Earth Observing System and forest management. *Journal of Forestry* 98 (6): 29–31.

Sader, S. A. 1995. Spatial characteristics of forest clearing and vegetation regrowth as detected by Landsat Thematic Mapper imagery. *Photogrammetric Engineering and Remote Sensing* 61 (9): 1145–51.

Sader, S. A., and A. T. Joyce. 1988. Deforestation rates and trends in Costa Rica, 1940–1983. *Biotropica* 20 (1): 11–19.

SBS (Sociedad Brasileira de Silvicultura). 2004. http://www.sbs.org.br. Accessed 14 July 2004.

Schlaepfer, R., and C. Elliott. 2000. Ecological and landscape considerations in forest management: The end of forestry? In *Sustainable forest management,* ed. K. von Gadow, T. Pukkala, and M. Tome, 1–67. Dordrecht, The Netherlands: Kluwer Academic.

Scholz, I. 1999. Protection and sustainable use of tropical forests: Points of departure in the Brazilian timber industry. Berlin: German Development Institute.

SGS (Société Générale de Surveillance). 2004. http://www.sgs.com. Accessed 14 July 2004.

Sherbinin, A., K. Kline, and K. Raustiala. 2002. Remote sensing data: Valuable support for environmental treaties. *Environment* 44 (1): 20–31.

Sist, P. 2000. Reduced-impact logging in the tropics: Objectives, principles and impacts. *International Forestry Review* 2 (1): 3–10.

Skole, D., and C. Tucker. 1993. Tropical deforestation and habitat fragmentation in the Amazon: Satellite data from 1978 to 1988. *Science* 260:1905–10.

Slik, F. J. W., R. W. Verburg, and P. J. A. Kebler. 2002. Effects of fire and selective logging on the tree species composition of lowland dipterocarp forest in East Kalimantan, Indonesia. *Biodiversity and Conservation* 11:85–98.

Souza, C. J., and P. Barreto. 2000. An alternative approach for detecting and monitoring selectively logged forests in the Amazon. *International Journal of Remote Sensing* 21 (1): 173–79.

Space Imaging. 2004. http://www.spaceimaging.com. Accessed 14 July 2004.

SPOT (Système Pour l'Observation de la Terre). 2004. http://www.spotimage.fr. Accessed 14 July 2004.

Steininger, M. K. 1996. Tropical secondary forest regrowth in the Amazon: Age, area and change estimation with Thematic Mapper data. *International Journal of Remote Sensing* 17 (1): 9–27.

St. Martin, K. 2001. Making space for community resource management in fisheries. *Annals of the Association of American Geographers* 9 (1): 122–42.

Stone, T. A., and P. Lefebvre. 1998. Using multi-temporal satellite data to evaluate selective logging in
Para, Brazil. *International Journal of Remote Sensing* 19 (13): 2517–26.

Sustainable Tree Crops Program (STCP). 2001. *Caracterização e avaliação da situação da tecnologia de proces-
samento das empresas madeireiras na Amazônia legal.* Vol. 1. Curitiba, Brazil: STCP.

Tanner, E. V. J., V. Kapos, and J. Adams. 1998. Tropical forests: Spatial pattern and change with time, as
assessed by remote sensing. In *Dynamics of tropical communities: The 37th symposium of the British Ecological Society,* ed. D. M. Newbery, H. H. T. Prins, and N. D. Brown, 599–615. Malden, MA: Blackwell
Science.

Thornber, K. 1999. Overview of global trends in FSC certificates. In *Instruments for sustainable private sector forestry series.* London: International Institute for Environment and Development.

Turner, M. D. 2003. Methodological reflections on the use of remote sensing and geographic information science in human ecological research. *Human Ecology* 31 (2): 255–79.

UNEP-WCMC (United Nations Environment Program–World Conservation Monitoring Centre), WWF
(World Wildlife Fund), FSC (Forest Stewardship Council), and GTZ (Deutsche Gesellschaft für Technische Zusammenarbeit). 2004. Information on certified forest sites endorsed by Forest Stewardship Council (FSC). http://www.certified-forests.org. Accessed 14 July 2004.

Ustin, S. L., and A. Trabucco. 2000. Using hyperspectral data to assess forest structure. *Journal of Forestry*
98 (6): 47–49.

Viana, V. M., P. May, L. Lago, O. Dubois, and M. Grieg-Gran. 2002. *Instruments for sustainable private sector
forestry in Brazil.* Instruments for Sustainable Private Sector Forestry series. London: International
Institute for Environment and Development.

Vogt, K. A., B. C. Larson, D. J. Vogt, J. C. Gordon, A. Fanzeres, J. L. O'Hara, and P. A. Palmiotto. 1999. Issues in forest certification. In *Forest certification: Roots, issues, challenges, and benefits,* ed. K. A. Vogt,
B. C. Larson, J. C. Gordon, D. J. Vogt and A. Fanzeres, 1–10. Boca Raton, FL: CRC Press LLC.

Whitmore, T. C. 1997. Tropical forest disturbance, disappearance, and species loss. In *Tropical forest remnants: Ecology, management, and conservation of fragmented communities,* ed. W. F. Laurance and R. O. J.
Bierregaard, 3–12. Chicago: University of Chicago Press.

Wilson, E. H., and S. A. Sader. 2002. Detection of forest harvest type using multiple dates of Landsat
TM imagery. *Remote Sensing of Environment* 80:385–96.

Wynne, R. H., and D. B. Carter. 1997. Will remote sensing live up to its promise for forest management?
Journal of Forestry 95 (10): 23–26.

Zimmerer, K. S. 2000. The reworking of conservation geographies: Nonequilibrium landscapes and
nature-society hybrids. *Annals of the Association of American Geographers* 90 (2): 356–69.

4 Productive Conservation and Its Representation: The Case of Beekeeping in the Brazilian Amazon

J. CHRISTOPHER BROWN

Productive conservation is one of the dominant conservation-with-development concepts in use today. Explained most fully in the Amazonian context by Anthony Hall (1997), productive conservation is a market-based effort to foster conservation and improve the livelihoods of local people—colonist farmers, indigenous peoples, rubber tappers, and *caboclos* (floodplain communities)—who make rain forest environments their home. Ideally, productive conservation efforts link local forest ecosystems, local resource users, and markets to bring higher incomes for local people and encourage forest conservation. International donor groups, including the World Bank, the Group of Seven (G7), and nongovernmental organizations (NGOs) such as the World Wildlife Fund and Friends of the Earth, have funneled millions of dollars of aid to local grassroots organizations in Amazonia to spur productive conservation programs.

This chapter is about beekeeping, one type of productive conservation activity that has received much funding in the Brazilian Amazon state of Rondônia (see fig. 4.1). An examination of how this activity came to be viewed and practiced as productive conservation in the Rondônian context reveals much about the complex spatialities of globalization and conservation this book addresses. By incorporating resource use and conservation of natural environments and biodiversity, productive conservation efforts like beekeeping can be described as part of the so-called second wave of conservation in that conservation goals are not viewed as incompatible with resource use and rising income levels (see Zimmerer, chap. 1 in this volume). Ideally, locally organized resource users would provide the market with a product that is derived from the existing local ecology deemed worthy of conservation. Increased income, in turn, provides local communities with the incentive required for sustainable use and conservation of the resource. As shown below, such logic was developed in the context of globalization of international con-

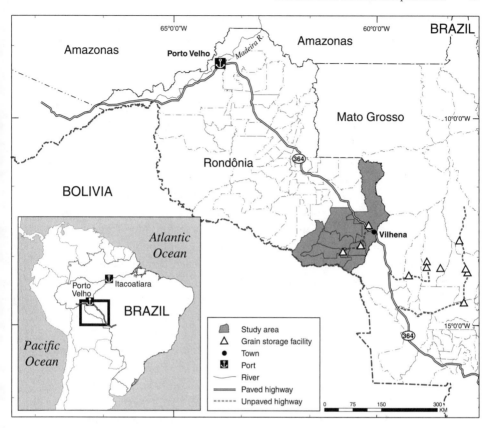

FIGURE 4.1. Map of study area: Rondônia, Brazil

servation policy and policymaking institutions since the 1970s (and even before), in association with the hegemony of neoliberal economic policies in
more recent years.

Moreover, productive conservation efforts in practice challenge the traditional concept that conservation ought to take place within narrowly demarcated territories with strict restrictions on activities within them. Indeed,
the spatialities of external, global-scale environments and social networks
appear to have become just as important as the spatialities of the conservation target zones themselves. This chapter explores the tensions between two
spatialities that are key to the practice of beekeeping as productive conservation. First, there is the spatiality of the social relations supporting beekeeping as an activity for development and conservation. These relations involve networks of local resource users, state actors, national agriculture
ministries, ecologists, international NGOs, and multilateral development
banks. Second, there is the spatiality of the production ecology of beekeep-

ing, one that is rooted in a conservation target zone with material processes linking people, bees, and the flowers and habitats that support honey production. Productive conservation efforts like beekeeping cannot exist without the two spatialities. Pursuing conservation with development in this way, however, comes with some familiar drawbacks of other development models—namely, the tendency to implement practices that are not well suited to local environments. Such a statement runs counter to conventional thinking, because productive conservation activities are known for their active involvement of local people and natural environments. What does and can go wrong? The goal of this chapter is to explain how the contradictions between the two spatialities (that of networks and that of production ecology) create fundamental challenges to achieving the goals of conservation. Only by becoming aware of these tensions, how they are created, and what role actors have in creating them, can these challenges be overcome.

This chapter is divided into four parts. The first places the development of the productive conservation concept in its historical political economic context. It is a sustainable development concept that arose as local NGOs began to partner with international actors to form novel networks that aimed to challenge the power national governments and their international financiers exerted over the development process. Development came to be practiced in a different way beginning in the 1990s in response to well-known development debacles of the 1970s and 1980s. Together with the neoliberal trend of seeking greater efficiency as a declared goal in development, international capital created a demand for local NGOs in the developing world to step forward to receive productive conservation funds.

The second part of the chapter uses beekeeping as a case study to explore how productive conservation activities came to fit the need of the rescaled development and conservation-funding agenda. I argue that the new networks used particular representations of beekeeping in their jockeying for power, at the expense of ensuring that the production ecology of beekeeping itself met the stated ideals of productive conservation. I show that discourses and narratives of beekeeping as some ideal production form for sustainable development change dramatically over time, indicative of efforts of actors within the network to gain control over development funds and practice.

Those dynamic changes in meaning and social relations are then contrasted in the third part with a discussion of the dynamics of biogeographic processes and human impacts that comprise the spatiality of the production ecology of beekeeping. Here, we may see what distance can develop in productive conservation initiatives between the meanings that support certain social relations and what actually happens among people, trees, and bees in conservation target zones. The fourth part reflects on the tensions that exist in productive conservation between the spatialities of networks and production ecology, and I offer some ways to resolve those tensions based on our understanding of the human and environmental dynamics of productive conservation.

Globalization and the Rescaling of Development and Conservation

Productive conservation did not arise in the Amazon in isolation from international political, economic, and environmental events (Klooster 1997). An increasingly global awareness of the social and environmental problems of development, coupled with the rise of neoliberalism and democratization, helped lay the foundation of the adoption of productive conservation ideals as a guide to achieve socially and ecologically sound development. Since the 1970s, wealth disparities between and within nations and concern about environmental degradation and human rights abuses in development sparked the convening of numerous international meetings to assess the state of the world and proposals for its improvement. The 1972 United Nations Conference on the Human Environment in Stockholm, the Brundtland Commission Report (World Commission on Environment and Development 1987), and the 1992 United Nations Conference on the Environment and Development in Rio de Janeiro all resulted in proposals that seem directed specifically at the problems of the Amazon, when indeed they were meant to apply to many regions of the world. Amid the problems of acid rain, global warming, and holes in the ozone layer, not to mention extinction crises wherever tropical forests were falling, the Amazon was merely one region of the world where development processes were raising awareness of the need for social and environmental justice.

Such global environmental awareness was raised most markedly with respect to the negative effects of national megadevelopment projects around the world that were financed by multilateral development banks (MDBs), fueling calls to change the status quo of centrally planned development. These negative effects of development, elaborated below for the case in Rondônia, Brazil, gave local and international environmental organizations reason to fight to change the development process, but it was the rise of democratization and the neoliberal development agenda that gave them the space to do so. Amid the economic recession and debt crisis of the early 1980s, it was clear to policymakers that the state-directed import substitution industrialization model of development had failed. As a result, the World Bank and the International Monetary Fund dictated structural adjustment programs (SAPs). These programs embraced the supposed power of a trimmed-down state and free markets to contribute toward export-oriented economic growth. By the end of the 1980s, these programs had been adopted by most indebted third world nations (Brohman 1990). With state expenditures slashed, little if any money remained in state coffers for basic social programs, let alone programs directed at environmental management. Thus, any centralized natural resource conservation policies of these governments were ineffective, and the decentralization trend that is a part of neoliberalism was seen as way to increase the efficiency of the delivery of services within social and environmental programs.

Finally, with the fall of the Soviet Union and the democratization of the late 1980s and 1990s, more space was created for the participation of ordinary citizens in development. In Brazil and elsewhere, this gave increased legitimacy to an organized civil society making demands on government. However, this came at a time when the state was not in any position financially to meet these demands. Development would have to be funded some other way. Here, the interests of multilateral agencies like the World Bank and governments converged. States needed money, and the World Bank needed a way to increase its legitimacy as an organization concerned about helping the poor directly and in an environmentally sound way. Adopting a participatory orientation to its development goals, the World Bank and states found a solution in strategies like productive conservation. This solution served as a way to maintain the flow of development project money and a way to maintain at least the appearance of repairing the damage of the 1970s and 1980s.

With a reason to challenge the status quo of development and the political space offered by democratization, international environmental organizations and local NGOs began to establish the spatiality of social relations that would support productive conservation. Scholars have come to describe these new configurations of social relations in many different ways. For example, Keck and Sikkink (1998) have described numerous cases of local NGOs fighting against the social and environmental abuses of development by pleading their cases at the international or global level, rather than approaching national- or state-level officials for remedies. This bypassing of the national level, termed the "boomerang" effect, brings greater attention to abuses at the local level, in effect challenging the power national governments have over the development process. In the geographic literature, Smith (1993) has called this "jumping scales," and the concept builds on a large theoretical literature in geographic political economy about the social construction of scale (Cerny 1995; Smith 1995; Swyngedouw 1997; Brenner 1999). Based on this literature, local NGOs who create ties with NGOs at the international or global scale are engaged in creating a new scale of human organization (neither local nor global) to gain access to power and resources in what could be called a process of "glocalization" (Swyngedouw 1997; Perreault 2003). Regardless of what the phenomenon is called, it is evidence of a real opportunity that globalization has presented to local communities organized around issues of environmental and social justice: they have used new social networks that are a product of globalization to challenge the power of the state, which has so long ignored their needs and shut them out of decision-making processes.

Statistical evidence underscores the importance of the trends discussed above. Keck and Sikkink (1998) report a fivefold increase in the number of NGOs that deal with social change at the international level between 1953 and 1993. Many of these international groups are the ones that helped elevate the concerns of local NGOs in regions like the Amazon to global importance, creating new funding channels for development and conservation. United Nations (UN), World Bank, and European Union (EU) development assistance is channeled increasingly through local NGOs, from 47 to 67 percent of all assistance between 1990 and 1994 alone ("Sins of the Secular Missionaries" 2000). More than 70 percent of World Bank development projects approved in 1999 involved NGOs in some way, indicating an increased willingness to treat both international and local NGOs as partners in development planning and implementation (World Bank 2000).

The local-international links that would eventually promote productive conservation were forged as NGOs at local and international scales combined forces to campaign against development practices of the Brazilian government and World Bank. But the formation of those alliances depended on existing networks of local organizations in Rondônia and the early action of NGOs in Europe and the United States. Many local NGOs in Rondônia count the efforts of the Catholic Church in the mid- to late 1970s as essential in encouraging local communities to take the first steps to help themselves when government programs failed. With respect to agricultural production, the Church helped sponsor the establishment of credit unions that stimulated investment and helped make cash more available; cooperatives facilitated the pooling of produce, especially coffee, to obtain higher prices. Northern NGOs had already been working since the early 1970s to educate the public about the necessity for social and environmental considerations in development. But the idea of productive conservation had yet to be achieved. It would arise as an idea generated within the local-international networks as a solution to the social and environmental crisis unfolding in places like Rondônia.

Before productive conservation could be presented as a solution, the NGO network first had to make people outside of Rondônia believe there was a crisis. First, the various institutions and individual persons of the network convinced the citizens of the developed world of the role they play in the functioning of the World Bank. A representative document from the NGO campaign, entitled "Bankrolling Disasters," outlined the important role of the U.S. government in World Bank affairs, and urged U.S. citizens to demand a say in how World Bank development funds were spent (Schwartzman 1986). The report denounced the environmental impacts of development projects, questioned the benefits of development spending to local populations, criticized the World Bank's inability to forecast social and environmental problems resulting from development projects, and challenged the unilateral imposition of first world economic and political priorities with limited input from governments in the borrowing countries.

In the United States NGOs took their case beyond ordinary citizens to influential members of Congress who sat on finance, environmental, and science committees and who had an interest in exposing the failures of the World Bank. Beginning in June of 1983 and continuing throughout the 1980s, over twenty hearings were held to address the NGOs' concerns about the practices of the multilateral development banks (Brown 1992). Several hearings dealt in part with development in Rondônia, and several expert witnesses focused their testimony on the failed rain forest colonization project in Rondônia called Polonoroeste. Upon examination of the evidence, several committees made strong recommendations to the U.S. Treasury and the MDBs. Those included giving greater attention to potential negative outcomes long before the project approval stage, hiring additional environmental staff, and including NGOs in the investigation of social and environmental problems and in project planning. The World Bank gave in to the pressure exerted by the NGO campaign, halting Polonoroeste project loan disbursements to the Brazilian government in April 1985 until the bank received renewed assurances Brazil would meet the social and environmental conditions of the loan agreement. Illustrating the impact that Polonoroeste and the MDB campaign had on the World Bank, then-president of the bank, Barber Conable, made specific reference to what had transpired in the rain forest frontier of Rondônia. He stated that Polonoroeste was a "sobering example of an environmentally sound effort which went wrong. The Bank misread the human, institutional and physical realities of the jungle and the frontier. . . . Protective measures to shelter fragile land and tribal people were included; they were not, however, carefully timed or adequately monitored" (quoted in a letter from Environmental Defense Fund to E. Patrick Coady, executive director, World Bank, 9 January 1990). Admitting that things could be done differently in the future, Conable added, "If the World Bank has been a part of the problem in the past, it can and will be a strong force in finding solutions in the future" (Rich 1989).

The World Bank also responded to pressure by announcing plans to include NGOs in the development process. Here, the Bank emphasized the importance of involving local communities in the preparation and implementation of projects that affect them (World Bank 1990). In addition, a series of "environmental projects" would be formulated specifically to improve environmental quality in rural and urban areas. Rondônia would be the target of one of these projects, called Planafloro (the Rondônia Natural Resources Management Project), which was designed to ameliorate the social and environmental problems of Polonoroeste. Loan disbursements for the project began in 1993. The project involved the promotion of improved natural resource management, better protection of indigenous, rubber tapper, and environmental reserves, socioeconomic and ecological macrozoning, development of smallholder agroforestry through research and rural credit, socioeconomic and service infrastructure development, and funding for administration, monitoring, evaluation, and technical assistance (Millikan 1997). Meeting the bank's earlier pledges, the project also included measures to ensure the formal participation of local communities and Rondônian NGOs in project planning, monitoring, and evaluation.

It did not take long before Planafloro became another controversial project. Within a year of the project's inception, a group of Rondônian NGOs, spurred on and funded by the World Wildlife Fund, called for Planafloro's suspension, citing numerous irregularities in its implementation, particularly a lack of NGO control over project implementation (Millikan 1995). After months of review, the World Bank chose to deny the NGOs' request. However, the NGOs did succeed in negotiating a deal with the Rondônian state government that sought to increase NGO participation in the project by allocating $22 million to special subprojects called PAICs (Community Initiative Assistance Projects). This decentralized development initiative gave local NGOs, representing colonist farmers, indigenous peoples, rubber tappers, and *ribeirinhos* (dwellers of the Amazon's floodplains), the opportunity to apply for up to US$150,000 to organize and implement their own development projects to better their communities (Planafloro 1997). Eligible projects included small-scale infrastructure development, environmental protection, environmental education, sustainable natural resource management and biodiversity conservation, and promotion of activities that fit the ideals of productive conservation.

In sum, the local-international environmental network successfully rescaled conservation. By most accounts, Rondônia was a development disaster, and the anti-MDB campaign convinced many decision makers that the MDBs had a role in causing the social and environmental ills of what were centralized, infrastructure-heavy, megadevelopment projects around the world. By decentralizing development in projects like Planafloro and involving local people in ecologically sound productive strategies carried out by communities themselves, productive conservation finally took shape as a concept around which the rescaled development and conservation networks could organize. In the end the policy allowed the World Bank to continue its most basic of operations, lending money, and NGOs in Brazil and in the North found themselves as new providers of political and economic power to the rural poor in Amazonia. As institutions, they developed their own economic needs, which were met by convincing taxpayers, government officials, and consumers that productive conservation deserved a chance. Money, in the form of government budget allocations, new World Bank loans, and profits from "rain forest crunch" products, began to flow into the production systems of Amazonia's rural poor so that they could stabilize and expand their efforts to save forests.

A Historical Geography of Beekeeping-as-Development and Conservation Narratives

Beekeeping was originally promoted through development projects that were internationally funded but nationally planned and executed. As such, some beekeeping programs suffered from familiar problems often attributed to the "top-down" approach of development programs. Through the struggle to control these development resources, beekeeping interests at the more local level would seek to rescale how beekeeping development would occur by casting beekeeping in the light of the ideals of productive conservation discussed above. I have examined narratives of beekeeping from a variety of sources as contained in the words of development specialists, anthropologists, government bureaucrats, NGO officials, and researchers from around the world in various situations. Here, I examine texts from conferences, mainly from the International Conferences on Apiculture in Tropical Climates, sponsored by the International Bee Research Association (IBRA) based in Cardiff, Wales; the texts reveal much about what beekeeping promoters want policymakers, development officials, potential donors, and financiers to know about this activity. These global narratives of beekeeping generated from the IBRA conferences are then compared with the narratives from proposals of local beekeeping cooperatives in Rondônia submitted to funding agencies in the 1990s to carry out development based on the ideals of productive conservation.

The link between U.K.-based IBRA and beekeeping development aid is clear once one realizes that the British were some of the first to promote beekeeping in the developing world as "development," as they did in east Africa. Arguably, colonial ties to the region had long exposed British bee enthusiasts to local beekeeping practices among tribesmen. Compared to so-called modern practitioners of beekeeping, the native peoples of east Africa used bees in a very extensive fashion for the provision of wax and honey, the latter being used mainly in the locally organized manufacture of fermented beverages such as honey wine, or mead. These practices persist today. A reasonable assumption is that the early British observers had an eye for exporting African honey to the tables of Europe, which would generate cash for exporters. To do this, however, the local beekeeping practices would have to be "improved." Quality control of the final product was necessary for export, and production levels had to be increased. The effort to improve and develop local beekeeping practices in east Africa was probably the first time the bee-human relationship found itself the object of policy based on the logic of modern development: to intensify commodity production to increase surplus, allowing for trade and export and the consequent generation of cash. As the decades passed, beekeeping would find itself the object of intervention based on even more notions about people, these insects, and their environment.

IBRA and Its Conferences on Apiculture in Tropical Climates
Based in Cardiff, Wales, and founded in 1949, IBRA is a not-for-profit organization that seeks to make people aware of the vital role bees play in agricultural and natural ecosystems. It also promotes the study of the conservation of bees. Every four years since 1976, IBRA has held conferences in cities throughout the world to bring government officials, bee researchers, beekeepers, and development aid organizations together to focus on issues of fostering beekeeping in the tropics.

National-International Scales of Beekeeping Development Programs
The goal of the first IBRA conference in 1976 at a very basic level was to ensure the establishment of particular social relations between national donor and recipient governments involving the flow of financial resources for development (Crane 1976). Eva Crane, director of IBRA, explicitly stated that its goal was for donors from the developed world and recipients from the developing world to meet and make the necessary social contact for the formulation of development projects between their respective countries. The first conference also established a consensus on the meaning of beekeeping that would support the formation of those social relations and justify the flow of financial resources between patron and client national governments. Tecwyn Jones, representing the United Kingdom, proclaimed that beekeeping as a practice fit the rural development needs of third world countries and the new funding mandates of first world government agencies, which were to target rural areas and peasant farmers, focusing on ways to directly alleviate rural poverty. For the first time, he formally declared the virtues of beekeeping: (a) it provides an extra nonperishable food for the peasant family; (b) it produces marketable, cash-generating crops of honey and wax; (c) it enables pollination of agricultural, horticultural, and tree crops; (d) it requires little capital; (e) it requires little land, and the quality of the land itself is not important; (f) it offers flexibility to the producer, as beekeepers may invest however much time they want into their operations, from low-tech, spare-time beekeeping, to more sophisticated, full-time operations; and (g) it does not compete for resources with any other agricultural activity—"it is a true and bountiful bonus" (Crane 1976, 4). These virtues, mainly of a socioeconomic nature, would be repeated time and again in the international beekeeping discourse.

Jones then commented on the favorable market conditions for honey and wax at the time, and that because of the introduction of modern technology and management techniques in place of extensive traditional management practices, beekeeping development projects would result in higher honey yields. He also emphasized the importance of the formation of cooperatives in beekeeping projects to provide processing, packing, and marketing centers for bee products. Where such projects had been implemented, Jones argued, honey production increased dramatically. The thinking behind early British efforts at beekeeping development in East Africa (identification of traditional beekeeping practices as a "problem" requiring the adoption of modern technology and management practices to generate cash income for peasants) is evident even in the words of speakers representing recipient countries who describe the traditional production systems as unproductive, requiring more "intelligent management" (Crane 1976, 13).

Consolidating the Local-International Scale with the Narrative of Productive Conservation

Beginning with the third conference in 1984, the narratives of beekeeping changed dramatically from conference to conference. The new narratives did not reflect significantly new knowledge about the spatiality of beekeeping as a social or ecological system. Rather, the narratives reflected efforts by those in the beekeeping development circles to rescale beekeeping development programs away from the national-international scale toward a more local-international scale in an effort to generate more interest in and financial support for beekeeping programs.

In the third conference in 1984, presenters began to discuss the decentralization of beekeeping development programs (IBRA 1985). Several participants challenged the status quo large-scale national beekeeping development programs in favor of those involving and empowering local people, especially women. Despite this growing awareness of challenges to the dominant national-international scale of programs, the overall interest of conference organizers remained ensuring the flow of resources to support beekeeping no matter where it was implemented or under what conditions. This is clear from repeated statements from participants stressing their view of beekeeping's virtuous social and environmental characteristics, irrespective of any social or environmental geography, local or global.

By the fourth conference in 1988, researchers who had worked in local communities involved in beekeeping appeared to take on a much more important role (IBRA 1989). They began to attribute new social and environmental characteristics to programs. B. Clauss and R. Zimba gave a presentation on the importance of native forests (*miombo*) in Zambia as especially good honey-producing areas. They praised traditional, "local" beekeeping; it is not primitive, they noted, but rather a highly skilled and appropriate technology, developed by people as an adaptation to very complex local conditions and resources. For the first time, researchers made the argument that in East Africa, where native woodlands dominate bee forage, beekeepers have an interest in protecting the environment, reasoning that woodland destruction would cause honey yields to drop. Their statements suggested a link between beekeeping and conservation, and other presenters emphasized beekeeping's importance through emphasis on the ecological service that bees provide, pollination, not only for cultivated plants, but for wild ones as well.

In the fifth IBRA conference in 1992, beekeeping promoters thrust beekeeping completely into discussions surrounding sustainable agricultural development and conservation in the developing world (IBRA 1994). conference papers explored the link beekeepers have to the environment, especially natural environments, through the bees that produce their honey. B. Clauss, again using examples from East Africa, claimed that beekeepers have an environmental vision different from others who exploit natural resources. Other cases presented from Sri Lanka and Belize cast beekeepers as protectors of forests in the developing world, but they did so without marshalling any systematic empirical evidence.

Throughout the papers of the fifth conference, there was clear dissatisfaction with the status quo beekeeping development programs because of the lack of participation of local people. Shu-Numfor Godlove Numfor from the Cameroon Association for Apiculture Development mentioned that governments were out to get financial assistance related to beekeeping without assessing already existing beekeeping resources in the country. The result is the importation of ideas and technology that may not be very appropriate for local beekeepers. He stated, "Gone are the days when we relied on funding agencies to develop our communities, for today funding agencies seem to be more interested in getting information for their dossiers than in funding programmes that come from communities." Obviously aware of the dangers of becoming reliant and overdependent on donor funds and the particular agendas associated with them, he emphatically stated, "Let the people hear of our projects and offer us assistance, rather than we looking for assistance to start projects" (IBRA 1994, 160).

This final conference evidenced an even stronger link beekeeping promoters make among the practice, broad goals of forest conservation, and traditional, local people. Researchers represented beekeeping as inherently conservationist, yet they never supplied systematic empirical evidence that such a universal property applies in all areas where bees and humans interact. There were calls for stronger local control over beekeeping projects, since past development planning and implementation have privileged the interests of national government recipients and national government donors. While certain case studies pointed to problems in the status quo promotion of bee-keeping, the international conference's purpose to keep development aid flowing from donor to recipient required global narratives around which every interest could organize: beekeeping and beekeepers are by their nature conservation oriented, and stronger local control of beekeeping planning and implementation brings about positive social and environmental outcomes.

A number of conclusions can be made at this point with respect to the ways in which the IBRA conferences reflected the rescaling of development and conservation from the 1970s to the 1990s discussed in the first section. First, though, beekeeping was promoted as an ideal rural development ac-tivity long before the notion of productive conservation would take shape in Amazonia as a response to human and environmental problems of develop-ment (Drescher and Crane 1982). Project planners and consultants had to jus-tify beekeeping development projects on some grounds, and at the first-ever IBRA conference, interests consolidated the basic tenets of a global narrative of how and why beekeeping should be promoted as a rural development strat-egy. Beekeeping as poverty alleviation would generate cash income, and the way to do this was to intensify already existing apicultural practices by local people via the introduction of modern technology and management prac-tices. Strikingly absent from the early discussions were references to ecologi-cal interactions among bees, flowers, and people. Rather, the presence of bees, beekeepers, and flowers in rural ecosystems, and some mutually beneficial relationship among them, were taken for granted. Conference participants portrayed beekeeping as simply taking advantage of already existing non-dynamic resources and generating cash from them. Perhaps this was why beekeeping seemed so attractive to development officials; unlike other agri-cultural development strategies, it seemingly required no alteration of the ecological system. Beekeeping was taking advantage of a "subsidy from na-ture" (Anderson, May, and Balick 1991).

The political economy of beekeeping development aid planning and im-plementation evident in the first conference maintained or strengthened the top-down approach of development. That approach involved donor nations' providing consultancy services to the agricultural ministries of developing countries to make beekeeping a more important part of their programs. It ap-pears that donor governments interested in funding development used bee-keeping as a way to channel funds to client governments. Client governments were not demanding this assistance; the social relations and meanings be-hind the IBRA meetings served as a way to spur demand. People in the devel-oping world took part in receiving beekeeping aid under the dominant de-velopment ideology expressed at the conference.

Beginning with the third IBRA conference, calls appeared to rescale bee-keeping development programs. Various speakers brought up the need to move beyond the national level and spur the growth of beekeeping. This was a call for a shift in power relations, and such a shift would be justified by discursively making beekeeping a different practice from what it was earlier. Beekeeping became a source of development that was culturally and technologically "appropriate" for local people, including women. Beekeeping also took on ecological qualities never mentioned before; beyond the pollination services bees provide to help conserve local plant species, the presence of beekeepers discourages overuse of pesticides that harm bee colonies. Beekeeping as a way to build social capital was even emphasized, and decentralized beekeeping development programs were promoted as being more efficient in their effects and delivery of services, a stated goal of neoliberalism. By the fourth conference in 1992, conference participants consolidated the full picture of beekeeping as an activity that brings development and conservation through the *international* funding of *local* beekeeping efforts. Strikingly, people who supposedly understand the most about the production ecology of beekeeping failed to present scientific evidence that beekeeping causes or brings about conservation.

In sum, calls for and shifts in power relations in beekeeping development programs occurred from the 1970s to the 1990s, paralleling the dynamics vis-à-vis the international political economy of development and conservation discussed in the first section. Beekeeping development planning and implementation shifted away from centralized, national programs involving technology transfer to one that attempted to put greater control in the hands of beekeepers themselves and recognize the value of traditional forms of beekeeping knowledge and practice. That such a shift would also conserve natural local environments was a final change in the meaning of beekeeping that supported the new local-international scale of beekeeping as productive conservation.

Beekeeping Representations in Funding Proposals from Rondônia
From the early to late 1990s, rural workers associations, unions, and cooperatives in Rondônia, Brazil, wrote proposals for productive conservation grants based on beekeeping. These project proposals were submitted to Planafloro and other granting agencies, including development and environmental organizations in North America and Europe, and large NGOs in Brazil. Some proposals were for values as high as US$250,000, while others were as humble as US$1,200. From 1995 to 1998, Planafloro alone distributed $1.7 million to thirty-eight local NGOs in Rondônia who proposed beekeeping as a significant component of their project, an incredibly large amount of money considering that the groups had never received direct development assistance for this activity (Brown 2001b). In this section I present and analyze the main elements of twelve grant proposals I obtained during field work in

Brazil. The names and locations of the local organizations have been with-held to protect their identities. What is so instructive about the narratives in these grant proposals is how they parallel the discussion in the later IBRA conferences of beekeeping in relation to development, conservation, and decentralization. My analysis suggests that the discourse of productive conservation had become a global one generated out of the spatiality of social relations that formed as worldwide development and conservation were rescaled. Similar to the murky descriptions of beekeeping's ecological link to conservation in the IBRA conferences, these project proposals pay little to no attention to the material production ecology of beekeeping in the places where it was to be practiced in these projects. The proposals assume that bee-keeping is "ecological" and that promoting it will lead to forest conservation and social justice.

There are two main types of project proposals: the initiation of a beekeep-ing project, and the continuation of work in progress. The former type of pro-posal often begins with a section that discusses the historical context of the local organization and its efforts. It tells the tragic story of the colonization of Rondônia: the government, calling the region a new El Dorado, stimulated colonization of the region, but because of poor planning and systematic mar-ginalization of the colonists, unsustainable agricultural was practiced, and, consequently, widespread deforestation took place. The organization laments the lack of government incentives, such as subsidized credit, for engaging in "ecological" production. Protests organized by rural workers in Porto Velho, Rondônia's capital, to call attention to these matters have only resulted in violence against the workers.

With the lack of schools, roads, transport, and health care in the country-side, life is already difficult for most who work the land, whether owners, day laborers, or sharecroppers. The environment as well suffers under these con-ditions. With the soils as poor as they are in Rondônia, farmers are forced to continue deforesting the land to produce food. Since more area each year has been deforested, then abandoned, a substantial proportion of deforested lands are in some form of regrowth (*capoeira*). The proposed project will help forge a resource development policy that will "fix" people on the land and avoid continued aggression against nature. Without this assistance, large landholders (cattle ranchers) will take over the land abandoned by small farmers, and even more deforestation will ensue. How can families produce in a way that will allow for environmental conservation and economic de-velopment? Beekeeping, along with the planting of agroforestry plots and the distribution of seedlings for reforestation, is the answer.

The typical project is based on the idea of demonstration areas, which will be used to train farmers and organizations in the region how to earn a liveli-hood based on ecologically and economically sustainable production activities. These new activities represent a wiser use of available resources, take advantage of available labor, and promise to raise family incomes. These proposals invariably speak of the establishment of a revolving fund: money granted to the organization will result in profits from the sale of production, which will be reinvested in future production. The grassroots nature of the organization is also emphasized in project proposals. One grassroots organization proposal made special mention that no outside group was imposing any agenda on it. One proposal argued directly for decentralization in development. "The idea of getting money directly into the hands of worker organizations gives credibility to producers who engage in activities that preserve the environment. This is better than giving money even to [international and even national] NGOs who really do not represent the interests of rural workers." In addition to equipment (ranging from basic beekeeping equipment, to desks and computers, to the motorcycle or Toyota Bandeirante 4×4 truck that is ever-present in proposals), the organization asks the donor to fund educational programs to convince rural workers that the adoption of "alternative agriculture" is a viable way to "guarantee the survival of man and nature." Further justifications for the project include projections of profits from beekeeping. Those profits are compared with projections of how much forest would have to be destroyed to plant an equivalent cash value of harvested corn. One project proposal claimed that thirty bee colonies produce 1,000 kilos of honey per year, equivalent to the value of corn produced on 10 ha of deforested land.

The second type of proposal, for projects already in progress, emphasizes how the recipient organization has already established successful alternative production practices thanks to seed money from other donors, and presses for more money in order to expand the activities of the organization. These include membership expansion, via implementation of demonstration areas in new regions and launching of new environmental education programs, and involvement of women in decision making and management of the organization. One proposal, for example, states that women, due to their dexterity, careful attention to detail, and expertise in cooking, will be exceptionally helpful in the weeding, transplanting of seedlings, and processing of fruit products from agroforestry plots. The organization may also claim that, having reached a certain level of production, it would benefit from the expansion of processing capacity. Here it may seek to acquire the industrial facilities necessary to receive honey and other agroforestry products from the countryside for processing and storage for later shipment to wholesalers. In addition to honey, these products include palm hearts, frozen fruit pulp, and fruit preserves. At this point, project budgets may rise into the hundreds of thousands of dollars. Such proposals often include computer systems, vehicles, large refrigeration and freezing equipment, and air conditioners.

The beekeeping program proposals from small farmer groups in Rondônia reveal much about the globalization of the productive conservation discourse up to the 1990s. The main elements of the proposals contain the very
same elements as the narratives that comprised the discourse evidenced
in the IBRA proceedings. First, the programs to which small farmer groups
wrote proposals were all local-international in their scale of organization. International capital was being funneled to local groups to carry out projects
local people wanted. Project proposals described beekeeping in socioeconomic and ecological terms that helped construct and reinforce this new
scale of organization that sought to wrest power from interests organized at
the national scale. The globalized narrative of beekeeping as productive conservation was simply reproduced by the farmers in their proposals of what
were supposedly local projects: large-scale, centralized development programs have created many social and environmental problems in the region;
the needs of local farmers have been ignored in development; money granted
to local farmer NGOs is the most efficient way to bring about change; decentralized programs help build social capital; and beekeeping is an ideal small-
scale, "ecological" activity that generates important cash income, unlike the
predominant unsustainable land uses of cattle ranching and annual cropping that cause deforestation. Similar to the narratives from the international meetings, the funding proposals reviewed above never presented systematic, empirical evidence of the link that exists between beekeeping and
forest conservation. It appears that for project donors, it was sufficient merely
to state that beekeeping was ecologically sustainable to ensure the flow of
funds.

Beekeeping's Material Production Ecology in the Amazonian Context
How well do the narratives above, which have supported the transfer of millions of dollars of development aid to local grassroots organizations in Amazonia over the last ten years, correspond to the actual production ecology
of beekeeping in Amazonia, particularly Rondônia? Does the idea that beekeeping leads to forest conservation, an idea that first developed in beekeeping as conservation in East Africa, apply in Amazonia? Recall that the logic of
beekeeping as productive conservation is as follows: bees depend on flowers
for nectar to make honey. Local people can make a living, or at least supplement their living, with the sale of honey. Therefore, beekeepers are environmental guardians of sorts because they ultimately depend on the maintenance of the same environment bees depend upon for their survival. This
environment is presumed to be a "natural," forested environment.

The global narratives about beekeeping that paralleled the general rescaling of conservation were both a product and a prop of the spatiality of social
relations emerging in the 1980s and 1990s in development and conservation.

Those narratives have masked, however, a contingent, spatial dynamic of the ecological biogeography of forests that supports the material production ecology of beekeeping, though this spatiality has emerged over a much longer time scale. The most important biogeographical reality masked by the narratives is that the bee species used for the productive conservation projects is not even native to the Amazon. The native honey-producing bees of the Amazon are stingless bees, so called because they evolved with a vestigial sting apparatus. Such bees are native to the tropics throughout the world, meaning they must have been widely present on Gondwanaland, the ancient southern supercontinent, which was composed of the present-day continents of Africa, South America, and Australia, before it began to break up through plate tectonics (Michener 1979). Indigenous peoples in the tropical Americas have known about and used stingless bees for thousands of years. Their cultural importance to Mayan peoples and some Amazonian groups is well studied (Kent 1984; Jara 1995; Schwarz 1948; Posey 1983). Indigenous and peasant peoples have come to use stingless bees in very sophisticated ways to maximize honey production and extend the life of colonies, and evidence exists that use of these bees has undergone periods of intensification to meet the honey and wax demands of both precolonial and colonial rulers. Of the 300 stingless bee species that exist in the neotropics, species of the genus *Melipona* are especially known and used for honey production. Rondônia contains at least seven *Melipona* species (Brown 2001a). *Melipona,* like all bee species that live in large colonies, are generalist foragers. Their nesting requirements in the wild, however, are rather specific—they require nesting in cavities of live trees in the forest (Camargo 1990). If enough forage is available, however, they can be removed from their live tree home and placed in man-made hives for management in deforested areas.

The common honeybee used so prominently in commercial beekeeping around the world is the cavity-nesting bee *Apis mellifera.* Unlike their stingless relatives, honeybees originated long after the Gondwanaland breakup. The group has an Old World origin (Michener 1979). Humans have introduced and used *A. mellifera* in Asia and the Americas far from where it originated. Since its origins, *A. mellifera* has evolved into numerous geographic races, or subspecies, within its vast natural range, which extends from the Cape of Good Hope to Scandinavia, from Dakar to the Ural mountains and to the eastern frontier of Iran (Crane 1976, 42). Like stingless bees, *A. mellifera* is a generalist forager. Unlike *Melipona,* however, *A. mellifera* is able to nest in a wide range of cavities, natural or man-made, and in the case of African races, it is able to nest in the open. Before colonial independence in Spanish and Portuguese America, the only bees available for honey production were stingless (Kent 1984). After independence, European races of *A. mellifera* were introduced at various times to various points in the Americas, from Mexico in the north, to southern South America. Crane (1992) places the date at 1839. These bees were never able to survive well in the intact humid tropical forests of the Amazon. Jesuit missionaries first introduced European races of *A. mellifera* to southeastern Brazil in the late 1830s. Later introductions by German colonists in Rio Grande do Sul, Santa Catarina, and Paraná followed in the 1840s (Nogueira-Neto 1972, 17–21). Introductions to the northern regions of Brazil occurred in the 1870s. By the mid-1950s, it appears that European bees were widespread in most regions of Brazil except for Amazonia. Researchers reported that few colonies of *A. mellifera,* if any, existed in the entire state of Amazonas (Kerr et al. 1967).

The state of *Apis* in the New World changed dramatically in 1957, when the Africanized "killer" bee was accidentally introduced to the New World in Rio Claro, São Paulo (Kerr 1967). In the mid-1950s, the Brazilian beekeeping community had reached a consensus that the European race of honeybee present in Brazil (*Apis mellifera mellifera*), though extremely gentle and easily managed, was poorly suited to the tropical and subtropical environment of Brazil. The poor adaptation of European bees to Brazil led to low honey production, high susceptibility to disease, and inefficient pollination, among other undesirable characteristics (De Jong 1996). High-level federal government agriculture department officials resolved to improve beekeeping in Brazil with the introduction of genes from African races of honeybees. In 1956, the Brazilian government dispatched Brazilian scientist Warwick Kerr to Sub-Saharan Africa to collect honeybee queens representing different subspecies of *A. mellifera.* In 1957, Kerr succeeded in introducing thirty-five African bee queens into European bee colonies in Brazil, and a few dozen eventually escaped, setting off one of the greatest ecological explosions of a nonnative species ever recorded. The African-European hybrids resulting from the introduction, currently referred to by most observers as the "Africanized honeybee" (AHB), colonized the South American continent at an average rate of 300 km/year, reaching Rondônia in the early 1970s when the first agricultural colonization projects were implemented (Taylor 1977, 1985).

No *A. mellifera* species is well suited to commercial honey production in intact humid tropical rain forest (O. Taylor, entomologist and Africanized honeybee expert at the University of Kansas, personal communication). Africanized bees, like all *A. mellifera,* are mostly adapted to open vegetation formations or edge areas where both forest vegetation and open vegetation formations area available. This may be due to their highly evolved, generalist foraging strategy that allows them to take advantage of so many different floral sources of nectar and pollen. Patchy environments, where both forest and nonforest vegetation are present, result in more diverse forage sources for bees over space and time. It may be that flowers blooming in the open areas (especially in weedy, abandoned areas) are important for the bees in one season, and in another they are able to find adequate forage in the tops of forest trees (Espina Pérez 1983). Humid tropical forest alone is not adequate habitat for bees (*A. mellifera*) or for a beekeeping operation. Even in forests that are drier than the Amazon, such as gallery forests in Rio Grande do Sul or areas of the Atlantic coastal forest, intact forests do not provide adequate forage for bees, and it is the more open *campos,* or grasslands, especially ones overrun with weeds, that provide the best forage. These areas are often quite large and uniform, yielding honeys that are clear and aromatic (Juliano 1972).

Due to the dynamics of development in Rondônia, Brazil, successive areas of forest were cleared, used for a brief time, and then left fallow as soil fertility levels declined. Such a process helped transform an intact humid tropical rain forest into a sun-drenched, weed-infested landscape that is ideal Africanized bee foraging habitat. Though able to nest in the open, Africanized bees found ideal and plentiful nesting sites in the dead wood littering the countryside. Ecological studies of honey production in Rondônia have confirmed that it is the weedy landscapes, especially those comprising a species of iron weed, *Vernonia polyanthes,* that makes commercial honey production in Rondônia a viable operation (Crane 1981; Condé 1989; Condé, Rezende, and Melo 1990). Other plants characteristic of successional growth also contribute to colony survival, especially during the wet season, by providing colonies with pollen (Marquez-Souza et al. 1993). In short, beekeeping, under its current configuration in Rondônia, would not be possible without the very process of deforestation that, under the pretext of productive conservation, it is supposed to halt.

Conclusion

In development discourses, beekeeping has always been grounded in the notion of mutual benefits to people and environment. The first narratives about beekeeping were based on early British efforts at intensifying beekeeping production with modern technology among traditional tribesmen of East Africa, with the goal of exporting honey and generating cash income among tribal peoples. Governments of donor countries worked with recipient countries' agricultural ministries to carry out beekeeping development based on its desirable socioeconomic characteristics. Later, however, beekeeping interests promoted beekeeping as an ideal activity for environmental reasons, and argued that local people, their traditional beekeeping practices, and their relationship to forest conservation initiatives were being ignored by beekeeping development programs. Existing information about the material production ecology of beekeeping in Rondônia, however, suggests that the narratives have developed with little empirical basis. Actors with a stake in productive conservation, however, promote false representations of beekeeping because it is in their interest to do so. At present, productive conservation appears to be more about contests for control over development planning and implementation and much less about producing anything in a way that is truly socially and ecologically sustainable.

People from different parts of the world, with different backgrounds, training, ideologies, and experiences, have held, and continue to hold, different notions as to what is important to encourage with respect to the relationship between people, bees, and flowers. In order for beekeeping as a practice to be promoted all over the world, its promoters were forced to take beekeeping out of its rootedness in place. They devised narratives to describe it in ways that have served the promoters' interests, no matter where beekeeping has been implemented as a development practice. When the provision of financial resources became key in social relations, as it has been with the promotion of beekeeping in productive conservation programs, then representations of beekeeping to transfer money out of donor countries to recipient countries have become more important than understanding the relationship between socioeconomic and ecological dynamics of places, which ultimately affect whether an activity succeeds or fails.

The globalization of conservation has presented us with opportunities and challenges. The increasing spatial interaction of globalization has allowed new spatialities of social relations to develop, giving a voice to previously powerless resource users such as indigenous peoples, rubber tappers, and colonist farmers. The global community now knows of their plight and has agreed that something has to be done about it. Creating such spatialities required a guiding, global concept of development with conservation, and the idea of productive conservation fills that role. There is also a material spatiality of the production systems that would be promoted as part of productive conservation efforts. How those spatialities unfold in particular places like Rondônia may actually be in conflict with the meanings that helped support the social networks that formed to promote the practice.

The challenge in the globalization of conservation is to shrink the "distance" that has developed between the spatialities. In the case of beekeeping, let us accept that the changes in social relations that occurred—the rescaling of development and conservation—continues as is. That leaves us with the possibility of either altering the meaning of beekeeping-as-productive conservation that guides or maintains those social relations, or changing the material production ecology of the practices so that they match better the ideals of productive conservation. To change the meaning would be to accept that beekeeping does not lead to conservation in the way it was originally proposed. The proper way to promote beekeeping as conservation is to promote it as an activity that takes advantage of already cleared and abandoned lands, in effect bringing them into production and hopefully making it possible for people to make a living without having to deforest new areas. Thus, beekeeping could be part of plans to *intensify* production. If one, however, wants to maintain the idea that bees are "natural" and require forests to survive, then the material production system must change. Africanized bees must be abandoned in beekeeping projects, and stingless bees of the genus *Melipona* must take their place. I make no claim that either of these strategies is free of contradictions or problems. I do claim that promoting Africanized beekeeping in Amazonia in the uncritical way it has been in the past as part of efforts to rescale development and conservation serves no one in the long term.

Until now, how productive conservation *activities* fit productive conservation *ideals* has not received much critical attention. This chapter has attempted to model what such a critical view should look like. Once done, we remember that productive conservation, like all ideas, operates in a political economy. For that reason, interests rely on representations (meanings) of the activities they promote to justify their control of resources. As long as those social relations are supported by representations that interests themselves create, and without any feedback from material, ecological processes, then our efforts at development with conservation in a globalized world merely involve shifts in power relations, which may or may not involve desired changes in the material landscape. If the beekeeping story is similar to that of fish farming, agroforestry, and other productive conservation activities among small farmers, indigenous peoples, and rubber tappers, then all we can say is that those groups have more control over development resources today than they did during the so-called Decade of Destruction of the 1980s (Cowell 1990). What those people do with that power—improving livelihoods and environments or not—is for us to investigate, not to assume.

References

Anderson, A. B., P. H. May, and M. J. Balick. 1991. *The subsidy from nature: Palm forests, peasantry, and development on the Amazon frontier.* New York: Columbia University Press.

Brenner, N. 1999. Globalisation as reterritorialization: The re-scaling of urban governance in the European Union. *Urban Studies* 36 (3): 431–51.

Brohman, J. 1990. *Popular development.* Cambridge, MA: Blackwell.

Brown, J. C. 1992. Development in Rondônia, Brazil 1980–1990: POLONOROESTE, nongovernmental organizations (NGOs) and the World Bank. MA thesis, University of Kansas.

———. 2001a. The effect of tropical deforestation on stingless bees of the genus *Melipona* (Insecta: Hymenoptera: Apidae: Meliponini) in central Rondônia, Brazil. *Journal of Biogeography* 28:623–34.

———. 2001b. Responding to deforestation: Productive conservation, the World Bank, and beekeeping in Rondônia, Brazil. *The Professional Geographer* 53 (1): 106–18.

Camargo, J. M. F. 1990. Stingless bees of the Amazon. In *Proceedings of the 11th International Congress of the International Union for the Study of Social Insects (IUSSI)*, ed. G. K. Veeresh, B. Mallik, and C. A. Viraktamath, 736–38. Bangalore, India: IUSSI.

Cerny, P. 1995. Globalization and the changing logic of collective action. *International Organization* 49 (4): 595–626.

Condé, P. A. A. 1989. Levantamento preliminar da flora apícola dos municípios de Ji-Paraná e Porto Velho, Rondônia. Porto Velho, Brazil: Secretaria de Est. de Indústria, Comércio, Ciência e Tecnologia. Unpublished manuscript.

Condé, P. A. A., H. U. Rezende, and O. H. Melo. 1990. Fluxo de néctar e pólen em onze colônias de *Apis mellifera* no município de Ji-Paraná, Rondônia. *Ciência e Cultura* 42 (11): 978–83.

Cowell, A. 1990. *The decade of destruction: The crusade to save the Amazon rain forest.* New York: Henry Holt.

Crane, E., ed. 1976. *Apiculture in tropical climates.* Full report of the first Conference on Apiculture in Tropical Climates. London: International Bee Research Association.

——. 1981. When important honey plants are invasive weeds. *Bee World* 62 (1): 28–30.

——. 1992. The past and present status of beekeeping with stingless bees. *Bee World* 73 (1): 29–42.

De Jong, D. 1996. Africanized honey bees in Brazil: Forty years of adaptation and success. *Bee World* 77 (2): 67–70.

Drescher, W. R., and E. Crane. 1982. *Technical cooperation activities: Beekeeping; A directory and guide.* Eschborn, Germany: Deutsche Gesellschaft für Technische Zusammenarbeit (GRZ) GmbH.

Espina Pérez, D. 1983. *Apicultura tropical.* Cartago, Costa Rica: Editorial Tecnológica de Costa Rica.

Hall, A. 1997. *Sustaining Amazonia: Grassroots action for productive conservation.* Manchester, UK: Manchester University Press.

IBRA (International Bee Research Association). 1985. *Proceedings of the third International Conference on Apiculture in Tropical Climates: Nairobi, Kenya, 5–9 November, 1984.* London: IBRA.

——. 1989. *Proceedings of the fourth International Conference on Apiculture in Tropical Climates, Cairo, Egypt, 1988.* London: IBRA.

——. 1994. *Fifth International Conference on Apiculture in Tropical Climates. Trinidad and Tobago, 7–12 September, 1992.* Cardiff: IBRA.

Jara, F. 1995. Bees and wasps: Ethno-entomological notions and myths among the Andoke of the Caqueta River in the Colombian Amazon. *Latin American Indian Literatures Journal* 11 (2): 148–65.

Juliano, J. C. 1972. Identificação de Espécies de Interesse Apícola da Flora do Rio Grande do Sul. In *II Congresso Brasileiro de Apicultura com Participação Internacional, Dias 4–6 de setembro de 1972,* 85–87. Sete Lagoas, Brazil: Congresso Brasileiro de Apicultura.

Keck, M., and K. Sikkink. 1998. *Activists beyond borders.* Ithaca, NY: Cornell University Press.

Kent, R. B. 1984. Mesoamerican stingless beekeeping. *Journal of Cultural Geography* 4 (2): 14–28.

Kerr, W. E. 1967. The history of the introduction of African bees to Brazil. *South African Bee Journal* 39:3–5.

Kerr, W. E., S. F. Sakagami, R. Zucchi, V. d. Portugal-Araújo, and J. M. F. Camargo. 1967. Observações sobre a arquitetura dos ninhos e comportamento de algumas espécies de abelhas sem ferrão das vizinhanças de Manaus, Amazonas (*Hymenoptera, Apoidea*). *Atas do Simpósio sobre a Biota Amazônica* 5:255–309.

Klooster, D. J. 1997. Conflict in the commons: Commercial forestry and conservation in Mexican indigenous communities. PhD diss., University of California, Los Angeles.

Marquez-Souza, A. C., M. L. Absy, P. A. A. Condé, and H. A. Coelho. 1993. Dados da obtenção do pólen por operárias de *Apis mellifera* no município de Ji-Paraná (RO), Brasil. *Acta Amazonica* 23 (1): 59–76.

Michener, C. D. 1979. Biogeography of the bees. *Annals of the Missouri Botanical Garden* 66:277–347.

Millikan, B. 1995. *Pedido de investigação apresentado ao Painel de Inspeção do Banco Mundial sobre o Planafloro.* Porto Velho, Brazil: Fórum das ONGs e Movimentos Sociais Que Atuam em Rondônia, Friends of the Earth.

——. 1997. *Análise crítica da implementação do Plano Agropecuário e Florestal de Rondônia um ano após o acrodo para sua reformulação.* São Paulo and Porto Velho, Brazil: Amigos da Terra Internacional; Programa Amazônia, Oxfam; Associação Recife-Oxford para a Cooperação ao Desenvolvimento.

Nogueira-Neto, P. 1972. Notas sobre a história da apicultura brasileira. In *Manual de apicultura,* ed. J. M. F. Camargo, 17–32. São Paulo, Brazil: Editora Agronômica Ceres Ltda.

Perreault, T. 2003. Making space. *Latin American Perspectives* 30 (128): 96–121.

Planafloro. 1997. *Programa de apoio às iniciativas comunitárias (PAIC): Cartilha informativa*. Pamphlet. Porto Velho, Rondônia: Secretaria de Estado do Plan, e Coord. Geral, Coord. Geral de Planafloro.

Posey, D. A. 1983. Folk apiculture of the Kayapó Indians of Brazil. *Biotropica* 15 (2): 154–58.

Rich, B. 1989. Funding deforestation: Conservation woes at the World Bank. *The Nation*, 23 January.

———. 1994. *Mortgaging the earth: The World Bank, environmental impoverishment, and the crisis of development*. Boston: Beacon Press.

Schwartzman, S. 1986. *Bankrolling disasters: A citizens' environmental guide to the World Bank and the regional multilateral development banks*. San Francisco, CA: Sierra Club.

Schwarz, H. F. 1948. Stingless bees (*Meliponidae*) of the Western Hemisphere. *Bulletin of the American Museum of Natural History* 90:1–546.

Sins of the secular missionaries. 2000. *The Economist*, 27 January.

Smith, N. 1993. Homeless/global: Scaling places. In *Mapping the futures: Local cultures, global change*, ed. J. Bird, 87–119. New York, Routledge.

———. 1995. Remaking scale: Competition and cooperation in prenational and postnational Europe. In *Competitive European peripheries*, ed. H. Eskelinen and F. Snickars, 59–74. Berlin: Springer Verlag.

Swyngedouw, E. 1997. Neither global nor local: "Glocalization" and the politics of scale. In *Spaces of globalization*, ed. K. Cox, 137–66. New York: Guilford.

Taylor, O. R. 1977. The past and possible future spread of Africanized honeybees in the Americas. *Bee World* 58:19–30.

———. 1985. African bees: Potential impact in the United States. *Bulletin of the Entomological Society of America* 31 (4): 15–24.

World Bank. 1990. *How the World Bank works with nongovernmental organizations*. Washington, DC: World Bank.

———. 2000. *The World Bank, NGOs and civil society*. Washington, DC: World Bank.

World Commission on Environment and Development (WCED). 1987. *Our common future*. Oxford: Oxford University Press.

PART II

Linking Scales in Livelihood Analysis and Global Environmental Science

Livelihoods, the ways people make a living, are central to sustainability strategies. Not surprisingly, livelihood analysis has become a core feature of the broadening concerns of global conservation. Agriculture, livestock raising, extraction of forest products, hunting, fishing, and other types of resource use make up the main livelihood styles that are spotlighted in the interface with global conservation and sustainability. Understanding the expanding interface requires building on the analysis of these resource-based livelihoods while, at the same time, incorporating the types of knowledge that tend to be the emphasis of global environmental science.

A number of interdisciplinary environmental approaches contribute the main threads to livelihood analysis in the chapters of this section. These approaches include both established fields and newer ones such as political ecology, human ecology, and cultural ecology. They are characterized by a level of interdisciplinary and specialized roots in the fields of geography, anthropology, sociology, biology, and environmental studies and sciences in general. The above approaches overlap, to an extent, with the global environmental sciences. One of the main contrasts is that the global environmental sciences tend to place emphasis on biogeophysical analysis, which is then framed according to new topical fields such as biodiversity science, conservation science, global change science, and human dimensions of global change. These types of expert knowledge are often merged, to varying degrees, and the chapters here faithfully reflect this trend by incorporating livelihood analysis and one or more of the global environmental sciences.

Scales of analysis are critically important to understanding livelihoods and the perspectives of the global environmental sciences. The paramount status of scale issues is vividly apparent in the biodiversity and conservation sciences, for example. For instance, there a scale emphasis is frequently applied to the spatial patterning of taxonomic diversity in wild plants and animals. By contrast, scale issues are much less understood in the application of the biodiversity and conservation sciences to agriculture and the spatial patterning of human-managed plants and animals that, taken in the broad sense, range from fully domesticated to the indirectly affected. At present, the biodiversity of agrarian landscapes is a particular interest in global conservation and sustainability circles, and it is a focus of two of the chapters of this section. First is a focus on the biodiversity of garden plants (chap. 5). Gardens are important sites of biodiversity worldwide. Garden biodiversity is especially pronounced in the house-lot gardens of developing countries of the tropics. These biodiverse gardens are commonly found in urban settings as well as rural locales. By their very ubiquity, house-lot gardens hold a great deal of potential interest for biodiversity conservation and livelihood security concerns that are increasingly prioritized in global sustainability circles. As a result of globalization, however, this potential importance is marked with a defining tension. Namely, the global tendencies toward urbanization, which are acute throughout the world, do seem to sever the life of cities and urban dwellers from the outlying rural areas. Yet, as chapter 5 demonstrates, urban house-lot gardens are highly dependent on the flows of seed stock and planting material that frequently occur via social networks that span the city and the countryside. As a result, we must analyze these important connections among key microenvironments at a scale across the urban-rural distinction.

A second locus for promoting the biodiversity of food plants is field agriculture (chap. 6), which is the single most widespread use of land resources worldwide. As a result, global conservation and sustainability interests are deeply concerned about the biodiversity of these agroecological habitats. In particular it is the biodiversity and conservation sciences that have examined the scientific basis of agrobiodiversity use in farm fields worldwide. The group of five food plants that form the mainstay of world nutrition—rice, wheat, maize, manioc, and potatoes—stand out as the most biologically varied since each crop complex (including semidomesticates and so-called wild crop relatives) comprises multiple species and thousands of varietal types. The management of biodiversity in these food plants has become a major concern of global conservation, food, and development organizations during the past decade. Building global interest has raised a basic tension, focused on scale issues, that exists in the prospects for agrobiodiversity conservation and sustainable use. Global conservation and sustainability organizations have thus far tended to adopt the scale of the rural community as the spatial foundation for spreading programs and projects worldwide. On the one hand the emphasis on the community scale must be seen as typical of globalization for several reasons. These factors include the inherent tendency of globalization efforts to adopt a single or certain constrained set of spatial models as well as the logistical ease and political credibility that are often gained in dealing directly with a rural community. On the other hand, however, various forms of land use activities involving crucial resources *may not* be scaled to the level of individual communities. Focusing on this point, chapter 6 evaluates how multiple scales and the predominance of the multicommunity scale are needed for the sustainability of the seed flows of agrobiodiversity in the Andean potatoes. Institutions and resource-user groups at the multicommunity scale are representative of an important challenge to global environmental science and sustainability policies.

Scale politics permeates issues beyond agricultural biodiversity, as demonstrated by chapter 7, on desertification, climate variability, range management, and livestock raising in the Sahel of Africa. This set of concerns has held an iconic sort of status in some of the most highly visible activities of global environmental science during the past three decades. The issues represented in the Sahel have been a particular concern of the global change and conservation sciences. These approaches have undergone formative scientific developments, particularly by integrating remote sensing (RS) and geographic information systems (GIS) analysis, which are increasingly central to presenting a picture of the Sahel as a global environmental problem. Scale is central to the analysis, interpretation, and policy guidance (e.g., scientific reports, recommendations, consultancies) that the global environmental sciences have created about the interactions among climate variability, range management, and livestock raising, and the possibility that desertification may result. One key role of scale is that it helps to focus our attention on a characteristic tension of globalization that is present in this analysis of the global environmental sciences. As chapter 7 demonstrates, global change and conservation sciences depend on those scales of analysis that typically cover large areas of the Sahel. It is such large areas that represent the sort of scale at which many data are collected, analyzed, and presented for policymakers in the global environmental sciences. Yet those scales are inadequate for the analysis of Sahelian livelihoods and resource use. The resulting tensions between the scale of global environmental science and that of existing livelihood activities has continued to become more apparent as a pulsating force of politics that both surrounds and infuses this environmental globalization.

5 Urban House-Lot Gardens and Agrodiversity in Santarém, Pará, Brazil: Spaces of Conservation That Link Urban with Rural

ANTOINETTE M. G. A. WINKLERPRINS

Dona Antonieta,[1] an avid gardener, has stated that "*uma casa sem jardim me deixa triste*"—"a house without a garden makes me sad." I once gave her a ride from the urban zone of Santarém to her rural-floodplain home on a boat that I had rented. The boatman had several potted plants on deck, one of them a little pepper plant, the fruit of which he used to spice up his daily fish. Being naturally attracted to plants, Dona Antonieta took a close look at the pepper plant while our three-hour trip unfolded. She decided that it was a variety she did not have and casually pocketed several peppers from the plant, "*para plantar no jardim*"—"to plant in the garden."

This vignette serves as an *entrée* into the world of house-lot gardens, agrodiversity, and urban-rural linkages. As globalization continues to redefine and reconfigure the world at all scales, the time has come to rethink the types of spaces in which conservation occurs. Urban areas in developing countries are not frequently considered when we think about biodiversity conservation. Yet when we broaden the definition of biodiversity to include agrodiversity, short for "agricultural biodiversity" (Brookfield 2001, 40–46), then we acknowledge that agriculture contains important material to be conserved, and the range of conservation spaces enlarges to encompass a broadly defined agriculture. Urban agriculture, including house-lot gardens, becomes a space in which important acts of conservation occur. In this chapter, through a case study on house-lot gardens in urban Amazonia, I address two interfaces that are key to the new geography of conservation (Zimmerer, chap. 1 in this volume): that between the urban and the rural, and that between conservation and agriculture.

House-lot gardens are sites of well-documented agrodiversity conservation. Indeed, much of the literature on homegardens has focused on their

1. All names are pseudonyms.

plant diversity and their spatial variation. From the original classics on home-gardens (e.g., Anderson 1954; Simoons 1965; Kimber 1966, 1973), to more recent publications (e.g., Gajaseni and Gajaseni 1999; Lamont, Eshbaugh, and Greenberg 1999; Agalet, Bonet, and Valles 2000; De Clerck and Negreros-Castillo 2000; Zaldivar et al. 2002) the garden literature is filled with accounts of the tremendous variety of plants within them. Although this plant diversity has been well documented, I find that few studies fully engage and theorize about the implications of this diversity for plant conservation. Gardens, including urban gardens, should be thought of as "new" spaces of conservation in the productive and dynamic landscape of the city. They can also be thought of as "vernacular" or "incidental" spaces of conservation, as their primary goal is not conservation but subsistence. Additionally, as this case study demonstrates, the garden conservation space extends beyond the urban garden and into a wide area of source material that includes rural gardens and other agricultural spaces. Therefore, the conservation space is multi-local and fluid and weaves together the city and its rural hinterland. This role of gardens as an interface between the urban and the rural has only recently been fully considered (Linares 1996; WinklerPrins 2002a).

House-lot gardens, also known as homegardens, housegardens, dooryard gardens, and kitchen gardens (Niñez 1984), have been defined as spaces around dwellings that are used to satisfy household needs for food, fiber, medicine, and construction materials, and provide places for recreation and aesthetic experiences (Kimber 1966). They can be found in both rural and urban areas, and, I argue, can be a medium through which what is "urban" and what is "rural" blur. House-lot gardens have traditionally been studied in rural areas where they are part of a continuum of cultivation areas, taking on many forms and functions throughout the world (Niñez 1984; Landauer and Brazil 1990; Netting 1993). Urban house-lot gardens bring the rural into the city ("ruralization"), the benefits and drawbacks of which are debated among urban planners (Sanyal 1985; Bibangambah 1992). In the urban zone, gardens often fall under the rubric of "urban agriculture," a term that also encompasses berm cultivation and other nongarden agricultural activities (e.g., Freeman 1991; Egziabher et al. 1994; Linares 1996; Madaleno 2000). I see urban gardens as crossover zones between rural and urban, places where householders can be both urban and rural at once as they make socioeconomic and cultural transition in their dynamic livelihoods. As such, gardens reflect the tension between rural and urban lifeways that many people and institutions confront in the increasingly globalized world of today.

I start this chapter with some comments on the scholarly literature on gardens as spaces of agrodiversity conservation, a description of the study area, and a note on the methods used for this research. I then concentrate on the empirical findings of a case study from the Brazilian Amazon, focusing my discussion on the flows of germplasm to and from gardens. I specifically consider how these flows link the urban to the rural, blurring the divide, highlighting the tension, and demonstrating the complex linkages between the two realms.

The Study of Gardens

Much attention in the literature on house-lot gardens has been on documenting their agrodiversity and physical structures (Niñez 1984; Landauer and Brazil 1990; Nair 2001). For many decades elegant descriptions have been published with accompanying diagrams and tables firmly documenting the extensive agrodiversity and spatial layout of homegardens (e.g., Esquivel and Hammer 1992; Gajaseni and Gajaseni 1999; Lamont, Eshbaugh, and Greenberg 1999; Agalet, Bonet, and Valles 2000; De Clerck and Negreros-Castillo 2000; Nair 2001; Zaldivar et al. 2002). Most authors list, document, and discuss the multitude of species found in gardens and how they are arranged, but few discuss the flows and networks that link gardens to the outside world. Some recent work, especially by geographers, is moving forward from the "landscape design" approach to a more political ecological approach to garden studies. This research more explicitly considers the sources of plant material and contextualizes gardens in their current socioeconomic situation (e.g., Brierley 1985; Thomasson 1994; Linares 1996; Lerch 1999; Greenberg 2003).

There are few published accounts of Amazonian house-lot gardens (e.g., Guillaumet et al. 1988; Padoch and de Jong 1991; Smith 1996; Lamont, Eshbaugh, and Greenberg 1999; Lima and Saragoussi 2000; Slinger 2000; Madaleno 2000; WinklerPrins 2002a; Murrieta and WinklerPrins 2003). This limited attention may be for two reasons. The first is that there has been an overwhelming research emphasis on "wild" vegetation in biomes such as forests and savannas in the region. Rural or urban gardens, as anthropogenic spaces, have been largely ignored until recently. Ironically, many of the "wild" forests of the Amazon that receive so much attention have been, and continue to be, managed and "gardened" by their keepers (e.g., Posey 1985; Balée 1989, 1994; Cleary 2001). A second reason for the lack of attention to gardens is due to the perceived absence of the land scarcity issue in Amazonia (Padoch and de Jong 1991). Much of the garden literature has come from areas in Asia (especially Java), parts of Africa, and the Caribbean islands, where there is either physical or structural land scarcity. Gardens are often conceptualized as a response to conditions of land scarcity through agricultural intensification (Netting 1993, 53–56); hence, their importance has not been considered in places dominated by extensive forms of land use, such as Amazonia.

The Amazon, however, is a changing place, where the impact of globalization on the environment is impressive. The population of the region, though still marginalized in terms of decision making, is profoundly affected by global, regional, and national economic forces and environmental policies. One of the consequences of this globalization of the Amazon is that, contrary to its popular image, the region is now predominately an urban place (Browder and Godfrey 1997). The most recent census data indicate that the region is now 70 percent urbanized (Instituto Brasileiro de Geografia e Estatistica 2002). Urbanization is most rapid in medium-sized cities such as Santarém, my study site.

Failed agricultural colonization projects, continued short-term boom-and-bust natural resource exploitation cycles, and the changing expectations of rural residents through access to television and other media have led to extensive rural-to-urban migration (Browder 2002). Unfortunately, economic development has not kept pace with urbanization, and employment opportunities are limited. This economic situation makes urban homegardens critical to sustaining urban life, and also continues to link urban newcomers to their rural homes. Urban homegardens offer a level of food security that is critical in an underdeveloped region such as the Amazon.

Study Area

The case study presented in this chapter is from Santarém, Brazil. This municipality (analogous to a county in the United States) is located in the western part of the state of Pará (fig. 5.1). While it is about the size of Maryland (24,154 km^2), less than 1 percent of the entire municipality (40 km^2) is defined as urban. Santarém is the fourth most populous municipality in the Amazon, with 262,672 residents in 2000 (Prefeitura Municipal de Santarém 2001). Of that total, 71 percent live in the official urban zone of Santarém (the "city") and the remainder in the rural area. Interestingly, and of importance to the topic of this chapter, is that people of Santarém are mainly "multilocal," making residence in both the city and the country and moving between the urban and rural homes with great fluidity and regularity. This makes census figures, which always attempt to fix a person in one place, a rather poor depiction of reality.

Urban growth in the municipality is due, in part, to regional economic changes similar to those that contribute to rural-to-urban migration elsewhere in the developing world. Changes in livelihood opportunities in rural zones, with accompanying changes in people's life expectations and attitudes, have resulted in rapid rural-to-urban migration. At the regional level, the collapse of jute cultivation has contributed to the flow of migrants from the floodplain region. Since the 1930s jute had been an important fiber crop grown in the rural floodplain area of Santarém and processed into sacking material in the urban zone (Gentil 1988; Homma 1998). Jute provided a livelihood for many locals and permitted year-round occupancy of the floodplain.

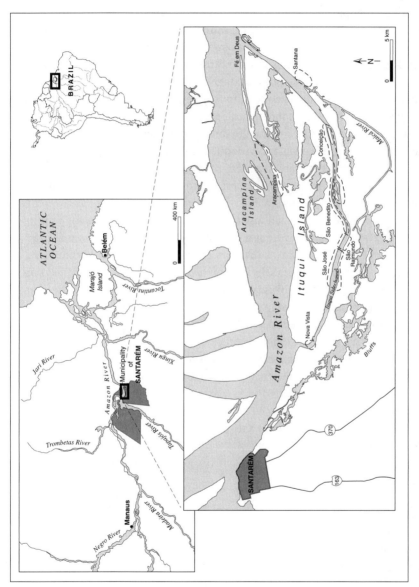

FIGURE 5.1. Map of study region (cartography by the Center for Remote Sensing and Geographic Information Science of Michigan State University)

Under most circumstances, year-round occupancy of the floodplain is challenging due to a high and unpredictable annual flood of the Amazon River, which prevents livelihood activity for five months out of the year. Jute, however, could be cultivated during the flood and for fifty years offered sustenance to floodplain dwellers in the season when all other activities are stagnant (WinklerPrins 1999). The Amazonian jute market collapsed around 1990 due to a bundle of mostly extralocal events: changes in commodity packaging, such as bulk handling and the substitution of synthetic for jute sacks to transport agricultural commodities; the removal of tariffs on imported jute; and the increasingly poor quality of jute produced in the region (Gentil 1988; Homma 1998). Since that time floodplain dwellers have been finding alternative livelihoods. In the rural floodplain zone there is increasing reliance on the marketing of agricultural products, fishing, ranching, and some timber extraction, but these are limited to the dry season. People have started migrating to upland areas on bluffs proximate to the floodplain during the flood season (WinklerPrins 2002b).

Part-time occupancy of a house in the urban zone is also increasingly attractive, and many from the rural area are seeking a house-lot in the "city." The average length of urban house-lot occupancy is nine years, which coincides with the timing of jute's decline. In the urban zone, fish and timber processing and several other minor industries offer some jobs, but most employment is anticipated and not real. Santarém is mostly a regional service town with limited substantial employment opportunities and is best thought of as "overurbanized" (Browder and Godfrey 1997, 13). The reality of much of the population that has migrated to the city is unemployment or underemployment. Under these circumstances most households become "split families with livelihood strategies straddling the rural and urban sectors" (Ellis 1998, 7). Having residences in both the urban and rural environments is very common in Santarém and allows the use of resources from both zones (Nugent 1993; WinklerPrins 2002a). House-lot gardens form a critical link between the urban and rural parts of households as germplasm and products are exchanged.

Methods
The empirical research on urban homegardens presented here builds on research conducted during 1995–96 in the rural floodplain zone of the municipality of Santarém (WinklerPrins 1999). The focus of subsequent visits has been continued research with the same kin and community groups in both the rural-floodplain and urban zones in order to document the transformation of a society as it urbanizes and becomes more integrated into the globalized world. The methodological approach has been qualitative and participatory, with a heavy emphasis on ethnographic interviews (Maxwell 1998).

Extensive fieldwork in the rural zone allowed social access to urban house-lot gardens. Two overlapping data sets are used in this case study, a preliminary study of twenty-one households from 2001 and a year-long investigation of twenty-five households from 2002–3. Initial fieldwork was timed to coincide with dominant urban occupancy during the flood season, permitting interviews and garden documentation in the urban zone. I used connections to families from the floodplain that I had worked with during my earlier work in the region. I also hired a local field assistant with whom I had previously worked. She knew the families that I wanted to study, and had an excellent appreciation for the urban-rural link. Her life story parallels the situation of so many families in the region: born and raised on the floodplain, but now recently "urbanized." She conducted the monthly interviews that were the central part of the year-long study.

Initially the objective of the research was to obtain a sense of the biological diversity of the urban gardens of formerly rural (or, as we found out, still partially rural) residents. Therefore, sampling was purposeful, yet it also targeted gardens of different sizes, household composition, and neighborhoods within the urban area. Interviews involved the completion of a questionnaire that served as a conversation guide, a tour of the garden with the person (usually female) who was primarily involved in maintaining it, and the completion of a list of all plants in the garden.[2] Interviews and related conversations were conducted in the house-lot garden. From the preliminary study it became clear that the urban households were still significantly connected to their rural antecedent. Therefore, the year-long follow-up study, from 2002 to 2003, focused on the urban-rural flows that the gardens enabled. This latter study involved an initial survey of the house-lot garden as well as monthly visits to document germplasm and product flows. The research presented here is based on evidence gathered in both the preliminary and full studies.

Garden Agrodiversity

The total number of plants (species and varieties) encountered in the urban house-lot gardens by category (fruit, medicinal, ornamental, vegetable, and other) is summarized in table 5.1. I documented a total of 182 species with 95 varieties for a total of 277 locally identified plants in the twenty-five urban gardens surveyed. This finding fits within a range of total plants species found in other studies of Amazonian homegardens: 61 (Guillaumet et al. 1988), 77 (Smith 1996), 136 (Lerch 1999), 161 (Lamont, Eshbaugh, and Greenberg 1999), 168 (Padoch and de Jong 1991), and 262 (Lima and Saragoussi 2000). This wide range of numbers is due, in part, to different sampling

2. Permission was obtained from all participants following Michigan State University's University Committee on Research Involving Human Subjects (UCRIHS) guidelines.

TABLE 5.1.   Enumeration of number of species in urban house-lot gardens in Santarém

Plant category	Total number of plants	Total number of species	Total number of varieties	Average number of plants per garden
Fruit	87 (31.5%)	41 (23%)	46	14
Medicinal	76 (27.5%)	65 (36%)	11	5
Ornamental	56 (20%)	36 (20%)	20	4
Vegetable	55 (20%)	37 (20%)	18	3
Other	3 (1%)	3 (1%)	0	<1
Total	277	182	95	25

Note: Not all plants have been positively identified and properly categorized.

strategies; for example, some researchers include ornamental and medicinal plants, while others do not.

Most plants in urban homegardens are classified as fruit trees and shrubs, followed closely by medicinal plants. Ornamental plants and vegetables make up smaller percentages of the plant composition. In fact, most plants categorized as vegetables are herbs and spices used to make condiments in small quantity. When considering total species, however, the medicinal plants dominate. There is an impressive variety of fruit plants: I documented ten banana varieties, seven mangos, five lime, and five *Inga*. In the other plant categories there were also impressive numbers of varieties, including eight peppers and even eight rose varieties. This demonstrates attentiveness to species variability and leads to a consideration of gardens as spaces of experimentation and perhaps incipient domestication (Clement 2001). It is also noteworthy that the species with most varieties (banana, mango, lemon) are all Old World plants.

The top ten plants in each category are listed in table 5.2. Best estimates of scientific names are included, but these are not based on voucher identification, only on common name interpretation. The historical origin of most of the medicinal and ornamental plants is not known. Of the fruit species in the top ten, the majority are Old World. For a complete listing of most of the plants, please consult WinklerPrins (2002b).

Flows of Germplasm: How Is Germplasm Obtained?

There are two relevant time frames when considering the source of germplasm used in gardens. The first is the historical time frame, the origins of plants now commonly grown in urban house-lots in Santarém. The second is the present time frame, and I consider the source areas and germplasm flows that supply urban gardens with new cuttings, seeds, seedlings, and so on. Historically, a considerable number of crops found in urban gardens of Amazonia originated in the Old World. Their arrival in the New World occurred

TABLE 5.2. The ten most important plant species by category found in urban house-lot gardens in Santarém, Pará, Brazil

Plant rank	Local name	English name	Scientific name
Fruit			
1	LIMÃO	LIME	*Citrus aurantifolia*
2	CÔCO	COCONUT	*Cocos nucifera*
3	Acerola	Barbados cherry	*Malpighia glabra*
4	Ata/Piña	Sweetsop	*Annona squamosa*
5	MANGA	MANGO	*Mangifera indica*
6	Goiaba	Guava	*Psidium guajava*
7	JAMBU	MALAY APPLE	*Eugenia malaccensis*
8	BANANA	BANANA	*Musa* spp.
9	Cupuaçu		*Theobroma grandiflorum*
10	LARANJA	ORANGE	*Citrus sinensis*
Medicinal			
1	Cidreira		*Lipia alba?*
2	Boldo (pequeno, grande)		*Perimus boldos*
3	Salva de Marajó		*Lipia grandis*
4	Marupazinho		*Simaruba amara?*
5	Arruda		*Ruta graveolens*
6	Folha grossa / malvarisco		*Plactantrus amboinicus*
7	Coramina		?
8	Manjericão		*Ocimum brasilicum*
9	Pião branco		*Jatropha curcas*
10	Vindicá		*Alpina rutans*
Ornamental			
1	Rosa	Rose	*Rosa* spp.
2	Generic "ornamental" plant[a]		
3	Tajá		*Caladium bicolor*
4	Samambaia		*Adiantum* spp.
5	Balão		?
6	Comigo ninguem pode		*Diffenbachia picta*
7	Laço de amor		?
8	Begonha	Begonia	*Begonia semperflorens*
9	Croto		*Dracaena fragens*
10	Papoula		*Hibiscus* ssp.
Vegetable			
1	Macaxeira	Sweet manioc	*Manihot esculenta*
2	CEBOLINHA	GREEN ONION	*Allium cepo*
3	Xicória		*Cichorium intybus*
4	Pimenta	Pepper (sweet)	*Capsicum* spp.
5	Alfavaca		*Ocimum micranthum*
6	Tomate	Tomato	*Lycopersicum esculentum*
7	COUVE COMUN	KALE (COMMON)	*Brassica oleracea*
8	Jerimum	Squash	*Cucubitae* spp.
9	Caruru		*Amaranthus flavus*
10	Pimenta malaguenta	Pepper (hot)	*Capsicum* spp.

(continued)

TABLE 5.2.   *(continued)*

Plant rank	Local name	English name	Scientific name
Other			
1	Seringa	Rubber	*Hevea brasiliensis*
2	Cuieira	Calabash gourd	*Crescentia cujete*

Notes: Plants listed in capital letters have known Old World origins. Voucher specimens of plants were not taken. Scientific names are interpreted from common name and are subject to change. Scientific names taken from Smith (1996, 1999) and Lima and Saragoussi (2000).
ᵃMany gardeners cannot identify every ornamental, and this is a generic category.

during the first wave of globalization that ensued after 1492, as part of the so-called Columbian Exchange of crops. Spanish and Portuguese colonists, as well as their imported slaves, brought over plants that dramatically changed the species composition of the Amazon Basin. The ubiquitous mango tree that graces the streets of many Amazonian cities and is extant in almost all of the gardens surveyed is not an Amazonian native (Smith 1999, 143). Approximately half of the total fruit and vegetables surveyed in the urban gardens are Old World natives, and these nonnative species dominate the top ten list (table 5.2). Historical germplasm flows, while ubiquitous in the Americas, are often forgotten when considering present day agriculture as the "new" crops have become so integrated and "normalized" (Crosby 1972; Zimmerer 2001a). Historical germplasm flow is more relevant in the context of agrodiversity (compared with conventional biodiversity), as human involvement in plant transfer is inherently stronger in agricultural than nonagricultural landscapes. Indeed, it would be difficult to imagine the region without bananas and mangos, yet they were introduced only about 500 years ago.

An often-forgotten transfer of germplasm, with accompanying knowledge system, was that brought by slaves from West Africa to the Americas (Carney 2001). In the Caribbean, where slaves could cultivate their own crops during weekends, slave "gardens were an amalgamation of tropical plants brought from Africa, Asia, plants borrowed from the surviving Carib and Arawak Indians, and some European plants adapted to tropical conditions" (Pulsipher 1990, 29–30). Descendents of those slave populations continue to cultivate this rich mixture of germplasm throughout the New World (Kimber 1966; Brierley 1985; Carney 2001). The persistence of an amalgam of plants of diverse origins also characterizes Amazonia, where people of African descent are a key component of the region's ethnicity.

Present-day germplasm is obtained predominantly through women's gift exchange networks among kin and acquaintances. Through these networks, households have access to different ecological niches: rural-floodplain, rural-upland, and urban. For example, kin connections between an urban household and one in the rural-upland zone may well supply an urban household with *cupuaçu (Theobroma grandiflorum)* seeds since these trees are quite numerous in the upland area. Frequent circulation of kin living in different zones make these exchanges easy and fluid. Friendship and kin socializing networks are critical to the circulation of agrodiversity. In Santarém gifts of seeds, pits, and cuttings circulate easily among female neighbors and kin. Gifts of germplasm are as key to the maintenance of social networks, especially women's networks, as are gifts of garden products themselves (Murrieta and WinklerPrins 2003).

Gift exchange networks are not barter networks. Barter is a direct exchange—a kilo of corn for a kilo of fish, for example—whereas within a gift exchange network value is "saved." In homegardens in the Caribbean island of Montserrat "exchange is so prominent that virtually all gardeners talked of the exchange role of their gardens and readily showed that what they were growing had exchange value" (Thomasson 1994, 26). This exchange value relates to both garden products and the genetic redistribution of the seeds and other plant "starts." Both of these are at work in Santarém and are crucial to the urban survival of the peasant class. Similar complex exchanges of germplasm have been documented elsewhere. In another part of the Amazon, Lerch (1999) found that most planting material is obtained through exchanges among neighbors and family. In Casamance, Senegal, Linares (1996) documented seed exchanges among kin groups who live in different parts of the rural-urban continuum. In Catalonia, Agalet, Bonet, and Valles (2000) traced complex networks to exchange medicinal plants among neighbors in rural communities and found that the significance of these networks frequently transcends their original purpose. Zimmerer's investigation of food plant networks in the Andean highlands illustrates the complexity of seed network geography: seeds are obtained through a multiscale procurement network from multiple environments, and from inter- as well as intracommunity exchanges (Zimmerer 2003).

In her investigation of immigrant Yucatec Maya house-lots near Cancun, Mexico, Greenberg found that planting material came from a variety of sources (Greenberg 2003). Plants were transferred from the original immigrant source village, a three-hour bus ride away, while some were received as gifts from neighbors or kin locally or in the source village. Still other plants were remnants of former forests that existed in the region, and lastly, some plants were volunteers from refuse piles. Greenberg's analysis clearly indicates the importance of germplasm flows between the source area of migrants and their destination. These plants help create miniature versions of the gardens from "back home" and as such are a critical space for the maintenance of ethnicity.

A second method of obtaining germplasm for gardens is through "self-provisioning" within an extended household. The majority of households surveyed are multi-local, occupying two to three places (urban, rural-floodplain, rural-upland) at once and self-provision between the locations. Germplasm is moved in small quantities from floodplain to city in plastic bags, coffee cans, and so on. For example, Dona Adela recently obtained an urban house-lot (fig. 5.2). She carefully chose her future urban house-lot so that it would have adequate sunlight and other conditions appropriate for the species she was planning to cultivate. Although her children were left in charge of building the new house and are now the ones occupying it (for schooling and employment purposes), she is in charge of planting the garden, and she arrives for her weekly visits with materials (including chickens) from her rural-floodplain garden.

Another form of self-provisioning within a household is the interface between cultivated fields and homegardens. Gardens are the location where field crop varieties are "stored" between seasons and during cycles when certain crops or varieties are not chosen to be part of the swidden system. My findings are similar to Lerch's in the Peruvian Amazon, where planting material sometimes comes from cultivated fields, and "one potential function of home gardens is as a reserve of cultivated plant diversity" (Lerch 1999, 57). Additionally, she documents experimentation with varieties and crops, a phenomenon of many gardens, including those in my study region (Murrieta and WinklerPrins 2003).

A third means of obtaining planting material is through purchase. In Santarém this is not an important source of planting material at all, perhaps a reflection of the still nonmonetized economy that my sampled gardens operate in. Purchases may become more important as households obtain more cash income. Elsewhere seed purchases are important. In the Peruvian Amazon, Lerch (1999) found that more than one-fifth of surveyed households purchased cultivated plants in nearby Iquitos. Much prestige and value is placed on "outside" germplasm; hence, sparse monetary resources are used to buy such planting stock. Another source of purchased germplasm is itinerant merchants who visit villages by boat. Whether conscious or unconscious, such openness to "outside" germplasm ensures continuation of diversity and a viable germplasm flow.

Lastly, in the context of conservation and globalization the role of outsiders bringing in germplasm cannot be discounted, especially as contacts between locals and outsiders increases with increasing globalization. Lerch (1999) documents numerous requests by locals of germplasm from researchers, missionaries, and others from far outside the neighbor/kin network, including herself. Balée (1994) recounts similar requests by the Ka'apor to expand their garden diversity with germplasm gifts from missionaries and anthropologists. Linares (1996) documents the importance of the introduc-

FIGURE 5.2. Dona Adela in her house-lot gardens: (*top*) with her daughter in the urban garden; (*bottom*) feeding chickens in her rural garden (photographs by the author, 2003)

tion of European varieties of crops, especially vegetables, to the cropping systems in and around Casamance, Senegal. In my study area, I also participated in this exchange network. I once received a bag of small mangos as a gift from a family in one of the floodplain communities I was studying. Upon arriving in another community I shared the bounty of the mangos (too many to eat alone) and was immediately asked for more of those because it was a "less prickly" variety and the people in the second community wanted to plant some. On a transnational scale, I have also participated in the gift network with packages of flower seeds bought in my local North American supermarket or garden shop. Also, responding to request, I have given packages of red cabbage seeds to the father of my field assistant in Santarém. These gifts from afar are highly valued.

"Outsider" diversity may seem trivial, but it demonstrates the great interest locals have in expanding the diversity of their gardens, especially with germplasm from far outside. The Columbian Exchange, as a form of early plant globalization, simply set in motion germplasm exchanges that continue to this day. Gardeners in the Amazon, as elsewhere, experiment with and rapidly adopt new germplasm. This has been demonstrated by the quick integration of Old World crops and other species that have been introduced since then into garden cultivation. It also demonstrates the very opportunistic nature of gardeners to seek and realize opportunities to obtain germplasm.

Flows of Garden Products: What Happens to Garden Products?

Approximately two-thirds of what is produced in urban gardens is for subsistence purposes. Those products not directly consumed by the producing household (about one-third of garden output) are given away through the vast gift exchange network. When asked about to whom the gifts were made, the response was overwhelmingly to family, neighbors, and friends. Few urban households in Santarém, only 2 percent, used products from their gardens to sell commercially, and again true barter was negligible. Food exchanges are an "important link in the food distribution system, effectively increasing the range of the food resource base and thereby enhancing . . . subsistence security" (Thomasson 1994, 26). Another important component that Thomasson notes is of gift giving and exchanges as a means of "storing credit." In other words, when my avocados are ripe, I give you several bags of them. Months later, when your pineapples are harvestable, you give some to me. This is also the essence of the garden product exchanges that occur in Santarém, and it requires a broadening of the idea of "subsistence." Garden production provides direct food provisions for householders but also permits access to products from other households through gift exchange, which contributes to meeting all of subsistence needs and contributes to urban food security. Garden products add to both the social and biological value of the city, comprising social and biological "capital" (Linares 1996).

It is also clear that the vast majority of households surveyed in the urban zone used the gift exchange network to obtain products not cultivated in gardens. Such products include other foods that could not be easily produced in the urban setting. Only 3 percent of households relied exclusively on purchased items to supplement what could be produced in the garden. Most notable was the finding that often there was a direct subsidy of the urban household by the rural household with products produced in the rural area (e.g., manioc, fish, corn for chickenfeed, beans; Nugent 1993; WinklerPrins 1999). Howorth, Convery, and O'Keefe (2001) documented a similar direct rural-to-urban subsidy within households in Dar es Salaam, Tanzania. Urban households are provisioned with rural products as needed and in season. In return, rural parts of households obtain products only produced in urban gardens, such as fruit from perennial trees, as needed and in season.

Exchanges go beyond goods to encompass services as well. For example, the husband of one of the participants in my survey was a bricklayer. His services were frequently exchanged for food products from other households. Also, access to government and other services was often facilitated through personal connections with a kin member (often a daughter or niece) who worked in one of the various bureaucracies. Given the long waits and stalemates possible at banks, social security offices, and other institutions, these connections genuinely improved the quality of urban life for locals and would be generously rewarded with a bag of whichever fruit was in season. "Service" exchanges have also been documented by Thomasson (1994) and Lerch (1999), especially in times of crisis such as hurricanes or floods.

Other Urban-Rural Links

Two other agricultural products flow between rural and urban zones of Santarém: manure and domestic animals. The dominant form of fertility management in urban gardens is the creation of a charcoal-rich *terra quiemada* (TQ)—burned earth, through frequent leaf litter sweeping and burning—which is applied to plants as a soil conditioner. However, manure is also used, either by itself or mixed in with TQ as an organic fertilizer and as a direct planting medium in raised beds for condiments. Chicken manure is frequently available within the garden itself or from a neighbor with chickens, part of the gift exchange network. Horse and cow manure are obtained from relatives or friends who have access to cattle and horses in rural areas, and are another example of the flows and links between urban and rural settings. Manure exchanges between rural source and urban sink are also documented in urban gardens in Senegal (Linares 1996).

Lastly, animals themselves, especially small fowl, flow between rural and urban zones. Chickens and ducks frequently form part of the house-lot. Chickens may be sent to live with relatives in the city during the flood season, when they are more difficult to keep on the floodplain. During the dry season these same chickens may be sent back to the floodplain to benefit from a larger area for foraging (fig. 5.2, bottom). Ducks, although they can swim and do well on the floodplain, are often fattened in the city where their movement is controlled. This is especially so during certain times of year when urban demand for duck meat is high, such as during religious festivals, in which regional duck dishes figure prominently.

Conclusions and Implications

House-lot gardens are places of extensive and dynamic agrodiversity. As such, they are clearly important sites for the conservation of agrodiversity. The use of gardens and the exchange of their various products maintain many species and varieties. In this chapter I have illustrated, by means of a case study from the Brazilian Amazon, the extensive informal-sector flows of plants and products between urban and rural areas that maintain house-lot gardens in the city. These gardens conserve agrodiversity while at the same time securing access to food in an overurbanized city. Thus, the gardens' social and ecological values are inextricably linked; as Linares (1996, 119) puts it: "Urban farming is one of the points where conservation interests and the interests of the cash-poor migrant begin to converge."

The overall number of species and varieties documented in the Santarém gardens is remarkable and echoes findings by others. The maintenance of garden agrodiversity is supported through women's gift and exchange networks that exist among kin, nonkin, and even outsiders in multiple ecological settings. These exchange networks yield different planting material and different foods during different times of the year, as well as offering access to services rendered by people within the network. There is great subtlety to germplasm exchanges: a few pocketed peppers, flowers received as a Mother's Day present, a mango pit from a friend, cuttings of favorite medicinals from a neighbor, even red cabbage seeds received as a gift from a researcher. Such things do not usually enter the realm of "business-as-usual" ecological or social science research, yet they actually play an important part in "incidental" conservation of agrodiversity and the sustainability of urban livelihoods. Urban home-gardens are "new," vernacular spaces for agrodiversity conservation as they are a type of "productive" or "utilized" landscape (Zimmerer and Young 1998; Zimmerer 2001b). They are worthy of investigation since "we . . . need an enlarged conception of [biodiversity], embracing commitment to the idea of preserving and enriching the complex ecosystems in which humans as well as other species live" (Linares 1996, 119).

The importance of homegardens to urban livelihoods has planning and policy implications. In many ways homegardens help ameliorate the cultural and livelihood shocks associated with the rural-to-urban movements endemic to globalization, and urban agriculture should be embraced by planners in regions with rapid urbanization. Gardens enable people to continue having access to a stable food supply outside of the cash economy, and therefore they enhance food security. They also offer occupational opportunities, and, most important, garden germplasm and products maintain critical kin and other social networks that ease the passage to becoming fully urban. They also offer psychological space, a zone of transition that offers a respite from the harsh realities of shifting regional economies. For women in particular, gardens help maintain a sense of aesthetic pride, an emotional link, and a psychological buffer as households move between rural and urban settings. Thus, the house-lot garden represents a crossover space that links rural and urban environments and reconciles traditional and modern lifeways.

To date there have been no deliberate development efforts (by governmental or nongovernmental offices) in Santarém targeting homegardens or any other sort of urban agriculture, and the practice is thus far rather invisible to government officials. In fact, one of the goals of my year-long research project was to make the municipal government aware of the importance of home gardening to urban quality of life and food security, so that gardens will not be legislated away, as has happened in many African cities (Sanyal 1985; Bibangambah 1992). Government officials, in their efforts to modernize the city, often enact legislation to exclude agricultural activities from cities. Outlawing cultivation of homegardens would have detrimental effects on urban residents, as their food security is dependent on it. Worldwide there have been much more extensive efforts at understanding and promoting research and information on urban agriculture. Canada's International Development Research Center (IDRC) has a program entitled Cities Feeding People that conducts research on and disseminates information about urban agriculture (IDRC 2004). The program also produces books and issues policy briefs on issues of urban agriculture (e.g., Egziabher et al. 1994).

Raising awareness of the importance of homegardens to local livelihoods has been considerable through my project. My yearlong research asked questions on a monthly basis that made women aware of their gardening in ways they never had been. Home gardening forms a part of everyday praxis and is often done with minimal self-consciousness. Over twelve months, however, my field assistant reported that women became much more aware of their work in the gardens and of the importance of their kin network. At the conclusion of the project we were thanked by many of the participants for conducting the study. Many women felt empowered by what they had learned of their own work, and hoped to use this knowledge to maintain the gardens they had created and to improve life in the city.

Acknowledgments
Research for this chapter was made possible by grants from the Center for Advanced Study of International Development, the Latin American and Caribbean Studies Program, and the Intramural Research Grant Program at Michigan State University, and support from Projeto Várzea (IPAM-Santarém).

I want to extend my sincere gratitude to Perpétuo Socorro de Sousa for her fantastic assistance in the field and to all those from Ituqui and "a cidade" for their patience and friendship. Many thanks also to Bilal Butt, Edna Wangui, Judy Hibbler, Christina Hupy, and João Mattos. This chapter was originally prepared for the EDARC workshop on 19 April 2002 in Madison, WI.

References

Agalet, A., M. A. Bonet, and J. Valles. 2000. Homegardens and their role as a main source of medicinal plants in mountain regions of Catalonia (Iberian Peninsula). *Economic Botany* 54 (3): 295–309.

Anderson, E. 1954. Reflection on certain Honduran gardens. *Landscape* 4:21–23.

Balée, W. 1989. The culture of Amazonian forests. *Advances in Economic Botany* 7:1–21.

———. 1994. *Footprints of the forest: Ka'apor ethnobotany; the historical ecology of plant utilization by an Amazonian people.* New York: Columbia University Press.

Bibangambah, J. 1992. Macro-level constraints and the growth of the informal sector in Uganda. In *The rural-urban interface in Africa: Expansion and adaptation,* ed. J. Baker and P. Petersen, 303–13. Copenhagen: Nordiska Afrikainstitutet.

Brierley, J. S. 1985. West Indian kitchen gardens: A historical perspective with current insights from Grenada. *Food and Nutrition Bulletin* 7 (3): 52–60.

Brookfield, H. C. 2001. *Exploring biodiversity.* New York: Columbia University Press.

Browder, J. O. 2002. The urban-rural interface: Urbanization and tropical forest cover change. *Urban Ecosystems* 6 (1/2): 21–41.

Browder, J. O., and B. J. Godfrey. 1997. *Rainforest cities: Urbanization, development, and globalization of the Brazilian Amazon.* New York: Columbia University Press.

Carney, J. A. 2001. *Black rice: The African origins of rice cultivation in the Americas.* Cambridge, MA: Harvard University Press.

Cleary, D. 2001. Towards an environmental history of the Amazon: From prehistory to the nineteenth century. *Latin American Research Review* 36 (2): 64–96.

Clement, C. R. 2001. Domestication of Amazonian fruit crops: Past, present, future. In *Diversidade biológica e cultural da Amazônia,* ed. I. C. G. Vieira, J. M. C. da Silva, D. C. Oren, and M. A. D'Incao, 347–67. Belém, Brazil: Museu Paraense Emílio Goeldi.

Crosby, A. W. 1972. *The Columbian exchange: Biological and cultural consequences of 1492.* Westport, CT: Greenwood Press.

De Clerck, F. A. J., and P. Negreros-Castillo. 2000. Plant species of traditional Mayan homegardens of Mexico as analogs for multistrata agroforests. *Agroforestry Systems* 48 (3): 303–17.

Egziabher, A. G., D. Lee-Smith, D. G. Maxwell, P. Ali Memon, L. J. A. Mougeot, and C. J. Sawio. 1994. *Cities feeding people: An examination of urban agriculture in East Africa.* Ottawa: International Development Research Centre.

Esquivel, M., and K. Hammer. 1992. The Cuban homegarden "*conuco*": A perspective environment for evolution and *in situ* conservation of plant genetic resources. *Genetic Resources and Crop Evolution* 39:9–22.

Ellis, F. 1998. Household strategies and rural livelihood diversification. *Journal of Development Studies* 35 (1): 1–38.

Freeman, D. B. 1991. *A city of farmers: Informal urban agriculture in the open spaces of Nairobi, Kenya.* Montreal: McGill-Queen's University Press.

Gajaseni, J., and N. Gajaseni. 1999. Ecological rationalities of the traditional homegarden system in the Chao Praya Basin, Thailand. *Agroforestry Systems* 46 (1): 3–23.

Gentil, J. M. L. 1988. A juta na agricultura de Várzea na área de Santarém: Médio Amazonas. *Boletim do Museu Paraense Emilio Goeldi, Antropologia* 4 (2): 118–99.

Greenberg, L. S. Z. 2003. Women in the garden and kitchen: The role of cuisine in the conservation of traditional house lot crops among Yucatec Mayan immigrants. In *Women and plants: Gender relations in biodiversity management and conservation*, ed. P. L. Howard, 51–65. London: Zed Books.

Guillaumet, J.-L., P. Grenard, S. Bahri, F. Grenard, M. Lourd, A. A. dos Santos, and A. Gely. 1988. Les jardins-vergers familiaux d'Amazonie Centrale: Un exemple d'utilisation de l'espace. *Turrialba* 40 (1): 63–81.

Homma, A. K. O. 1998. A civilação da juta na Amazonia: Expansão e declino. In *Amazônia: Meio ambiente e desenvolvimento agrícola*, ed. A. K. O. Homma, 33–60. Brasília, Brazil: Empresa Brasileira de Pesquisas Agropecuárias (EMBRAPA).

Howorth, C., I. Convery, and P. O'Keefe. 2001. Gardening to reduce hazard: Urban agriculture in Tanzania. *Land Degradation and Development* 12:285–91.

IDRC (International Development Research Center). 2004. Cities Feeding People. http://web.idrc.ca/en/ev-5911-201-1-DO_TOPIC.html. Accessed 5 August 2004.

Instituto Brasileiro de Geografia e Estatistica (IBGE). 2002. Tabulação Avançada do Censo Demografico 2000: Resultados Preliminares da Amostra. Rio de Janeiro: IBGE.

Kimber, C. 1966. Dooryard gardens of Martinique. *Yearbook of the Pacific Coast geographers* 28:97–118.

———. 1973. Spatial patterning in the dooryard gardens of Puerto Rico. *Geographical Review* 63:6–26.

Lamont, S. R., W. Hardy Eshbaugh, and A. M. Greenberg. 1999. Speices composition, diversity, and use of homegardens among three Amazonian villages. *Economic Botany* 53 (3): 312–26.

Landauer, K., and M. Brazil, eds. 1990. *Tropical home gardens.* Tokyo: United Nations University Press.

Lerch, N. C. 1999. Home gardens, cultivated plant diversity, and exchange of planting material in the Pacaya-Samira National Reserve Area, Northeastern Peruvian Amazon. MA thesis, McGill University, Montreal.

Lima, R. M. B. de, and M. Saragoussi. 2000. Floodplain homegardens on the Central Amazon Floodplain. In *The Central Amazon Floodplain: Actual use and options for sustainable management*, ed. W. J. Junk, J. J. Ohly, M. T. F. Piedade, and M. G. M. Soares, 243–68. Leiden, the Netherlands: Backhuys Publishers.

Linares, O. F. 1996. Cultivating biological and cultural diversity: Urban farming in Casamance, Senegal. *Africa* 66 (1): 104–21.

Madaleno, I. 2000. Urban agriculture in Belém. *Cities* 17 (1): 73–77.

Maxwell, J. A. 1998. Designing a qualitative study. In *Handbook of applied social research methods*, ed. L. Bickwan and D. J. Rog, 69–100. London: Thousand Oaks Press.

Murrieta, R. S. S., and A. M. G. A. WinklerPrins. 2003. Flowers of water: Homegardens and gender roles in a riverine caboclo community in the Lower Amazon, Brazil. *Culture and Agriculture* 25 (1): 35–47.

Nair, P. K. R. 2001. Do tropical homegardens elude science, or is it the other way round? *Agroforestry Systems* 53:239–45.

Netting, R. M. 1993. *Smallholders, householders: Farm families and the ecology of intensive, sustainable agriculture.* Stanford, CA: Stanford University Press.

Niñez, V. K. 1984. *Household gardens: Theoretical considerations on an old survival strategy.* Lima, Peru: International Potato Center.

Nugent, S. 1993. *Amazonian caboclo society: An essay on invisibility and peasant economy.* Providence, RI: Berg Publishers.

Padoch, C., and W. de Jong. 1991. The house gardens of Santa Rosa: Diversity and variability in an Amazonian agricultural system. *Economic Botany* 45 (2): 166–75.

Posey, D. A. 1985. Indigenous management of tropical forest ecosystems: The case of the Kayapó Indians of the Brazilian Amazon. *Agroforestry Systems* 3:139–58.

Prefeitura Municipal de Santarém. 2001. Diagnóstico institucional preliminar para apoio à elaboração do Plano Estratégico Municipal para Assentamentos Subnormais-Pemas. Santarém, Brazil: Prefeitura Municipal de Santarém.

Pulsipher, L. M. 1990. They have Saturdays and Sundays to feed themselves: Slave gardens in the Caribbean. *Expedition* 32 (2): 24–33.

Sanyal, B. 1985. Urban agriculture: Who cultivates and why? A case study of Lusaka, Zambia. *Food and Nutrition Bulletin* 7 (3): 15–24.

Simoons, F. J. 1965. Two Ethiopian gardens. *Landscape* 14:14–20.

Slinger, V. A. 2000. Peri-urban agroforestry in the Brazilian Amazon. *Geographical Review* 90 (2): 177–90.

Smith, N. J. H. 1996. Home gardens as a springboard for agroforestry development in Amazonia. *International Tree Crops Journal* 9 (1): 11–30.

———. 1999. *The Amazon river forest: A natural history of plants, animals, and people.* New York: Oxford University Press.

Thomasson, D. A. 1994. Montserrat kitchen gardens: Social functions and development potential. *Caribbean Geography* 5 (1): 20–31.

WinklerPrins, A. M. G. A. 1999. Between the floods: Soils and agriculture on the lower Amazon floodplain, Brazil. PhD diss., University of Wisconsin, Madison.

———. 2002a. House-lot gardens in Santarém, Pará, Brazil: Linking rural with urban. *Urban Ecosystems* 6 (1/2): 43–65.

———. 2002b. Recent seasonal floodplain-upland migration along the lower Amazon River, Brazil. *Geographical Review* 92 (3): 415–31.

Zaldivar, M. E, O. J. Ocho, E. Castro, and R. Barrantes. 2002. Species diversity of edible plants grown in homegardens of Chibchan Amerindians from Costa Rica. *Human Ecology* 30 (3): 301–16.

Zimmerer, K. S. 2001a. Report on geography and the new ethnobiology. *Geographical Review* 91 (4): 725–34.

———. 2001b. The reworking of conservation geographies: Nonequilibrium landscapes and nature-society hybrids. *Annals of the Association of American Geographers* 90 (2): 356–69.

———. 2003. Geographies of seed networks for food plants (potato, ulluco) and approaches to agrobiodiversity conservation in the Andean countries. *Society and Natural Resources* 16:583–601.

Zimmerer, K. S., and K. Young. 1998. Introduction: The geographical nature of landscape change. In *Nature's geography: New lessons for conservation in developing countries,* ed. K. S. Zimmerer and K. Young, 3–34. Madison: University of Wisconsin Press.

6 Multilevel Geographies of Seed Networks and Seed Use in Relation to Agrobiodiversity Conservation in the Andean Countries

KARL S. ZIMMERER

Introduction: Agrobiodiversity and Globalization

In "Return of the Native Seeds," a feature article of *The Ecologist* magazine, the authors paint a partly optimistic if ironic picture of the fate of agrobiodiversity amid globalization (Nellithanam, Nellithanam, and Samiti 1998). Focusing on the diverse food plants of India, these socially committed agronomists describe how the local impacts of the globalization of agriculture have led farmers to return to using their local varieties of rice, millet, lentils, and various other biologically diverse crops. Propelling the "return of the native seeds" is a confluence of forces that includes the deepening farm crisis, which is particularly acute for saturated markets of imported high-yield varieties (HYVs), and steep cutbacks in government support for agriculture. Agrobiodiversity cultivation, and hence de facto conservation, is thus sketched as a local response to these globalization pressures. As a result, agroecological sustainability is potentially compatible with economic globalization, at least in the example of India they emphasize. This observation highlights the newly complex and often contradictory relation, one that may be both antagonistic and enabling, between on-farm (in situ) conservation of agrobiodiversity and the forces of economic, political, and cultural globalization (Brookfield 2001; Brush 2000, 2004; Gade 1999; Louette, Charrier, and Berthaud 1997; Padoch and Sears 2005; Rhoades and Nazarea 1999; Zimmerer 1996, 2003c).

The expanded role of various global conservation, environment, and development organizations is also leading to the increasingly prevalent framing of agriculture as a conservation issue. Conservation organizations with a global perspective have invested a high level of interest, unprecedented for these groups, in sustainability initiatives and policies related to agriculture, livelihoods, and land use. Examples include major nongovernmental organizations (NGOs) that were originally concerned nearly exclusively with wilderness preservation, such as the World Wildlife Fund (WWF, also referred to as the Worldwide Fund for Nature), the International Union for the Conservation of Nature (IUCN; also known as the World Conservation Union), and Conservation International. In addition, a number of major global and international agricultural development agencies, including the main centers of the Consultative Group for International Agricultural Research (the CGIAR), now devote a growing share of their resources to agriculture as seen through the lens of conservation. Prominent examples include the International Potato Center (Centro Internacional de la Papa, or CIP) in Peru and the International Center for Tropical Agriculture (Centro Internacional Para la Agricultura Tropical, or CIAT) in Colombia (Ashby and Sperling 1995). Once in the vanguard of the Green Revolution and prominent players in the spread of farming systems based on HYV crops from the 1960s through the 1980s, the global agricultural centers are now severely underfunded due to reduced subsidies, the privatization of agricultural research and development, and the general shift of plant breeding to corporate agribusiness. In response, these global agricultural institutions are generating a new "green" image of their activities in developing countries. In these attempts at a pro-environment and pro-conservation makeover, the global centers of crop development have turned their missions, at least partially, to sustainable development and conservation that includes an emphasis on agrobiodiversity.

The various global institutions that are framing agriculture as a major conservation issue have been influenced by the so-called biodiversity phase of environmentalism (McNeely and Scherr 2003; Thrupp 1998; Western and Wright 1994; Wood and Lenné 1997). Their interests tend to place emphasis on the biological diversity component of the complex farming systems of developing countries and those in the tropics in particular. Various other environmental perspectives also influence the newer green view of agriculture, as demonstrated by alternative environmental and socioeconomic models that are based on organic and low-input farming systems, small-scale agrodiversity, and agroecology. Such models are increasingly framed with reference to global environmental change and the global environmental sciences (Altieri 1992, 1995; Brookfield and Padoch 1994; Brookfield and Stocking 1999; Carroll, Vandermeer, and Rosset 1990; Carruthers 1997; Gliessman 2001; Gliessman, Engles, and Krieger 2000; Hecht 1995; Rosset and Altieri 1997; Salick and Merrick 1990; Tenberg et al. 1998; Thrupp 1998; Vandermeer et al. 1998). While the perspective of these models may be equally important to agricultural sustainability, the agrobiodiversity emphasis has gained the most noticeable prominence among the global environmental organizations. Indeed, the biodiversity conservation emphasis has been increasingly applied during the past ten to fifteen years to the diverse biotic components of agroecosystems: crop plants (major and minor, food and nonfood), wild crop relatives, domestic livestock, soils, and noncrop biota.

Two heated and highly global debates are actively shaping the role of agrobiodiversity in global environmental conservation. The first involves the legal status of intellectual property rights over the domesticated, semidomesticated, and otherwise directly or indirectly managed organisms of agriculture (Brush 1991, 2004; Juma 1989; Kloppenburg 1988; Richards 1996; Shiva 1997). The preferences of farmers or countries seeking or holding proprietary rights to agrobiodiversity stand in stark contrast to the dominant common-pool approach that is used by global crop centers, breeding interests, agribusiness corporations, and the intermediate "farmers' rights" approach. The second debate concerns the spread of genetically engineered crops. Scientific works couched in terms of global environmental science and environmental activism have expressed concern about the spread of genetically modified organisms (GMOs) as a threat to agrobiodiversity and agroecosystems (McAfee 2003; Quist and Chapela 2001). In general it is interesting to note how the explicitly political, economic, and technological nature of the agrobiodiversity debates contrasts with some of the more visible controversies over the conservation of forests and wildlife. In the latter, "anthropogenic" and "pristine" are polarizing categories of controversies over landscape quality and conservation value, and much debate over conservation strategy revolves around the relative merits of "sustainability" versus "strict protection" models (Denevan 2001; Redford and Sanderson 2000; Schwartzman, Moreira, and Nepstad 2000; Stevens 1997; Zerner 2000). This broader conservation debate stemming from forest and wildlife conservation is important to agrobiodiversity since it emphasizes landscapes and livelihoods in a way that is often precluded in the focus on intellectual property rights and GMO diffusion.

Seed Supply and Agrobiodiversity in the Globalization of Conservation and Development

Seed supply has gained the status of a concrete issue as global conservation, agricultural, and development organizations direct attention to the farmers and farms responsible for agrobiodiversity (Almekinders et al. 1994; Almekinders and Elings 2001; Almekinders and Louette 2000; Almekinders and Louwaars 1999; Bentley and Vasques 1998; Cleveland, Soleri, and Smith 1994; Friis-Hansen and Sthapit 2000b; Thiele 1999; Witcombe, Virk, and Farrington 1998; Zimmerer 2002b). Among such global organizations and their national and local counterparts, programs in support of seed supply are typically designed to maintain and improve the biodiversity, quality, and affordability of planting materials of the diverse food plants. Development concerns are frequently central to these initiatives since seeds are a crucial and productive component, yet also often a limiting factor, in the farming systems of smallholder, peasant, and indigenous farmers. The importance of seed supply to development change is redoubled since it also helps provide a mainstay of the diets of many rural people in developing countries. Seed supply is similarly considered paramount to issues of intellectual property rights since it is the point where agrobiodiversity originates and hence the first crucial link in the chain of ownership, control, and accessibility. And, most recently, seed supply is regarded as crucial to concerns over GMO spread and contamination. Seed supply in this case refers *both* to the diffusion of the GMO crops themselves, such as the spread of GMO maize in Mexico, *and* to the geographic flow patterns of agrobiodiversity, which strongly shape the exposure and vulnerability of these diverse crops.

Specific initiatives on seed supply and agrobiodiversity are divided here for the purpose of introduction into three general areas: (a) special seed projects and farmer-organization certified seed programs involving agrobiodiversity; (b) participatory crop improvement (PCI) and participatory plant breeding (PPB) programs; and (c) management plans in the context of protected-area (PA) conservation. Each area reflects the entwining of the globalization of conservation, agriculture, and development.[1] Special seed projects include a growing suite of innovative approaches to local conservation and development that support seed loan programs through the payback of farmer production, community seed-bank storage facilities, seed fairs, and pro-conservation farmer networks (Bauman 1987; Friis-Hansen and Sthapit 2000b; Gonzáles 2000; Huamán 2000; Tapia 2000; Wildfong 1999). Such projects have become the basis of many new attempts to combine the support of seed supply with agrobiodiversity conservation.

Certified seed programs are increasingly run through farmer organizations, rather than government agencies or private enterprises. These programs are partly a response to the widespread farm crises of developing countries that especially beset those with smaller size farms, including many producers who are peasants and/or indigenous peoples. In the Andean countries of South America (Peru, Bolivia, and Ecuador), for example, certified seeds of potatoes and other important crops are provided through farmer-run research committees (such as the Comítes de Investigación Agropecuria Local [CIALS] in Peru), farmer organizations that produce seed, and NGOs that support them (Bentley, Tripp, and Delgado de la Flor 2001; Bentley and Vasques 1998; Thiele 1999). These institutions have become common and widespread because they attempt to fill an important void: previous governmen-

---

1. These projects and programs do not include the expanded activities of private seed companies, whose usual efforts neither involve the biologically diverse crops of small farmers nor promote the growing agrobiodiversity. (In fact, almost all the activities of private seed companies are associated with the production of HYVs and genetically modified crops [GMOs] that are not biologically diverse and that often displace the cultivation of agrobiodiversity.)

tal or "formal" seed programs in Peru, Bolivia, and Ecuador, though only minimally successful in contributing to overall seed production, have now been mostly eliminated by the cutbacks of neoliberal structural adjustment programs (SAPs; Thiele 1999). Global conservation organizations are already offering support to special seed projects, and they would be well advised to consider backing seed programs certified by farmer organizations, since these groups represent an existing type of institution that is almost uniquely capable of contributing to agrobiodiversity conservation and development.

A second area of specific initiatives on seed supply and agrobiodiversity is the PCI or PPB approach, which combines selected elements of scientific plant breeding and seed production with the participation and knowledge of farmers (Almekinders and Elings 2001; Cleveland and Soleri 2002; Friis-Hansen and Sthapit 2000a; McGuire, Manicad, and Sperling 1999; Thiele, van de Fliert, and Campilan 2001). In the area of seed supply the PCI/PPB approach seeks to merge so-called informal or local farmer-based seed production with formal or mainstream institutional seed production (Almekinders and Louwaars 1999; Almekinders, Louwaars, and de Bruijn 1994; Witcombe, Virk, and Farrington 1998). Formal seed systems alone have suffered from administrative difficulties, high expenses, and limited success in serving small farmers, and they were eliminated or severely curtailed in neoliberal reforms of the past decade (Thiele 1999). This merging of informal and formal systems of seed supply in PCI/PPB and related approaches is widely agreed to depend upon farmers' existing repertoires of skill, knowledge, and expertise in seed management (Sperling and Loevinsohn 1992).

Global agricultural centers and development organizations are the main supporters of the PCI/PPB approach. In the Andean countries, major institutional actors in PCI/PPB projects include the CIAT in Cali, Colombia, and the International Potato Center, headquartered in Lima, Peru, which are part of the global Consultative Group for International Agricultural Research (CGIAR) system (McGuire, Manicad, and Sperling 1999; Thiele, van de Fliert, and Campilan 2001). Other major global institutions that support PCI/PPB projects in the Andean countries are the United Nations' Development and Environment Program (PNUMA in Latin America) and the International Plant Genetic Resources Institute (IPGRI, a division of the Food and Agriculture Organization [FAO] of the United Nations, which sponsors the International Undertaking on Plant Genetic Resources for Food and Agriculture; see CIP and PNUMA 2000; Iriarte, Terrazas, and Aguirre 1998; Iriarte et al. 1999).

Seed supply is central to the PCI/PPB approach's priority of varietal selection and evaluation within important food plants. This priority emphasizes the incorporation of a greater degree of the diversity of farmers' existing varietal management (Cleveland and Soleri 2002; Cleveland, Soleri, and Smith 2000; Louette and Smale 2000; Smale, Bellon, and Gómez 2001; Soleri, Smith, and Cleveland 2000; Thiele et al. 1997), which depends in turn on the nature of seed acquisition and provisioning networks and the variation that is supplied among farmers across diverse geographical scales. As suggested in the names "participatory crop improvement" and "participatory plant breeding," these approaches prioritize the participation of farmers in agricultural research and development, thus adopting an emphasis similar, in general outline, to the participatory approaches that are also favored among global conservation organizations (Ribot 1999; Wells and Brandon 1993; Western and Wright 1994). To a growing extent the PCI/PPB approach also combines agroenvironmental and conservation goals, including particular concern for the continued use and in situ conservation of agrobiodiversity (McGuire, Manicad, and Sperling 1999).

Yet, at least thus far, many specific PCI/PPB initiatives are composed of design features and resource commitments that tend to incorporate farmers as a means of testing variety-level preferences and varietal management. The participation of farmers in these initiatives consists of a greater farmer-based role in the evaluation of new varietal releases, which may include biologically diverse ones although the agrobiodiversity emphasis is highly varied. In the case of the participatory breeding of seed tubers of the Andean potato and ulluco crops, for example, the preliminary PCI/PPB plans focus on such goals as incorporating farmers into the production of varietal types possessing favorable characteristics, including disease resistance, faster maturation time, and preferred eating qualities (CIP and PNUMA 2000; Iriarte et al. 1999; Iriarte, Terrazas, and Aguirre 1998; Thiele et al. 1997). Another potential avenue is for PCI/PPB to engage in agroecological "fine-tuning," so that varieties gain beneficial new traits while still representing diverse biological populations (Eyzaguirre and Iwanaga 1996). Both varietal selection and agroecological fine-tuning as described above will depend heavily upon proper understandings of the patterns and processes of supplying seed (Zimmerer 1998).

A third area of major interest in seed supply and agrobiodiversity is the incorporation of farmers, their resources, and their activities into conservation PAs. This incorporation of agriculture into conservation territories is highlighted in such designations as the buffer zones and transition zones of globally recognized biosphere reserves and the expansive conservation corridors that are being promoted by global conservation organizations. This area of interest in seed supply and agrobiodiversity, when compared to the preceding two, is potentially most directly related to global environmental conservation and may affect the landscapes and livelihoods of farmers most powerfully. This area also is in the most formative stage of development, so discussion of it here is focused on the specific initiatives that do exist. In the case of the Andean countries, for example, the designation of PAs that incorporate farmlands, such as biosphere reserves and corridors, has more than doubled during the past decade (Zimmerer and Carter 2002; for an Andean case study, see Rhoades 2001). Official designation in May 2002 of the 300,000 km² Vilcabamba-Amboró Corridor promises a vastly expanded incorporation of Andean farmland since it connects agricultural landscapes of Bolivia and Peru into a protected area that is coordinated through Conservation International. In situ agrobiodiversity conservation is likely a high-priority goal of incorporating farmlands into new and existing protected areas, so that agrobiodiversity-rich landscapes are treated not as set-asides but rather as opportunities for productivist in situ conservation (Brush 1995).

The conservation of agrobiodiversity is clearly a growing emphasis within seed supply initiatives, special seed projects, farmer-organization seed certification programs, PCI and PPB programs, and protected-area management. Initiatives and growing interest in these areas offer potentially promising benefits to agrobiodiversity conservation. Yet, as I will argue below, the geographic dynamics of existing seed flows and seed use may be negatively affected through the unintended consequences of these approaches. The following sections employ a case study analysis to evaluate the existing patterns and processes of seed flows and thus to contribute to the revision and enhancement of strategies for agricultural sustainability and agrobiodiversity conservation.

Case Study Design, Methodological Approach, and Theoretical Background

The case study is based on my field research with farmers and farmer organizations in eastern Cuzco in Peru and the Cochabamba region of central Bolivia. The Quechua and mestizo farmers of these places, and their 4–6 million counterparts that also undertake agriculture in the Andean countries, have helped create and maintain extremely diverse arrays of food plants, including Andean potatoes, minor Andean tubers (ulluco and oca, for example), maize, and quinoa (Gonzáles 2000). Numerous places in eastern Cuzco and Cochabamba have been designated as sites for agrobiodiversity conservation, involving selected seed projects, PCI/PPB programs, and agrobiodiversity conservation within the context of protected-area management. Several places are located within the buffer zones of major biosphere reserves, particularly the Manu Biosphere Reserve of southern Peru, one of the world's largest protected areas that encompass both tropical lowland and tropical mountain environments. The farming communities in both the Cuzco and Cochabamba sites are located squarely within the center of the Vilcabamba-Amboró conservation corridor project (or the Andean Mega-Corridor, as it is also known).

This chapter is oriented theoretically along four axes that are used to frame the analysis of farm-level dynamics of agrobiodiversity management, local spatial organization of agriculture and land use units, scaling of seed flows, and agroecological adaptation. The first piece of the framework, which is adapted from political and cultural ecology, is the emphasis on causal connection between macro-level conditions, such as farm policies, and micro-level effects, such as farmer resource levels that exert direct influences on existing agrobiodiversity practices and the dynamics of conservation projects (Bellon 1991; Bianco and Sachs 1998; Brookfield 2001; Brush 2004; Brush, Taylor, and Bellon 1991; Zimmerer 1991c, 1996). Such influences are likely to be more complex, resilient, and historically prone to change than simpler models would predict. The second piece of the framework consists of a spatioenvironmental, process-based view of the land use organization and ethnogeographies of rural production that are likely to influence agrobiodiversity use, including seed flows (Mayer 2001; Zimmerer 2001, 2003a). Third, socioenvironmental scaling, occurring at multiple levels with interactions among individuals, social groups, institutions, and the environment, leads to the preeminence of certain scales and people at the expense of others (Swyngedouw 1997; Turner 1999; Zimmerer and Bassett 2003). The fourth main piece of this chapter's framework is developed from the ecological and biogeographic idea that seed scaling, through movements across the landscape, is both a cause and an effect of agroecological adaptation in diverse food plants (Almekinders and Louwaars 1999; Altieri 1995; Zimmerer 1998). Mutually interacting influences thus relate the nature of these plants to the patterning of seed flows.

In this study I present and discuss findings on the geographic dynamics of seed networks and seed use, with emphasis on the agrobiodiverse Andean potato crop that is widely grown among individual farmers of twelve communities in eastern Cuzco (fig. 6.1; see also Zimmerer 2003a, 2003b; the Cochabamba studies are presented separately in Zimmerer 2004). Study methods and research design are chosen in order to investigate the social-environmental components of seed flows among the farm-level producers of Andean potatoes. The social component is based on a household-level survey of the asset portfolios (resource types, use, and employment) and agricultural seed provisioning and acquisition of forty-five farm families that reside during at least a main part of the year in the study area. This survey is designed to address the seed transfers that farmers direct to and from each of their fields (quantity, date of transfer, and location of destination or source), which typically number between five and twenty in the eastern Cuzco area. The agroecological component has two parts. First, it involves multiyear field sampling of crop and variety types, location, and field-level agroenvironments (climate, soils, vegetation; Zimmerer 2003c). Field sampling is coordinated with the household survey on seed provisioning and acquisition practices for multiple crops (Andean potatoes, which are the focus of this study,

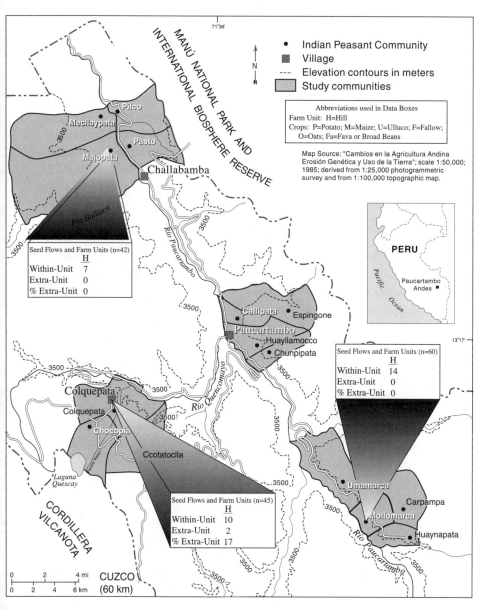

FIGURE 6.1. Multicommunity study areas of eastern Cuzco and seed flows of agrobiodiverse Andean potatoes

as well as ulluco, maize, and quinoa; see Zimmerer 2003a, 2004). Second, the agroecological component entails detailed benchmark sampling methods that evaluate spatial-environmental distribution patterns at the within-field and across-landscape levels (Zimmerer 1991a, 1996; Zimmerer and Douches 1991). The design of the study covers farmers in multiple communities, both multiple adjacent locales as well as clusters in separate areas of the study region. This research is being completed with the consent and cooperation of community authorities and a regional farmers' organization, which have aided in the design of the project, the implementation of field study methods, and the dissemination of results thus far.

Mindful of the opening account of "Return of the Native Seeds," which as discussed is both provocative and problematic, the following sections are intended to show how the existing practices of agrobiodiversity seed supply and seed use underpin the viability of these food plant types amid the responses of farmers to globalization-led changes. In addition, these sections aim to show how existing practices must be accounted for in the new sustainability policies that often contain the imprint, to widely varying extents, of global or international agendas.

Seed Flows and Use in the Andean Agricultural Landscape

Seed flows to and from the fields of Andean potatoes at upper elevations are critical for agrobiodiversity conservation in Andean countries. Higher-elevation fields contain the greatest concentration of agrobiodiversity in the agricultural landscape, based on the numbers of farmer varieties of the Andean potatoes and minor tuber crops such as the locally important ulluco (Brush 1992; Zimmerer 1991b). Estimated variety-level diversity of these food plants in the eastern Cuzco region is over 200 potato varieties and more than ten ulluco varieties (Zimmerer 1996). The significant contribution to seed flows and agrobiodiversity of farming at the upper-elevation sites must be seen as disproportionately large relative to the overall farm landscape. Although the number of fields of this type is slightly less than one-fifth the total of growing sites that are typically cultivated, the upper-elevation farm landscape is especially important to local agrobiodiversity production. For this reason the summary below is focused on this particular geographic element within the Andean agricultural landscape, a unit that in the local ethnogeographies of Peruvian farmers is designated as the "Hill" (loma) landscape.

The Hill landscape contains a fairly broad spectrum of planting environments that are traversed by seed flows since farmers do not differentiate among within-Hill field habitats for the purpose of seed suitability. Among the study communities of eastern Cuzco, the growing sites at upper elevations encompass habitats as varied as deep-soil Andean plateau surfaces; steep, arid-tending and thin-soil slopes; and wetland patches of various small and medium sizes (typically less than 0.25 ha). The elevation range of the upper-elevation field sites is typically 300 meters or more within individual communities. The substantial range of field elevations is associated with a significant variation in climate, soils, and topography. Since the farmers of eastern Cuzco do not identify subareas within the upper-elevation growing area for their seed management, the diverse varieties of Andean potatoes are typically spread across a characteristically wide range of field environments at upper elevations.

The agrobiodiversity of Andean potato fields in the upper-elevation Hill landscape typically depends on extrafield seed, that is, the seed tubers that derive from cultivated parcels other than the planting site itself. Indeed, in the 1997 planting season only 3 percent of these fields were grown using seed tubers that had been produced in the same planting site (fig. 6.2a). The prevalence of extrafield provisioning is due to the widespread knowledge and effectiveness of this technique as a means of lessening the yield reducing loads of soil-borne nematodes and diseases of both the potato and ulluco crops (Pestalozzi 2000). By contrast, the most common sort of regular seed flow, accounting for 90.9 percent in the study year, is the transfer of seed that takes place from a different upper-elevation field within the same community. These different fields belong in approximately equal fractions either to the same household or to different families, with the latter situation characterized by exchange mechanisms such as sale, barter, and labor-for-seed swaps (Zimmerer 2003b). The extrafield sourcing of Andean potato is closely related also to multi-crop rotation sequences and fallow periods (Zimmerer 2002b). As a result of customary rotation and fallow practices, each of the many field sites within the upper-elevation area is rarely sown with the same crop type during successive years.

Women farmers generally manage household seed supply and conduct the majority of extrahousehold seed transactions. In the case of extrahousehold transactions involving seed tubers of Andean potatoes, women's control over seed procurement predominates in regular planting cycles, although it is especially prevalent in those exchanges that are associated with initial sourcing (fig. 6.2c). *Initial sourcing* refers to the procurement of seed for the start of expansion of production into new fields, such as occurs when a newly married couple establishes their farm household. Women farmers also hold the primary responsibility for recent acquisitions (fig. 6.2b), which are defined as occurring when seed is obtained outside the household for the continued planting of existing field sites. Recent acquisitions may occur regularly, and for the Andean potatoes they are associated with the periodic "freshening" of seed, which involves obtaining seed, typically from more distant sources, that carries a lesser load of the soil-borne nematodes and diseases that reduce yields. A substantial number of other recent acquisitions

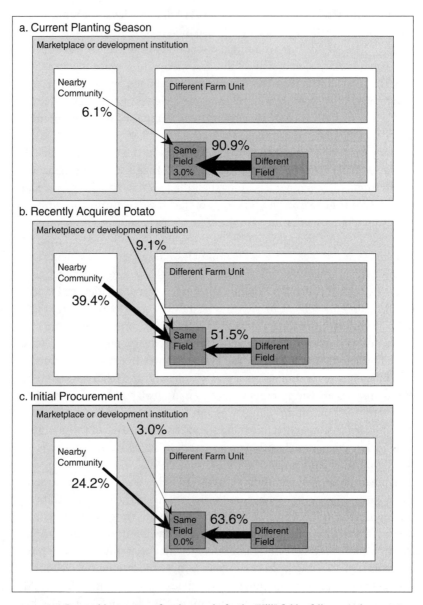

FIGURE 6.2. Geographic structure of seed networks for the "Hill" fields of diverse Andean potatoes (eastern Cuzco, 1998)

occur on an irregular basis, as when crops are damaged as the result of agro-environmental hazards common to Andean mountain agriculture, such as frost, hail, and drought.

Seed obtained from neighboring communities is important, to varying degrees, in regular annual planting, initial procurement, and recent acquisitions (fig. 6.2). The seed of Andean potatoes is sourced from neighboring communities on a regular basis in annual planting to 6.1 percent of fields. The frequency of extracommunity seed acquisitions was significantly higher in the initial planting (24.2 percent), as when a recently established household of newlyweds obtains seed from in-laws that reside in a neighboring community. Extracommunity seed flows are still more important in the case of recent acquisitions (39.4 percent), such as following localized, weather-related crop damage. The extracommunity seed flows of upper-elevation fields are notable in that they typically involve networks of farmers among several clustered communities, rather than strictly paired or primarily intercommunity networks. In the case of the Andean potatoes the extracommunity clustering of seed flows seems to take place among several (four to ten) communities that center on large-size areas of upper-elevation agricultural landscapes, where seed provisioning is a local specialty in farm household strategies. In the case of other agrobiodiverse crops, which in the Andes include maize and quinoa, the clustering of multicommunity seed flows appears to occur across scales and agroenvironments within the farm landscape, making these seed geographies significantly different from that of the Andean potatoes (Zimmerer 1991c, 2004).

In multicommunity seed flows, women and men are involved to a similar degree. As a result, the gendering of extracommunity procurements of potato and ulluco seeds tends to differ from that of procurement work occurring predominantly within the community, since men are more involved in the extracommunity transactions. Both male and female farmers rely extensively on social networks for seed procurement. Such networks frequently involve individuals who recognize one another as social kin (*compadrazgo*), which is an extremely important social relation in transactions that involve trust and continuity. This social relation is especially vital to seed flows at the multicommunity level, where there are fewer simple acquaintances than are located in the home community. Seed flows at the multicommunity level, as well as those that are more locally restricted, are enabled through cash transactions, barter, and the exchange of seed for labor, often in harvesting. The use of bartering is not uncommon in the acquisition of seed through multicommunity networks, since the Andean potatoes are often traded for maize, vegetable crops, or field access at sites lower than those in the upper elevation communities.

Seed sourcing through extracommunity networks, which includes procurement from more distant sources such as local or regional marketplaces and development institutions, comprises slightly more than one-half of the most recent acquisitions of seed tubers for the upper-elevation sites of Andean potatoes. Extracommunity sources are also important in the "Ox Area" fields, another local geographic unit where many diverse Andean potatoes

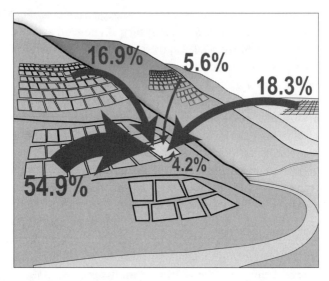

FIGURE 6.3. Seed networks for the current season plantings of "Ox Area" fields of the diverse Andean potatoes, showing seed sourcing from the same field (4.2 percent); from the same farm unit within the community (54.9 percent); from a different farm unit of the community, principally the "Hill" unit (16.9 percent); from a nearby community (5.6 percent); and from a marketplace or development institution (18.3 percent)

are grown (Brush 1992; Zimmerer 2003b). This type of seed-provisioning network in "Ox Area" fields is important both to current-season plantings, shown in figure 6.3, and to recent seed acquisitions. Farmers often choose to use these extracommunity sources as a means of periodically freshening their seed when they make these acquisitions and in response to local weather-related crop losses. In sum, the growing of agrodiverse potato and ulluco types in upper-elevation fields is dependent upon varied seed flows that can be ranked in this order of importance: farmers' own production, exchanges between farmers, and flows from marketplaces and development institutions. In eastern Cuzco, the latter includes the Centro de Estudios y Servicios Agropecuarios (CESA), an agricultural NGO with seed projects throughout the Andes and a presence of more than ten years in the area, and the Centro para la Investigación de Cultivos Andinos (CICA), a unit of the local university that has coordinated community-based seed projects. Geographically, the seed flows of diverse potato and ulluco varieties are made up of environmental and spatial networks that traverse moderate to large gradients of habitats and farmers in eastern Cuzco.

The recent decrease in the extent of upper-elevation agriculture is of special concern to agrobiodiversity conservation in general and among eastern Cuzco farmers in particular. The number and area of these fields, which are the uppermost cultivation sites of local agriculture, have diminished by as much as 30–40 percent during the past twenty to twenty-five years (Zimmerer 2003a). This decrease of upper-elevation farming is due to concomitant changes that combine disintensification from above and intensification of other farming systems from below. Disintensification pressure emanates from labor-time reallocation to off-farm work, which deprioritizes farming at the highest elevations (Pestalozzi 2000; Reardon, Berdegué, and Escobar 2001). Intensification pressure stems from the expansion of malt barley that local farmers produce under agribusiness-based contracts (Mayer 2002; Mayer and Glave 1999). Reduction of the areas of upper-elevation farming appears to be widespread in eastern Cuzco. This reduction adds further to the frequency of procurement transactions that extend beyond small or strictly local areas due the shrinking number of seed producers, and thus it tends to expand the spatial extent of seed flows.

The first conclusion to draw from these findings is that extracommunity flows are important in each type of seed movement that was investigated. Seeds flows are not neatly circumscribed at the household and community levels. The actual importance of extracommunity flows is varied according to the type of procurement and the unit of farming. Generally, the extracommunity scale is more important in the procurement types associated with recent acquisitions and initial sourcing (compared to regular seasonal plantings that depend heavily on the self-provisioning of seed). Nearly one-half of the recent acquisitions of diverse Andean potatoes were supplied through extracommunity channels. Extracommunity seed flow is clearly significant in upper-elevation farming, which provides the local microcenters of agrobiodiversity. In addition, seed flow to units other than the upper-elevation area depends to a great degree on markets as seed sources. Extracommunity flows to upper-elevation fields (the most important to agrobiodiversity), by contrast, rely chiefly on networks to neighboring communities.

The second point to draw here is that contributions of both men and women are important to the multilevel procurement of potato and ulluco seed. Gendering of seed procurement tends to vary according to geographic scale. Women farmers are responsible for the great majority of seed procurement from sources within the community, especially for meeting the planting needs of the upper-elevation fields. Men and women participate almost equally as seed procurement agents at the extracommunity level. Women's role is reduced, although still significant, in procurement from marketplaces and development institutions. Recognizing that the contribution of women farmers varies according to geographical scale, rather than being uniform and invariable, is an important step in "de-romanticizing" their potential role in sustainability and conservation programs (Howard 2003; Jewitt 2000).

Third, seed movements of both potatoes and ulluco take place within an ample range of agroecological habitats that are encompassed in the cultivation area that is described here as the upper-elevation unit. This Hill unit shows substantial agroecological variation. At the same time, the local Hill unit is used in farmers' management to guide a characteristic spatial patterning of seed flow. Seed flow within the bounds of the unit is most characteristic of farming in upper-elevation areas. This seed is primarily grown in, sourced from, and dispersed to other fields within the upper-elevation unit of agriculture. In sum, these results show that a geographic model of broadbased adaptation, rather than microenvironment-level organization, is evident in seed production and procurement of the agrodiverse Andean potatoes and ulluco. Seed flows also demonstrate the important role of farmers' own designation of production spaces (e.g., upper-elevation planting) in their procurement and provisioning decisions.

Discussion: Multilevel Seed Flows and Geographic Issues of Agrobiodiversity Conservation

The various approaches to supporting seed supply outlined above have become central to the prospects of in situ agrobiodiversity conservation, which is an increasingly major goal of conservation, development, and agricultural organizations, at global, national, and local scales. While many plans are currently in an early stage of design and preproject or policy analysis, such initial attempts at agrobiodiversity conservation must be open to critical revisions and case-study analysis (Agrawal and Gibson 1999; Cleveland and Soleri 2002; Zerner 2000). These initial attempts must also pay careful attention to existing research on agriculture and resource use that is designed specifically to address the potential for in situ agrobiodiversity conservation and to dialogue with the rapidly expanding array of proposed approaches. Since these goals are a primary purpose of this chapter, this discussion and conclusion are framed in order to focus fully on the findings of multilevel seed flows and the significance of these socioenvironmental networks for agrobiodiversity conservation.

In the case of the farmers and agricultural landscapes of eastern Cuzco, the seed flows that currently exist are relevant to the preproject and prepolicy analysis of food plant agrobiodiversity conservation involving the highly diverse Andean potatoes. The geography of these seed flows is characterized as multi-environment and, frequently, multicommunity. One main concern is that this type of seed flow geography may not be accounted for in the main spatial foundations of expanding conservation, at least in the case of the Andean potatoes. In fact, the geographic nature of seed flow documented in this case study offers a contrast to a pair of environmental and social models that are currently central to agrobiodiversity conservation. In particular, the geographies of multilevel seed flows contrast with what has been suggested as the basis of a "second paradigm" of conservation-based, post–Green Revolution agricultural development. This "second paradigm" is premised on agro-biodiversity's serving as the basis of narrow genotype-by-environment inter-action (G × E), also referred to as *yield stability*, in variety selection and evalua-tion (Eyzaguirre and Iwanaga 1996, 33). Yet, as seen in this chapter, a narrow G × E interaction model runs counter to the existing spatial and environ-mental patterning of seed flow networks for diverse Andean potato varieties in eastern Cuzco. This chapter and other recent publications suggest that PCI/PPB must pay close attention to the full scope of seed flows that create a range of adaptability and serve to limit the degree of agroecological special-ization, *not* eliminate it (see also Cleveland 2001).

As a result of these findings on seed flow geographies it is advisable that global conservation, development, and agricultural interests seek to analyze the specific seed geographies of food plant complexes under consideration for in situ agrobiodiversity conservation. Varietal types may display a broad style of G × E, which is evident in the staple food plants of the diverse Andean potato varieties, as well as the local ulluco varieties (on ulluco, see Zimmerer 2003a). This style of agroenvironmental interaction may become more vital as the result of development change, especially with predicted increases in the frequency and spatial extent of seed flows due to the contraction of farming areas at upper elevations in many Andean regions (such as eastern Cuzco). It is likely that broad-style G × E interaction is similarly associated with multilevel seed geographies and perhaps with certain types of agroeco-logical settings in particular (for example, tropical mountain environments such as the eastern Andean slopes). This reasoning does not necessarily apply to the seed-flow geographies of other food plants and farming regions, al-though it is predicted that similar sorts of seed flow are likely in various other complexes of agrodiverse food plants (Joshi, Sthapit, and Witcombe 2001).

One additional agroenvironmental aspect is also worth reexamining in order to best inform agrobiodiversity conservation based on the findings of this chapter. To date, the agroenvironmental planning classifications used in PCI/PPB have tended to adopt a system of zones based on climate and soil parameters (Witcombe, Virk, and Farrington 1998). Climate-soil zonation is commonly used to guide the distribution of on-farm trials that involve the testing and evaluation of varietal materials, which thus far has usually been a chief form of farmer participation in PCI/PPB. Yet climate-soil zonation, by itself, does not necessarily offer an adequate geographic model for the exist-ing management of seed flows and hence the dispersal of varietal types. For example, the application of this type of zonation planning to the diverse An-dean potato varieties of eastern Cuzco would be misguided if the design of zones were based solely on soil-climate parameters and thus did not account for the existing agroecological variation of upper-elevation potato and ulluco farming. Indeed, potential PCI/PPB efforts that are designed strictly on the basis of climate-soils zonation would likely pose the risk of limited adoption, due to lack of social usefulness, among those farmers and farm communities whose existing field sites and seed-flow geographies were significantly different from the zonal design. Due to conflicts with seed-flow practices, this problem of limited social usefulness would jeopardize the potential "scaling up" of participation and benefits beyond those farmers that happen to possess the field and farm types that match the climate-soils zones of PCI/PPB templates.

Agrobiodiversity conservation in the domain of protected areas is faced with a similar challenge for zonation-based agroecological management. The commitment of resources to in situ agrobiodiversity conservation within protected areas is growing in conjunction with the emphasis on sustainability and sound land use management. Not long ago it was only the latter that had been the main goal of zone-based planning in protected areas (on the Andes, see Young 1993; Zimmerer 2003c). Agrobiodiversity conservation would be impeded, however, through the designation of types of management zones within protected areas that were too narrowly defined relative to existing seed-flow geographies. Restrictive farm zonation could easily hinder seed transfers, especially among small farmers lacking in the number of fields and the range of seed sources that are typically utilized among their better-off neighbors. Inappropriate zonation could thus cause a shortage of seed, a common factor that has often led small farmers to reduce or eliminate the production of diverse food plants. In general, careful consideration must be given to a concept of zonation that is suited to existing farm management and seed flows and that would support in situ conservation and the local value of agrobiodiversity.

Special seed projects deserve a particularly high degree of consideration since they are increasingly a mainstay strategy of sustainability in protected areas and are commonly associated with PCI/PPB. Programs such as community seed banks, community seed fairs, and "peasant conservationist" associations are promising and growing rapidly in size and number. Dozens of these programs already exist in the Andean countries, where they are organized through the financing and efforts of local, regional, and national NGOs (Friis-Hansen and Sthapit 2000b; Huamán 2000; Iriarte et al. 1999; Iriarte, Terrazas, and Aguirre 1998; Tapia 2000). These NGOs tend to work in cooperation—to varying degrees—with the governments of Peru, Bolivia, and Ecuador. The national ministries of agriculture of the Andean countries have offered at least some cooperation with the PCI/PPB initiatives. At the same time, many of the PCI/PPB programs are funded or supported through global agricultural and development organizations or international agencies, and thus far these programs have proven generally compatible with the predominant processes of neoliberal restructuring and decentralization (though PCI/PPB is not necessarily allied to neoliberal policies per se).

Still, the convergence of interest in and support for special seed programs must be accompanied by greater attention to the geographic design of these undertakings. This study highlights the potential problems that would plague those approaches to in situ agrobiodiversity conservation that rely on single-location sites. The standard geographic design of the single-location model is based on one element such as the community garden, the community field, or the diversity-rich parcel of the "peasant conservationist." Contrary to the case with the single-site model, the findings of this chapter indicate the general importance of cross-site geography, ample agroecology, and supporting socioenvironmental networks of existing seed flows. These findings are notably at odds with single-site seed production in a spatial model, usually unarticulated, that favors the generation of seed types within an agroecological range that is narrower than would be useful to many local farmers. In the farm communities of eastern Cuzco, for example, effects of the long-term deployment of single-site seed production would be especially adverse to poorer farmers whose seed use is likely to depend on multi-environment and cross-unit flows (for an analysis of important similarities and differences in West African rice, see Richards 1995). Plans for special seed projects, especially those incorporating PCI/PPB and the possibility of sustainable management in protected areas, need to pay close attention to the overall spatial patterning of seed flows both *to* and *from* such key sites as the community garden and the diversity-rich plot of the "peasant conservationist." Such strategies should encourage the coordinated use of multiple sites, within single communities and among multiple communities (discussed further below), and the establishment of cross-site flows of seed that resemble the spatial and environmental patterns of existing management.

The social model of agrobiodiversity conservation is also in need of reconsideration given the findings of this chapter. In particular, special seed projects as well as PCI/PPB and protected-area management are prone to adopt a spatial framework of community-based conservation. The findings of this chapter, by contrast, demonstrate that the individual rural community is only a partial representation of the existing patterning of seed flow in the Andean potatoes. Indeed, various other cultural and agroecological aspects of the seeds of food plants, such as varietal naming, management techniques, people-plant interactions, and origin stories are often associated with multi-community or region-scale spatialities, both in the Andean countries and in other world regions (Gonzáles 2000; Nazarea 1998; Richards 1995, 1996; Salick, Cellinese, and Knapp 1997; Steinberg 1999; Zimmerer 1996). In eastern Cuzco, farmers within a particular community typically depend on seed procurement of diverse Andean potatoes and ulluco through the socioenvironmental networks created with their counterparts that reside in five or more outside communities. While the act of procurement is conducted among individual farmers, the prevalence and spatiality of these extracommunity flows indicates the importance of the multicommunity level in seed provisioning. Indeed, this finding is among the most important of this chapter since it calls for more extensive planning for multicommunity involvement in agrobiodiversity conservation. Specifically, this chapter recommends that special seed projects, PCI/PPB, and protected-area efforts be designed to build upon existing multicommunity organizations that already operate in the seed flows of diverse food plants.

Multicommunity involvement in agrobiodiversity conservation holds particular promise in places of eastern Cuzco and other areas of the Andean countries that support an existing form of coordinated land use that is known as common field agriculture or sectoral field agriculture (Mayer 2002; Pestalozzi 2000; Zimmerer 2002a). Common field agriculture is of considerable promise to conservation efforts that include agrobiodiversity as well as range management. It is a community-based institution that combines coordinated cropping with grazing fallow. Importantly, the cultivated area that is designated through common field agriculture typically comprises fields in upper-elevation areas of agriculture, which is the farm unit of greatest value to agrobiodiversity conservation in the Andean countries. While common field agriculture is community based and the governing institutions exist within individual communities, the actual practice of common field agriculture is usually found among neighboring communities that informally coordinate their adjoining cultivation systems at upper elevations (Zimmerer 2002a). This de facto coordination serves to facilitate multicommunity seed flows in places that are characterized by the practice of common field agriculture, since upper-elevation farmers can easily gain or provision seed across community boundaries. The existing multicommunity networks of seed flows and common field agriculture make up a sort of socioenvironmental network that could feasibly become a key feature of expanded, multilevel designs for in situ agrobiodiversity conservation.

References

Agrawal, A., and C. C. Gibson. 1999. Enchantment and disenchantment: The role of community in natural resource conservation. *World Development* 27 (4): 629–49.

Almekinders, C. J. M., and A. Elings, eds. 2001. Collaboration of farmers and breeders: Participatory crop improvement in perspective. *Euphytica* 122 (3): 425–38.

Almekinders, C. J. M., and D. Louette. 2000. Examples of innovations in local seed systems in Mesoamerica. In *Encouraging diversity: The conservation and development of plant genetic resources,* ed. C. Almekinders and W. Boef, 219–22. London: Intermediate Technology Publications.

Almekinders, C. J. M., and N. Louwaars. 1999. *Farmers' seed production: New approaches and practices.* London: Intermediate Technology Publications.

Almekinders, C. J. M., N. P. Louwaars, and G. H. de Bruijn. 1994. Local seed systems and their importance for an improved seed supply in developing countries. *Euphytica* 78:207–16.

Altieri, M. A. 1992. Sustainable agricultural development in Latin America: Exploring the possibilities. *Agriculture, Ecosystems, and Environment* 39 (1): 1–21.

———. 1995. *Agroecology: The science of sustainable agriculture.* 2nd ed. Boulder, CO: Westview.

Ashby, J. A., and L. Sperling. 1995. Institutionalizing participatory, client-driven research and technology development in agriculture. *Development and Change* 16 (4): 753–70.

Bauman, D. 1987. El programa de semilla del Nor Yauyos. In *Sistemas agrarios en el Perú,* ed. E. Malpartida and H. Poupon, 147–53. Lima, Peru: Universidad Nacional Agraria La Moliza (UNALM) and l'Office de la Recherche Scientifique et Technique Outre-Mer (ORSTOM).

Bellon, M. 1991. The ethnoecology of maize variety management: A case study from Mexico. *Human Ecology* 19 (3): 389–418.

Bentley, J. W., R. Tripp, and R. Delgado de la Flor. 2001. Liberalization of Peru's formal seed sector. *Agriculture and Human Values* 18:319–31.

Bentley, J. W., and D. Vasques. 1998. *The seed potato system in Bolivia: Organisational growth and missing links.* London: Overseas Development Institute Agricultural Research and Extension.

Bianco, M., and C. Sachs. 1998. Growing oca, ulluco, and mashua in the Andes: Socioeconomic differences in cropping practices. *Agriculture and Human Values* 15:267–80.

Brookfield, H. 2001. *Exploring agrodiversity.* New York: Columbia University Press.

Brookfield, H., and C. Padoch. 1994. Appreciating agrodiversity: A look at the dynamism and diversity of indigenous farming practices. *Environment* 36:7–11, 37–45.

Brookfield, H., and M. Stocking. 1999. Agrodiversity: Definition, description, and design. *Global Environmental Change* 9:77–80.

Brush, S. B. 1991. A farmer-based approach to conserving crop germplasm. *Economic Botany* 39:310–25.

———. 1992. Ethnoecology, biodiversity, and modernization in Andean potato agriculture. *Journal of Ethnobiology* 12 (2): 161–85.

———. 1995. In situ conservation of landraces in centers of crop diversity. *Crop Science* 35:346–54.

———, ed. 2000. *Genes in the field: On-farm conservation of crop diversity.* Rome: International Plant Genetic Resources Institute.

———. 2004. *Farmer's bounty: Locating crop diversity in the contemporary world.* New Haven, CT: Yale University Press.

Brush, S. B., J. E. Taylor, and M. Bellon. 1991. Technology adoption and biological diversity in Andean potato agriculture. *Journal of Development Economics* 39:365–87.

"Cambios en la agricultura andina, erosión genética, y uso de la tierra." 1985. Unpublished project map of C. Fonseca, S. B. Brush, and E. Mayer.

Carroll, C. R., J. H. Vandermeer, and P. Rosset. 1990. *Agroecology.* New York: McGraw-Hill.

Carruthers, D. V. 1997. Agroecology in Mexico: Linking environmental and indigenous struggles. *Society and Natural Resources* 10 (3): 259–72.

CIP (Centro Internacional de la Papa) and PNUMA (Programa de las Naciones Unidas para el Medio Ambiente). 2000. *Efectividad de las estrategias de conservación i situ y el conocimiento campesino en el manejo y uso de la biodiversidad.* Lima, Peru: CIP.

Cleveland, D. A. 2001. Is plant breeding science objective truth or social construction? The case of yield stability. *Agriculture and Human Values* 18:251–70.

Cleveland, D. A., and D. Soleri, eds. 2002. *Farmers, scientists, and plant breeding: Integrating knowledge and practice.* Wallingford, CT: CAB International.

Cleveland, D. A., Soleri D., and S. E. Smith. 1994. Do folk crop varieties have a role in sustainable agriculture? *BioScience* 44:740–51.

———. 2000. A biological framework for understanding farmers' plant breeding. *Economic Botany* 53 (4): 377–94.

Denevan, W. M. 2001. *Cultivated landscapes of native Amazonia and the Andes.* Oxford: Oxford University Press.

Eyzaguirre, P., and M. Iwanaga. 1996. *Participatory plant breeding.* Rome: International Plant Genetic Resources Institute.

Friis-Hansen, E., and B. Sthapit. 2000a. *Participatory approaches to the conservation and use of plant genetic resources.* Rome: International Plant Genetic Resources Institute.

———. 2000b. Overview: Participatory approaches for establishing community seed banks and improving local seed systems. In *Participatory approaches to the conservation and use of plant genetic resources,* 140–63. Rome: International Plant Genetic Resources Institute.

Gade, D. W. 1999. *Nature and culture convergent: Geography, historical ecology, and ethnobiology in the Andes.* Madison: University of Wisconsin Press.

Gliessman, S. R. 2001. *Agroecosystem sustainability: Developing practical strategies.* Boca Raton, FL: CRC Press.

Gliessman, S. R., E. Engles, and R. Krieger. 2000. *Agroecology: Ecological processes in sustainable agriculture.* Boca Raton, FL: Lewis.

Gonzáles, T. A. 2000. The cultures of the seed in the Peruvian Andes. In *Genes in the field: On-farm conservation of crop diversity,* ed. S. B. Brush, 193–216. Rome: International Plant Genetic Resources Institute.

Hecht, S. B. 1995. The evolution of agroecological thought. In *Agroecology: The science of sustainable agriculture,* ed. M. A. Altieri, 1–19. Boulder, CO: Westview.

Howard, P. L. 2003. *Women and plants: Gender relations in biodiversity management and conservation.* London: ZED.

Huamán, Z. 2000. Semilleros comunales de papas nativas del Perú. *Revista AgroNoticias* (Lima) 251: 28–31.

Iriarte, V., L. Lazarte, J. Franco, and D. Fernández. 1999. *El rol del género en la conservación, localización, y manejo de la diversidad de papa, tarwi, y maíz.* Cochabamba, Bolivia: Lauro.

Iriarte, V., F. Terrazas, and G. Aguirre. 1998. *Memoria: Primer encuentro taller sobre el mantenimiento de la diversidad de tuberculos andinos.* Cochabamba: Poligraf.

Jewitt, S. 2000. Unequal knowledge in Jharkhand, India: De-romanticizing women's agroecological expertise. *Development and Change* 31:961–85.

Joshi, K. D., B. R. Sthapit, and J. R. Witcombe. 2001. How narrowly adapted are the products of decentralized breeding? The spread of rice varieties from a participatory plant breeding programme in Nepal. *Euphytica* 122 (3): 589–97.

Juma, C. 1989. *The gene hunters: Biotechnology and the scramble for seeds.* Princeton, NJ: Princeton University Press.

Kloppenburg, J. R., Jr. 1988. *First the seed: The political economy of plant biotechnology, 1492–2000.* New York: Cambridge University Press.

Louette, D., A. Charrier, and J. Berthaud. 1997. In situ conservation of maize in Mexico: Genetic diversity and maize seed management in a traditional community. *Economic Botany* 51 (1): 20–38.

Louette, D., and M. Smale. 2000. Farmers' seed selection practices and traditional maize varieties in Cuzalapa, Mexico. *Euphytica* 113 (1): 25–41.

Mayer, E. 2002. *The articulated peasant: Household economies in the Andes.* Boulder, CO: Westview.

Mayer, E., and M. Glave. 1999. Alguito para ganar (a little something to earn): Profits and losses in peasant economies. *American Ethnologist* 26 (2): 344–69.

McAfee, K. 2003. Corn culture and dangerous DNA: Real and imagined consequences of maize transgene flow in Oaxaca. *Journal of Latin American Geography* 2 (1): 18–42.

McGuire, S., G. Manicad, and L. Sperling 1999. *Technical and institutional issues in participatory planting breeding.* Cali, Colombia: Centro Internacional Para la Agricultura Tropical (CIAT).

McNeely, J. A., and S. J. Scherr. 2003. *Ecoagriculture: Strategies to feed the world and save biodiversity.* Washington, DC: Island.

Nazarea, V. D. 1998. *Cultural memory and biodiversity.* Tucson: University of Arizona Press.

Nellithanam, R., J. Nellithanam, and S. S. Samiti. 1998. Return of the native seeds. *The Ecologist* 28 (1): 29–33.

Padoch, C., and R. R. Sears. 2005. Conserving concepts: In praise of sustainability. *Conservation Biology* 19 (1): 39–41.

Pestalozzi, H. 2000. Sectoral fallow systems and the management of soil fertility: Indigenous knowledge in the Andes of Bolivia. *Mountain Research and Development* 20 (1): 64–71.

Quist, D., and I. H. Chapela. 2001. Transgenic DNA introgressed into traditional maize landraces in Oaxaca, Mexico. *Nature* 414:541–43.

Reardon, T., J. Berdegué, and G. Escobar. 2001. Rural nonfarm employment and incomes in Latin America: Overview and policy implications. *World Development* 29 (3): 395–409.

Redford, K. H., and S. E. Sanderson. 2000. Extracting humans from nature. *Conservation Biology* 14 (5): 1362–64.

Rhoades, R. E. 2001. *Bridging human and ecological landscapes: Participatory research and sustainable development in an Andean agricultural frontier.* Dubuque: Kendall/Hunt.

Rhoades, R. E., and V. D. Nazarea. 1999. Local management of biodiversity in traditional agroecosystems. In *Biodiversity in agroecosystems,* ed. W. W. Collins and C. O. Qualset, 215–36. Boca Raton, FL: CRC Press.

Ribot, J. C. 1999. Decentralisation, participation, and accountability in Sahelian forestry: Legal instruments of political-administrative control. *Africa* 69 (1): 23–65.

Richards, P. 1995. The versatility of the poor: Indigenous wetland management systems in Sierra Leone. *Geo-Journal* 35 (2): 197–203.

——. 1996. Culture and community values in the selection and maintenance of African rice. In *Valuing local knowledge: Indigenous people and intellectual property rights,* ed. S. B. Brush and D. Sabinskey, 209–29. Washington, DC: Island.

Rosset, P. M., and M. A. Altieri. 1997. Agroecology versus input substitution: A fundamental contradiction of sustainable agriculture. *Society and Natural Resources* 10 (3): 283–95.

Salick, J., N. Cellinese, and S. Knapp. 1997. Indigenous diversity of cassava: Generation, maintenance, use and loss among the Amuesha, Peruvian Upper Amazon. *Economic Botany* 51 (1): 6–19.

Salick, J., and L. C. Merrick. 1990. Use and maintenance of genetic resources: Crops and their wild relatives. In *Agroecology,* ed. C. R. Carroll, J. H. Vandermeer, and P. Rosset, 517–48. New York: McGraw-Hill.

Schwartzman, S., A. Moreira, and D. Nepstad. 2000. Rethinking tropical forest conservation: Perils in parks. *Conservation Biology* 14 (4): 1351–57.

Shiva, V. 1997. *The enclosure and recovery of the commons: Biodiversity, indigenous knowledge, and intellectual property rights.* New Delhi: Research Foundation for Science, Technology, and Ecology.

Smale, M., M. R. Bellon, and J. A. A. Gómez. 2001. Maize diversity, variety attributes, and farmers' choices in southeastern Guanajuato, Mexico. *Economic Development and Cultural Change* 50:201–25.

Soleri, D., S. E. Smith, and D. A. Cleveland. 2000. Evaluating the potential for farmer and plant breeder collaboration: A case study of farmer maize selection in Oaxaca, Mexico. *Euphytica* 116 (1): 41–57.

Sperling, L., and M. E. Loevinsohn. 1992. The dynamics of adoption, distribution, and mortality of bean varieties among small farmers in Rwanda. *Agricultural Systems* 34:441–53.

Steinberg, M. K. 1999. Maize diversity and cultural change in a Maya agroecological landscape. *Journal of Ethnobiology* 19 (1): 127–39.

Stevens, S., ed. 1997. *Conservation through survival: Indigenous peoples and protected areas.* Washington, DC: Island.

Swyngedouw, E. 1997. Neither global nor local: "Glocalization" and the politics of scale. In *Spaces of globalization: Reasserting the power of the local,* ed. K. R. Cox,. 137–66. New York: Guilford.

Tapia, M. E. 2000. Mountain agrobiodiversity in Peru: Seed fairs, seed banks, and mountain-to-mountain exchange. *Mountain Research and Development* 20 (3): 220–25.

Tenberg, A., J. Ellis-Jones, R. Kiome, and M. Stocking. 1998. Applying the concept of agrodiversity to indigenous soil and water conservation in eastern Kenya. *Agriculture, Ecosystems, and Environment* 70:259–72.

Thiele, G. 1999. Informal potato seed systems in the Andes: Why are they important and what should we do with them? *World Development* 27 (1): 83–99.

Thiele, G., E. van de Fliert, and D. Campilan. 2001. What happened to participatory research at the International Potato Center? *Agriculture and Human Values* 18:429–46.

Thiele, G., G. Gardner, R. Torrez, and J. Gabriel. 1997. Farmer involvement in selecting new varieties: Potatoes in Bolivia. *Experimental Agriculture* 33:275–90.

Thrupp, L. A. 1998. *Cultivating diversity: Agrodiversity and food security.* Washington, DC: Worldwatch Institute.

Turner, M. D. 1999. Merging local and regional analyses of land-use change: The case of livestock in the Sahel. *Annals of the Association of American Geographers* 89 (2): 191–219.

Vandermeer J., M. van Noordwijk, J. Anderson, C. Ong, and I. Perfecto. 1998. Global change and multispecies agroecosystems: Concepts and issues. *Agriculture, Ecosystems, and Environment* 67 (1): 1–22.

Wells, M. P., and K. E. Brandon. 1993. The principles and practice of buffer zones and local participation in biodiversity conservation. *Ambio* 22 (2–3): 157–62.

Western, D., and R. M. Wright, eds. 1994. *Natural connections: Perspectives in community-based conservation.* Washington, DC: Island.

Wildfong B. 1999. Savings seeds: Farmers and gardeners are the best hope for protecting what remains of food plant diversity. *Alternatives Journal* 25 (1): 12–17.

Witcombe, J., D. Virk, and J. Farrington, eds. 1998. *Seeds of choice: Making the most of new varieties for small farmers.* London: Intermediate Technology Publications.

Wood, D., and J. M. Lenné. 1997. The conservation of agrobiodiversity on-farm: Questioning the emerging paradigm. *Biodiversity and Conservation* 6:109–129.

Young, K. R. 1993. National park protection in relation to the ecological zonation of a neighboring human community: An example from northern Peru. *Mountain Research and Development* 13 (3): 267–80.

Zerner, C. 2000. *People, plants, and justice: The politics of nature conservation.* New York: Columbia University Press.

Zimmerer, K. S. 1991a. Labor shortages and crop diversity in the southern Peruvian sierra. *Geographical Review* 81 (4): 414–432.

———. 1991b. Managing diversity in potato and maize fields of the Peruvian Andes. *Journal of Ethnobiology* 11 (1): 23–49.

———. 1991c. The regional biogeography of native potato cultivars in highland Peru. *Journal of Biogeography* 18:165–78.

———. 1996. *Changing fortunes: Biodiversity and peasant livelihood in the Peruvian Andes.* Berkeley and Los Angeles: University of California.

———. 1998. The ecogeography of Andean potatoes: Versatility in farm regions and fields can aid sustainable development. *BioScience* 48 (6): 445–54.

———. 2001. The common ground of geography and the new ethnobiology: Links to a framework of ethno-landscape ecology. *Geographical Review* 91 (4): 725–34.

———. 2002a. Common field agriculture in the geographical development of resource use. *Journal of Cultural Geography* 19 (2): 37–63.

———. 2002b. Social and agroecological variability of seed production and the potential collaborative breeding of potatoes in the Andean countries. In *Farmers, Scientists, and Plant Breeding: Integrating knowledge and practice,* eds. D. A. Cleveland and D. Soleri, pp. 83–107. Wallingham: CABI International.

——. 2003a. Environmental zonation and mountain agriculture in Peru and Bolivia: Toward a model of overlapping patchworks and agrobiodiversity conservation. In *Political ecology: An integrative approach to geography and environment-development studies,* ed. K. S. Zimmerer and T. J. Bassett, 141–69. New York: Guilford.

——. 2003b. Geographies of seed networks for food plants and approaches to agrobiodiversity conservation. *Society & Natural Resources* 16 (3): 583–601.

——. 2003c. Just small potatoes (and ulluco)? The use of seed-size variation in "native commercialized" agriculture and agrobiodiversity conservation among Peruvian farmers. *Agriculture and Human Values* 20:107–23.

——. 2004. The geographic structure of seed supply for agrobiodiversity of Andean maize and quinoa in Cuzco, Peru, and Cochabamba, Bolivia. University of Wisconsin–Madison, Department of Geography. Manuscript, August.

Zimmerer, K. S., and T. J. Bassett. 2003. Approaching political ecology: Society, nature, and scale in human-environment studies. In *Political ecology: An integrative approach to geography and environment-development studies,* ed. K. S. Zimmerer and T. J. Bassett, 1–28. New York: Guilford.

Zimmerer, K. S., and E. D. Carter. 2002. Conservation and sustainability in Latin America. *Yearbook of the Conference of Latin Americanist Geographers* 27:22–43.

Zimmerer, K. S., and D. S. Douches. 1991. Geographical approaches to crop conservation: The patterning of genetic diversity in Andean potatoes. *Economic Botany* 45 (2): 176–89.

7 Shifting Scales, Lines, and Lives: The Politics of Conservation Science and Development in the Sahel

MATTHEW D. TURNER

The case of environmental and social change in the Sudano-Sahelian region of West Africa has, since the early 1970s, strongly influenced development and conservation thought and practice. It has done so by serving as an early environmental calamity attracting global concern (Buttel and Taylor 1994) and a benchmark to which other areas are compared (Middleton and Thomas 1997). In images, reports, and articles the Sahel is portrayed as a region experiencing economic stagnation, chronic malnutrition, and environmental degradation. These portrayals, in juxtaposition, played an instrumental role in the emergence of various conservation-with-development ideas in the 1980s and 1990s (Adams 1990; Lélé 1991; Warren 1995). These ideas all sought underlying reasons for the creation of a semi-arid landscape seemingly denuded by expanding human demands for food, forage, fiber, and fuel. Population, rural poverty, inappropriate resource management institutions, mismanagement, environmental degradation, and decline in resource endowments are seen to form causal links in a downward spiral leading to both environmental and humanitarian disasters.

Conservation and development interest in the Sahel has declined since the mid-1990s, as the post–cold war policy calculus has drawn shrinking development funds away from the region. Moreover, the new emphasis on biodiversity protection, described by WinklerPrins, Zimmerer, and Sierra (chaps. 5, 6, and 9, respectively, in this volume), has also worked to guide conservation investments away from biosparse regions such as the Sahel (Zerner 1996; Bowker 2000). Conservation and development funding for the region has declined but not disappeared. The contemporary approach to conservation-with-development in the region can best be described as enviro-economic stabilization: an approach that seeks to limit the potential for widespread crisis and political destabilization. In this holding pattern of support, the region has become the commonly used example of an economic, ecological, and climate-change endpoint.

Lowered interest in the Sahel can be seen as an opportunity to move beyond the failures of large-scale, crisis-driven, technocratic development projects of the past toward programs that are more closely tied to the ecological, political, and cultural realities of the Sahel (Baker 1984; Ferguson 1994). Without question, past conservation programs have met with almost ubiquitous failure. It is now recognized that blame rests less with the character of the recipients of donors' largesse and more with failures in conservation approach. These failures largely stem from a shockingly poor understanding of the ecological and social contexts of extant livelihood systems in the Sahel. Poor environmental science coupled with profound naïveté about or lack of interest in the social relations that animate production systems led to misdiagnoses of environmental degradation and to inappropriate social policy to correct these "problems." The dilution of the crisis mentality could provide an expanded space for novel conservation approaches to develop in the region.

With this opportunity in mind, I will relate new approaches in environmental assessment and conservation in the Sahel to more contemporary understandings of ecology and livelihood strategies in the region. In so doing, I will argue that contemporary approaches to environmental monitoring and conservation are not fully informed by contemporary understandings and that this results from the persistence of outmoded methodological features that are consistent with globalized visions of conservation science and programs. In particular, I evaluate the choice of scales and boundaries by conservation scientists and practitioners alike. I will show that these seemingly innocuous choices have real political implications for the subjects of conservation programs and result from the confluence of multiple spheres of often inward-looking scientific and bureaucratic politics along with the more obvious politics of state and local interests. While much of this chapter focuses on the methods, scales, and boundaries adopted by conservation scientists and practitioners, these choices reflect more than good and bad science. They reflect a subtle and complex politics of scale and categorization that works to reorder the political and material lives of those that find themselves within a conservation area.

Spatial Scaling and Boundary Drawing in Conservation
Environmental conservation seeks to change the relation of people to their physical environment. The distributions of the things that conservationists seek to protect (such as biodiversity, vulnerable soils, or particular species) are not regularly distributed on the landscape. Nor are rural populations and their productive activities. Therefore, the scale at which resources and pro-

ductive activities are monitored and sampled will affect one's evaluation of conservation status and strategy. Technology and labor constraints affect what type of environmental changes can be effectively monitored at a particular scale. It is difficult, for example, to monitor soil fertility changes over broad spatial scales (>1 km^2). Therefore, the spatial scale chosen will influence what kind of environmental changes one can effectively monitor.

Because of these sampling problems, spatial abstractions are a necessary part of conservation planning and are sprinkled throughout the analyses found within conservation planning or monitoring texts. The statement "the density of wildlife species X is Y while the human population density is presently Z and rapidly expanding" is a common form of analytical statement in conservation documents. Such statements are made to provide quantitative measures of both the status of the conservation object (in this case, wildlife species X) and the potential human threat to that object. One could question the reliance on population densities as indicators of the conservation status of wildlife species X and human threats, but more important to this discussion is that such statements are often sensitive to the particular area over which average densities are calculated.

Relating such statements to on-the-ground realities reveals a host of complications. Wildlife species are often mobile, and therefore the stated density in a particular area is strongly influenced by the season and even the time of day of the animal census. The decision to include or exclude an important watering hole in a dryland savanna area may very much influence the area's stated wildlife density. In that same savanna area, human settlements are irregularly distributed—the inclusion or exclusion of a large town at the edge of the area will often significantly affect the stated human population density for an area. Certainly, this suggests that such data can be easily manipulated to suit the interests of the analyst, but more commonly analysts do not deliberate on these scaling issues and instead allow the data to guide them in making their scaling choices—by using data at the level of the administrative district, for example. Such data-driven choices are far from neutral—they rely on unexamined inclusions and exclusions of humans and conservation objects from analyses that are used to support the causal claims that guide conservation strategy and practice.

The major problem with such analyses is not necessarily that they rely on spatial abstractions but the unreflective way in which they perform them. These quantitative measures are simply proxies for what is of most interest—the interaction of human activities with the life cycle of wildlife species X. Choosing the appropriate scale of analysis necessarily involves an understanding of the human activities that most influence wildlife species X and at one point in its seasonal and life cycles. The spatial area that encompasses the species range (at the season and life-cycle stage of highest vulnerability) and the human activities that most influence the wildlife species approximate the appropriate spatial scale for monitoring. Obviously, the actual choice of the monitoring scale and area is far from an exact science and necessarily involves deviating from the ideal, but it is important to use available process information to think seriously about the appropriate scale. At the very least, one would be better positioned to understand how one's data-driven choice of scale may bias one's analysis.

The environmental science that supports conservation planning is scale sensitive. The choice of scale necessarily draws boundaries through ecological and social landscapes, which leads to exclusions and inclusions. Conservation practice is very much about changing how people interact in a material way with the objects of conservation. For that reason, a large part of conservation is about inscribing social boundaries on the landscape and, in so doing, defining spatial scales of political institutions. Either through rule setting, institution building, technology introduction, or education, conservationists are often interested in changing human-environment relationships in a very spatially defined way. This is most obvious in the case of protected-area management, where social boundaries that are inscribed on the landscape explicitly bound human activities (Vandergeest 1996). The spatial explicitness of preservation as a social project also holds true for conservation on more humanized landscapes. On more humanized landscapes, conservation is also concerned with *how* productive activities are performed. Therefore, the changes promoted by conservationists can take a wide range of forms (e.g., low-till farming, rotational grazing, and community forestry). These forms are still likely to involve the promotion of rules, sanctions, or benefits that are spatially differentiated and thus work to define social boundaries. The degree to which such boundaries and their associated spatial scales are appropriate to the productive ecology and extant political structures of an area is a major determinant of the success or failure of the conservation strategy.

A major premise of this chapter and the other chapters in this section (see WinklerPrins, chap. 5, and Zimmerer, chap. 6 in this volume) is that both the implicit and explicit boundaries drawn around human activities by conservation programs and by the scalar choices made by conservation science have played often unrecognized roles in shaping the political and ecological outcomes of the larger conservation project. Despite the promise of *new* approaches for environmental monitoring and conservation practice that have developed in the Sahel over the past decade, a major failing is their inattention to these boundary and scaling issues. I will argue that many of the scaling and boundary choices made are inappropriate for the Sahel. The mismatch is caused by a combination of the persistent influence of past approaches, the political economy of conservation in the region, and globalized modes of conservation practice that have been wrongly imposed upon the ecological and social realities of the Sahel.

I will first provide an overview of current understandings of Sahelian land-use ecology followed by a delineation of how, since the early colonial period, environmental scientists have typically engaged with this complex reality to monitor environmental change. I will then outline two major innovations in environmental monitoring and conservation practice that have developed over the past fifteen years. Conservation professionals and their hired consultants have increasingly used satellite imagery data to monitor environmental change. This has led to a broadening of the scales at which the environment can be monitored. At the same time, there has been a reorientation in conservation practice toward community-based conservation, with the spatial scope of conservation attention being the village territory. While both are advancements over environmental monitoring and conservation approaches of the past, they can be strengthened by greater attention to issues of scale and boundaries. I will critically review each of these innovations and suggest reasons for the persistence of certain features of both monitoring and conservation practice that severely limit their success in a Sahelian context. I will conclude by outlining alternative multiscaled approaches for monitoring and conservation in the Sahel.

Environmental Scientific Engagements with the Socioecological Complexity of the Sahel

The Sahel is defined here as the strip of land lying south of the Sahara desert that receives a long-term average of 100–600 mm of rain falling sometime from June through September each year (see fig. 7.1). It is an agropastoral region, where both livestock raising and dryland farming are practiced.[1] Vegetative structure grades from steppe to dry savanna from north to south. Herbaceous vegetation is dominated by annual grasses and forbs. The productivity of pasture and cropland is limited by both sparse rainfall and low soil fertility. Patches of higher pasture and cropland productivity shift from year to year as a result of the confluence of higher soil fertility and well-distributed rainfall.

Due to the high spatiotemporal variability of vegetative productivity, the livelihood strategies of farmers and herders incorporate a large degree of spatial flexibility. Farmers will often seed a larger number of fields over a broad area (often outside the "territory" of their village) than they will weed or harvest. Herders have historically conducted long-range north-south seasonal

[1]. Both livestock production and millet farming are practiced in the region (most farming activity falling south of the 300 mm isohyete). While ethnic and caste identities in the region are often still tied to particular livelihoods, many "farmers" own livestock and many "pastoralists" farm.

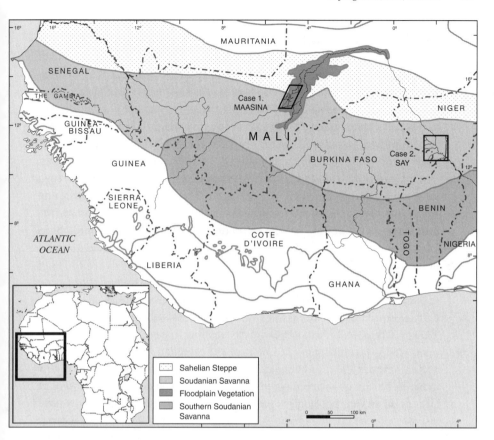

FIGURE 7.1. Map of study area

movements to take advantage of high-quality but sparse pastures to the north
during the rainy season and returning south to the cropping zone during the
dry season. The specific north-south path taken and where herders pasture
their animals at the rainy- and dry-season destination areas will vary from
year to year depending in large part on rainfall distribution. In short, the
high spatiotemporal variability of rainfall leads to *both* farming and herding
strategies that encompass broad spatial scales—transcending the village territory and facilitated by extravillage social networks. The importance of extralocal social networks for Sahelian livelihoods mirrors their importance
for in situ agrodiversity conservation as described by WinklerPrins and Zimmerer (chaps. 5 and 6 in this volume).

The large interannual swings in vegetative cover, species composition, and
cropland and pasture productivity in the Sahel are caused primarily by
changes in the pattern and magnitude of rainfall. Many of the impacts of
human land uses are not only obscured by high interannual variability of

rainfall but are actually mediated (qualitatively and quantitatively) by rainfall variability. This complicates the study of human-induced environmental change. Human-induced changes in vegetation and soil productivity do not leave "fingerprints" that can be easily identified through casual inspection.[2] Many changes that are visibly apparent (species composition of herbaceous layer, vegetative cover, standing biomass) are actually ephemeral and are not necessarily related to changes in the productive potential of the land. More persistent human-induced changes become evident over many years and require study of the structure and fertility of soils. To understand the complex web of causation behind human-induced ecological change, there is a strong need to link human resource uses to the physical environment in space and time. Under such circumstances where rainfall, vegetation, and human production practices display significant spatiotemporal variability, observations of land use and the environment are strongly shaped by the spatial and temporal scales at which they are made.

Since early in the colonial period (early 1900s), the tremendous variation in the environmental conditions of the West African Sahel has attracted attention and concern among environmental scientists (Gritzner 1988; Grove 1994; Warren 1995; Swift 1996). There was much speculation and concern about the growing impoverishment of the Sahelian environment (Hubert 1920; Chevalier 1934; Stebbing 1935; Boudet 1972). Confronted by the highly dynamic nature of both ecological and land-use systems, scientific observers (agronomists, range ecologists, geographers, and others) relied on short-term visual observations, the quality and spatial breadth of which were severely limited by conceptual, infrastructural, and political constraints. We now recognize that these were spatially and temporally circumscribed observations of often-ephemeral conditions interpreted through inappropriate models of ecology and society (Behnke, Scoones and Kerven 1993; Turner 1993).[3]

Despite increasing sophistication in the applied ecological sciences, meth-

2. A major exception is the felling of trees for fuelwood and cultivation. The cause of tree loss is readily apparent on the ground although its longer-term implications for the biological productivity are not.

3. A major conceptual adherence, most evident among range management experts, but also among agronomists was equilibrium-based notions of ecology (Braun-Blanquet 1932; Boudet 1975; Stoddart, Smith, and Box 1975; Breman and de Ridder 1991; Niemeijer 1996). For example, in considering the potential for grazing-induced environmental change, range systems were seen as naturally prone to evolve toward an internally regulated equilibrium or "climax" type. The goal of range management was simply to identify the deviations from the desired community type (may differ from the climax) and recommend changes in stocking rate that would move the system in the desired direction. The beauty of this framework was its seemingly facile application to different socioecological situations—management prescriptions could be made without performing controlled studies on grazing impact and without understanding existing livestock management. As a result, only a hand-

ods for environmental monitoring in the Sahelian region retained certain common methodological features during much of the twentieth century, which together can be termed an environmental monitoring tradition (Turner 2003). The environmental monitoring tradition in the Sahel has relied on spatially and temporally circumscribed visual descriptions of ecological state with little understanding or reference to past or present human uses of the observed patch. Early claims of desert expansion were based on regional generalizations of highly localized observations (Stebbing 1935; Stamp 1940; Swift 1996). Such visual descriptions provide very little insight into the causes behind changes in vegetative parameters and why these changes persist. Human-induced degradation was often inferred from the spatial proximity of signs of vegetative deterioration and signs of human use. Such analyses have a high potential for circular argument—human signs and signs of degradation are often one and the same.

The legacy of this environmental monitoring tradition is a large body of ecological information on vegetative conditions at particular points in time with very little understanding of how these conditions were affected by human activities. Within this void, the specter of human-induced degradation that provokes and informs conservation efforts is spatially vague and abstracted since it results less from empirical study and more from the joint persuasiveness of the independent assessments of two seemingly separate groups of dispassionate observers: environmental scientists describing poor vegetative conditions and social scientists advocating various social models of overuse: overpopulation, tragedy of the commons, cultural attachments to livestock, poverty-induced shortening of planning horizons, and so on. In this way, these somewhat independent systems of knowledge and discourse, by producing seemingly independent and consistent but incomplete and unverified environmental assessments, support a common narrative of human-induced degradation. As a result, there has existed a situation in the Sahel that would seem curious to any outside observer: conservation and development practitioners are convinced that human-induced degradation is widespread, but are not able to point to any spatially specific study that documents such degradation.

With the growing recognition that much of the interannual variation in vegetative parameters is caused by rainfall fluctuations and not human activities, the equilibrium assumptions that supported the environmental monitoring tradition in the Sahel came under attack in the 1990s (Ellis and

---

ful of controlled or semicontrolled grazing studies have been performed in the Sahelian region, and the relevance of stocking rate prescriptions to mobile livestock production systems were not scrutinized, until recently (Behnke, Scoones, and Kerven 1993; Turner 1993).

Swift 1988; Westoby, Walker, and Noy-Meir 1989; Behnke, Scoones, and Kerven 1993). There is now recognition among most scientists that much of the environmental change that was described in the past as land degradation was in fact ephemeral and that short-term visual observations alone provide insufficient evidence for land degradation (Tucker, Dregne, and Newcomb 1991; Thomas 1993; Middleton and Thomas 1997). To understand how human productive activities affect the Sahelian environment, process-oriented research, informed by the spatial and historical contexts of climate and human land uses, is needed.

Revisionist views of ecological dynamics have led to increased questioning of conventional conservation programs that generally sought to transform malfunctioning rural institutions and production strategies by the infusion of Western technology and social organization (Baker 1984). No longer can it be assumed that poor vegetation condition is improved by reducing human pressure on the land. In this way, the relevance of social models that delineate causal pathways leading to human "overuse" have been questioned. Attention has shifted away from a preoccupation with the magnitude of human demands on resources to include careful considerations of *how* resources are managed. While in the past mobility was linked with wastefulness and the lack of property institutions, there is now renewed appreciation for the ecological rationality behind the mobility and spatial flexibility of Sahelian livelihoods. Therefore, rather than attempting to transform these strategies in the name of conservation, there are now calls to reinvigorate them in ways that are appropriate to present socioecological contexts (Vedeld 1992; Scoones 1994; Niamir-Fuller 1999).

New Approaches to Environmental Monitoring and Conservation

During the period of mounting criticism of the environmental scientific and conservation traditions in dryland Africa (mid-1980s to present), funding for conservation and development work in the Sahel has declined. As argued earlier in this chapter, this situation presents an opportunity to develop, in a more deliberate, measured, and participatory fashion, conservation approaches across these humanized landscapes that are more fitting to their ecological and social realities. With these opportunities in mind, I will consider the new emphases in environmental monitoring and conservation practice in the Sahel that have developed since the mid-1980s, paying particular attention to spatial scaling and boundary issues. Environmental monitoring and conservation practice have moved in opposite directions with respect to the spatial scales at which they work. Environmental monitoring has "scaled up"—measuring environmental status at broader spatial scales through the increased reliance on satellite data, which have spatial breadths of thousands of square kilometers and maximum effective resolutions of 0.2

square kilometers.[4] Environmental conservation practice, on the other hand, has scaled down, working at spatial scales of the village territory as a result of the growing reliance on village-based resource management projects, which have spatial breadths of tens of square kilometers and resolutions of 0.01 square kilometers. These changes represent improvements over past approaches to environmental monitoring and conservation practice. Still, they do not address many of the concerns concurrently raised by the revisionist critics of past monitoring and conservation approaches. Moreover, there is a mismatch between the scales at which the environment is monitored and at which conservation is practiced. In the following sections, I will briefly outline the ways in which these new approaches do not address more contemporary understandings of Sahelian ecology and society. In so doing, I will argue that their promotion and diffusion since the 1980s is animated less by new understandings about the Sahelian region than might be expected. Instead, these new approaches reflect both the persistent influence of old approaches and the emergence of new, more globalized conventions in conservation today—each disturbingly removed from Sahelian realities.

Environmental Monitoring through Remotely Sensed Data
Concurrent with the growing revisionist critique of the environmental monitoring tradition in the Sahel, the availability of satellite imagery rapidly increased. In fact, the analysis of remotely sensed imagery played an important role in exposing the spatial and temporal bias and inappropriate generalizations of the environmental monitoring tradition. For example, satellite imagery data were used effectively to counter claims of the marching desert—claims that were generated from site-specific studies along an ill-defined desert edge (Hellden 1991; Tucker, Dregne, and Newcomb 1991). Still, the widespread adoption of these technologies by the conservation and development communities has not addressed many of the problems associated with the environmental monitoring tradition in the Sahel and in fact has worked to reinforce some of its most problematic features.

Remotely sensed data are recordings of the reflectance of visible and invisible wavelengths of energy from the land to an airborne camera or digital wave-band sensor. Satellite imagery data became generally available in the mid-1980s from LANDSAT MSS and TM, SPOT-XS, and NOAA-AVHRR sensors. The availability of these data dramatically increased analysts' ability to monitor changes in the pattern and extent of cultivated areas and vegetative cover over broader areas and time frames than in the past. This contribution

---

4. By "effective resolution," I am not referring to the technical resolution of the sensor but to the scale at which features can be readily distinguished and measured on the satellite imagery (feature plus sufficient matrix), often representing tens of pixels in length and width on such arid landscapes.

is especially important in arid regions such as the Sahel where population is sparse, land use is extensive, and vegetative cover is spatiotemporally variable. Therefore, uses of satellite imagery data have gone a long way toward addressing the problems of spatial and temporal bias that have plagued the environmental monitoring tradition. This is a very important contribution.

In the 1990s, environmental analysts increasingly relied on remotely sensed data for monitoring of environmental change in semi-arid areas such as the Sahel. I am not referring to cutting-edge science here; instead, I refer to the environmental analyses and assessments performed to support development and conservation planning that, despite their lack of sophistication, represent the bulk of the environmental assessment in the Sahel. Remotely sensed data have proven to be attractive to donors and analysts alike because of the spatial breadth they provide compared to field measures, the temporal depth provided through repeat coverages, their lower cost compared to field-work, and their connection to what is seen as unbiased measures of land cover. Moreover, these data contributed to the expanding influence of the global environmental change community (Buttel and Taylor 1994), which uses these data as an important scientific resource to garner funding and to lend scientific legitimacy to its claims about the nature and causes of environmental change.

Despite the real advances that remotely sensed data offer for the monitoring of land-cover change, a number of problems are revealed in their use for environmental assessment in the Sahel. First, the promise that satellite sensors can reveal more than can be observed by the human eye because of their wider spectral breadth (features such as above-ground biomass, crop yield, soil moisture, and species composition of vegetation) has not been realized in the Sahel. This is due not only to limited infrastructure and financial resources but also to the technical problems caused by the considerable spectral heterogeneity that results from the combination of sparse vegetation cover and variable soil background (Beck, Hutchinson, and Zauderer 1990; Huete and Tucker 1991; Leprieur, Kerr, and Pichon 1996).[5] The power of remotely sensed data rests largely in their form. Satellite data are uniform, repeatable quantitative measures of reflectance in a spatial grid, a form that is much more analytically useful than visual descriptions of an on-the-ground observer (Bastin et al. 1993). Still, they remain largely visual assess-

---

5. Moreover, rigorous ground-truthing is rarely performed by development and conservation agencies or their consultants in the Sahel. Remotely sensed data are widely available, but collecting on-the-ground data remains difficult and expensive. Armchair environmental assessment performed halfway around the world has become increasingly common for the Sahel. Given the limited Sahelian field experience of many of these distant analysts, the accuracy of much of the land cover classifications is questionable.

ments of changes in vegetative cover. While such assessments have proven useful for tracking patterns of tropical deforestation, they provide surprisingly little information about the persistence or mechanisms behind changes in vegetative cover in the Sahel.

Similar to the environmental monitoring tradition in the Sahel, understandings of environmental change supplied by remotely sensed data are visually descriptive (of land-use or vegetative cover) and alone provide little insight about the prevalence and causes of land degradation in the Sahel. Especially for data of poor spatial resolution (NOAA AVHRR, Landsat MSS), it is often difficult to identify uniquely human signatures on the dryland landscape. With data of higher resolution, the human signature is land cover change—therefore, human activities are seen as cropped areas and devegetated zones. Again consistent with the environmental monitoring tradition, there is a strong tendency to equate human activities with measures of environmental change. Such depictions, if not downright teleological, facilitate very reduced forms of socioecological analysis. More sophisticated uses attempt to infer human impacts on vegetative cover from spatial correlations between land-cover change and social variables. The social data used are usually simply human or animal population data, because of the lack of other data at broad spatial scales. Causal inference from spatial correlations among a reduced set of variables (e.g., population density and land cover) is hazardous, given the high autocorrelation of human ecological data sets (Turner 2003).

In sum, the analytical uses of remotely sensed data have greatly increased the temporal and spatial breadth of conventional environmental assessment in the Sahel. At the same time, the uses of remotely sensed data show strong similarities with the environmental monitoring tradition in the Sahel, a tradition that was being roundly critiqued during the same period that remotely sensed data were increasingly used. In both cases, visual descriptive analyses are produced with little connection to human context and ecological process. The continuation of this mode of analysis is not simply a legacy of the past. In the contemporary funding context, monitoring through remote sensing is seen as more cost-effective and appropriate for arid, resource-sparse regions such as the Sahel. In addition, it reflects new political struggles within international conservation and global environmental change communities over what is legitimate environmental assessment. The global environmental change community has adopted remote sensing as an indispensable tool and in so doing has linked an upward shift in spatial scales with data-intensive empiricism and technically sophisticated science (Turner 1999b). In contrast, both the social scientist talking to people and the field ecologist walking a transect are seen as old-fashioned and ineffective. In this context, the scientific goal of generalization has curiously been reworked

178   MATTHEW D. TURNER

from the classic focus on understanding process and rejection of description, to one based on spatial scaling criteria. In the applied environmental scientific context, studies that describe change and pattern over broad scales are often seen as being more general and scientific than local studies of social and ecological process (Turner 2003).

Conservation through Village-Based Resource Management Projects
The past fifteen years has also seen a significant reorientation of international conservation practice from top-down technocratic programs toward political decentralization in resource management (Agrawal and Ribot 1999; Ribot 1999) with a heavy reliance on community-based projects (Western and Wright 1994; Brosius, Tsing, and Zerner 1998; Agrawal and Gibson 1999; Kellert et al. 2000). In the Sahel, community-based conservation has largely taken the form of the *gestion des terroirs villageois* (GTV), or village lands management, approach (Barrier 1990; Toulmin 1993; Painter, Sumberg, and Price 1994; Batterbury 1997; Gray 1999; Gray, chap. 12 in this volume). Although implementation of GTV varies, one can identify four novel features associated with this approach: (a) devolution of resource management authority to rural communities, in practice defined as villages; (b) the use of outside facilitation, often performed by nongovernmental organizations (NGOs), to ensure broad-based community involvement and problem solving; (c) an interest in increasing land tenure security by clarifying access rules and increasing the exclusionary powers of community members; and (d) a reliance on land-use zoning that divides the village territory into areas reserved for agriculture, grazing, and wild product collection.

A major feature of the GTV approach, as in most community-based conservation projects, is the intention to devolve resource management authority to local communities. This innovation is particularly welcomed, given the highly coercive and extractive history of state-sponsored conservation in the Sudano-Sahelian region. Much of the literature and debate about GTV and community-based conservation in general has concerned the promise and pitfalls of decentralization and devolution processes. It is unclear whether governments will transfer real authority over natural resources to local communities or use these projects as an opportunity to tighten their control. In the Sahel, where local communities have often effectively used state agents' ignorance of local conditions to their advantage, the GTV approach, by documenting local resources, tenure, and conflicts, may be seen as a threat to local autonomy, despite its devolution veneer. Moreover, unchecked devolution to local communities has been questioned due to the nonrepresentative form of village-based governance in many areas (Engberg-Pedersen 1995; Ribot 1996).

These are important concerns, as the promise of the GTV approach very much rests on rearrangements of power and interest between and among local and state institutions. I would, however, like to focus here on the choice of the "village" as the primary target for devolved authority over resource management in the GTV approach. Is village-based authority the appropriate locus for resource management decision-making and is the village territory the appropriate scale for resource management? Misunderstandings of "community" have long plagued community-based conservation (Agrawal and Gibson 1999). In community-based conservation, the "community" should include the social networks that mediate and govern human use of the resource(s) of concern. Because of the complexity of such networks (Berry 1989), community-based conservation programs most typically choose human settlements as a proxy for "community." At first glance, this may seem to be an appropriate proxy in the Sahel. Agricultural usufruct is controlled by village chiefs, and supravillage indigenous institutions governing resource use are either weak or nonexistent. Still, the production strategies of farmers and livestock managers necessarily transcend often ill-defined boundaries between villages because of the high spatiotemporal variability of productive resources. The spatially restricted focus on the village and its "territory" follows a long political tradition in French West Africa of elevating village-based indigenous polities over less territorial political networks as well as those covering broader areas. As has been argued elsewhere (Bonnet 1990; Marty 1993; Painter, Sumberg, and Price 1994; Turner 1999a), the village territory cannot contain the range of resources on which farmers and herders depend to meet subsistence needs in the Sahel. From this perspective, the present lack of supravillage institutions does not arise from a lack of need for them but from a colonial and postcolonial history in which they were eroded by state actions.

The present GTV focus on the village territory, a veritable sandbox within the broader productive system described earlier, would likely have inadvertent negative consequences on resource management if it were to reduce the mobility and flexibility of Sahelian livelihoods. As described earlier, farmers and herders must be able to adjust their production efforts in response to the changing mosaic of productive resources. There is a geography to the social relations governing access. A farmer cannot travel hundreds of kilometers to tend a field outside his village territory—these fields must be within a reasonable distance from his home (about 20 km). To get from point A to point B, a herder must pass through a specific geographical area. Therefore, decisions that effectively restrict resource access to outsiders within a particular village territory *will* reduce mobility and may exacerbate extraction pressures elsewhere. The classic example of this is the vulnerability of transhumance corridors (paths) on which herders rely to move through agricultural areas. Ineffective protection of the integrity of these corridors by the state along with the erosion of supravillage institutions governing pastoral production has led to the fragmentation of many of these paths in much of the Sudano-Sahelian region.

The spatial scale and implicit boundaries erected by GTV projects are causes for concern. In addition, the nature of the social boundaries that are erected and enforced by these community-based projects need to scrutinized. As argued earlier, the broad spatial scales at which much environmental monitoring is done today provide very little information about the causes behind changes in vegetative cover: localized drought, wood collection, overgrazing, and overfarming could all be implicated. It is in this void that conservation practitioners find themselves. Most are convinced that environmental degradation is widespread, but they just do not have access to any empirical information that would allow them to even pinpoint the livelihood activities that are causing the problem, let alone the underlying social factors contributing to mismanagement. Therefore, they work within the community-based conservation framework that they have inherited as GTV without having the opportunity to seriously consider how some its underlying premises match Sahelian realities.

As discussed in more detail elsewhere (Turner 1999a), the model of mismanagement that underlies the GTV approach is that mismanagement stems from institutional failure—particularly those institutions that govern access to common property. In this way, GTV prescriptions can be tied to a major metanarrative that shapes the social policies of international conservation: namely, that local institutions governing common property resources need to be reengineered to make the rules of exclusion clearer and more enforceable (Goldman 1998). GTV projects often prescribe participatory mapping in order to delineate different sections of the village territory that are restricted to cropping, grazing, or wild resource extraction. By conceptualizing land as *the* resource to manage, the GTV approach advocates land-use zoning at a micro scale. Land, however, is only one of a number of biophysical resources utilized by cropping and livestock husbandry. The location of other resources such as rainfall, manure, and vegetation is ephemeral. Dissecting the rural landscape with socially rigid boundaries may reduce the mobility of agropastoral production, thereby effectively reducing farmers' and herders' access to spatially variable resources. In this environment, a reduction in resource access may not only increase the vulnerability of rural producers but also exacerbate environmental problems by limiting the adjustment of extraction pressures to changing resource availabilities.

Multiscale Complementarities and Socially Adjudicated Boundaries
In this chapter, I have highlighted the relationships between spatial scale and boundaries in the context of environmental monitoring and conservation planning. The scale chosen by the environmental scientist, planner, and prac-

titioner will affect what he or she sees and the social and ecological outcomes of his or her work. The scales at which we monitor human-environment relations and the nature of the social boundaries that we erect across conservation landscapes should not follow from convention but should be deduced by a joint consideration of the spatialities of production, institutions of governance, and ecological variability. Improper impositions of globally promulgated spatial scales and boundaries will not only lead to misreadings of environmental change but could unknowingly exacerbate social and ecological vulnerabilities.

I used the West African Sahel as my case to explore these issues. Two recent innovations in conservation practice in the Sahel have been the increased use of remotely sensed imagery for environmental monitoring and village-based resource management. Each of these innovations holds significant promise for improving the effectiveness of conservation in the region. Each, however, has failed to fully realize its potential because of inattention to scaling and boundary issues. With respect to spatial scaling, these innovations represent movements in opposite directions from past approaches. The increased reliance on satellite imagery has led to the monitoring of environmental change over much broader areas than in the past, while the increased popularity of community-based conservation has led to the spread of village-based conservation in the Sahel with conservation interventions encompassing only the village's territory.

Nothing is inherently wrong with either of these methodological and scalar movements—in fact, they represent significant improvements over past approaches. The major problem is that conservation scientists and practitioners have recognized the advantages but not the disadvantages of these scalar movements and the methods associated with them. In the case of satellite imagery, a broadening of spatial scales is associated with the invigoration of highly criticized methods for environmental monitoring that focus on visual changes with limited knowledge of human activities and management. The village land management approach to conservation has reduced the size of the social group affected by a conservation program, thereby improving the prospects for meaningful participation. However, it has also reduced the spatial scales of conservation attention below that which most rural producers operate, and has introduced rigid land-use restrictions that may reduce the spatial flexibility that have allowed farmers and herders to persist in this highly variable environment.

The reasons for these methodological problems, many of which have long been identified, are complex and cannot be explained by ignorance, lack of interest, or simple mediations of political interest. Moreover, they are not simply determined by technology or underlying theories of ecology or society. Instead, I have attempted to outline how these problems have emerged from multiple spheres of political contestation as shaped by technology, theory, and the unique prerogatives of conservation. For example, visual measures of environmental change have reemerged, despite their inappropriateness for monitoring land degradation in the Sahel. This resurgence results from the growing importance of satellite technology, its cost-effectiveness for measuring change over broad areas, and its popularity with the global change community, which uses remote sensing to elevate its global research agenda over highly contextualized ecological and social studies. The GTV approach's adoption of land-use zoning at the micro scale in this highly variable environment suggests an accommodation between the interests of village-based elites and unarticulated conservation prerogatives that favor rules and fixed boundaries over the messiness of political mediation and flexible boundaries (Ferguson 1994; Scott 1998; Ribot 1999).

In short, I argue that these deficiencies are not determined and are amenable to corrective adjustment if conservation players display sufficient insight and political will. In the case of environmental monitoring, multiscaled approaches need to be developed that are tied to the geographies of livelihoods. Such multiscaled engagements would undoubtedly include satellite imagery coupled with process-specific monitoring at key sites chosen for their vulnerability or representativeness. Given the limited conservation and monitoring resources for such detailed work, such multiscaled work is likely to include the monitoring of resource management project areas through joint efforts of project staff, local residents, and resource scientists. Such work may therefore involve blurring the distinctions between scientific, conservation, and development work as well as disordering standard project phases (e.g., resource assessment, problem identification, and project design and implementation). Ecological monitoring would be improved by having greater access to information about on-the-ground ecological and social realities and resource management programs would benefit from a greater understanding of the human role behind environmental change in their project areas.

Resource management programs would benefit additionally from a more serious consideration of the spatial scales at which livelihood activities and political institutions operate. Of particular importance is the form which social boundaries erected or supported by these programs should take. As I have argued elsewhere (Turner 1999a), replacing the political processes through which farmers and herders gain access to resources over a broad area with formal rules circumscribing the pool of outside candidates and narrowing their channels of access will increase both order and rigidity. In many cases, strengthening procedural rules for negotiating access may be more effective than specifying rights to particular resources. Before pushing the community to clarify and formalize boundaries, conservationists should reflect more broadly about informal mechanisms on which the resiliency of agropastoral systems depend. Without such reflection, they run the risk of promoting zoning schemes that increase both social and ecological vulnerabilities in the region.

References

Adams, W. W. 1990. *Green development: Environment and sustainability in the third world.* London and New York: Routledge.

Agrawal, A., and C. Gibson. 1999. Enchantment and disenchantment: The role of community in natural resource conservation. *World Development* 27:629–49.

Agrawal, A., and J. Ribot. 1999. Accountability in decentralization: A framework with South Asian and West African cases. *Journal of Developing Areas* 33:473–502.

Baker, R. 1984. Protecting the environment against the poor: The historical roots of soil erosion orthodoxy in the third world. *The Ecologist* 14:53–60.

Barrier, C. 1990. Développement rural en Afrique de l'ouest soudano sahélienne: Premier bilan sur l'approche gestion de terroirs villageois. *Les Cahiers de la Recherche Développement* 25:33–42.

Bastin, G. N., G. Pickup, V. H. Chewing, and C. Pearce. 1993. Land degradation assessment in central Australia using a grazing gradient method. *Rangeland Journal* 15:190–216.

Batterbury, S. P. J. 1997. The political ecology of environmental management in semi-arid West Africa: Case studies from the Central Plateau, Burkina Faso. PhD diss., Clark University, Worcester, MA.

Beck, L. R., C. F. Hutchinson, and J. Zauderer. 1990. A comparison of greenness measures in two semi-arid grasslands. *Climatic Change* 17:287–303.

Behnke, R. H., I. Scoones, and C. Kerven, eds. 1993. *Range ecology at disequilibrium.* London: Overseas Development Institute.

Berry, S. 1989. Social institutions and access to resources in Africa. *Africa* 59:41–55.

Bonnet, B. 1990. Elevage et gestion de terroirs en zone soudanienne. *Les Cahiers de la Recherche Développement* 25:43–67.

Boudet, G. 1972. Désertification de l'Afrique tropicale sèche. *Adansonia,* series 2 (12): 505–24.

———. 1975. *Manuel sur les paturages tropicaux et les cultures fourragères.* Paris: Institut d'Élevage et de Médecine Vétérinaire des Pays Tropicaux (IEMVT).

Bowker, G. 2000. Biodiversity and datadiversity. *Social Studies of Science* 30:643–83.

Braun-Blanquet, J. 1932. *Plant sociology: The study of plant communities.* New York: McGraw-Hill.

Breman, H., and N. de Ridder. 1991. *Manuel sur les paturages des pays sahéliens.* Paris, France, and Wageningen, the Netherlands: Editions Karthala and Technical Centre for Agricultural and Rural Cooperation.

Brosius, J. P., A. L. Tsing, and C. Zerner. 1998. Representing communities: Histories and politics of community-based natural resource management. *Society and Natural Resources* 11 (2): 157–69.

Buttel, F., and P. Taylor. 1994. Environmental sociology and global environmental change: A critical assessment. In *Social theory and the global environment,* ed. M. Redclift and T. Benton, 228–55. New York: Routledge.

Chevalier, A. 1934. Les places dépourvues de végétation dans le Sahara et leur cause sous le rapport de l'écologie végétale. *Comptes rendus de l'Academie des Sciences* (Paris) 194:480–82.

Ellis, J. E., and D. M. Swift. 1988. Stability of African pastoral ecosystems: Alternate paradigms and implications for development. *Journal of Range Management* 41:450–59.

Engberg-Pedersen, L. 1995. *Creating local democratic politics from above: The "gestion de terroirs" approach in Burkina Faso.* IIED Issue Paper no. 54. London: International Institute of Environment and Development.

Ferguson, J. 1994. *The anti-politics machine: "Development," depoliticization and bureaucratic power in Lesotho.* Minneapolis: University of Minnesota Press.

Goldman, M., ed. 1998. *Privatizing nature: Political struggles for the global commons.* New Brunswick, NJ: Rutgers University Press.

Gray, L. C. 1999. Is land being degraded? A multi-scale investigation of landscape change in southwestern Burkina Faso. *Land Degradation and Development* 10:329–43.

Gritzner, J. 1988. The West African Sahel: Human agency and environmental change. Geography Research Paper no. 226. University of Chicago.

Grove, R. H. 1994. A historical review of early institutional and conservationist responses to fears of artificially induced global climate change: The deforestation-desiccation discourse 1500–1860. *Chemosphere* 29:1001–13.

Hellden, U. 1991. Desertification: Time for an assessment. *Ambio* 20:372–83.

Hubert, H. 1920. Le desséchement progressif en Afrique Occidentale. *Bulletin du Comité D'Études Historiques et Scientifiques de l'Afrique Occidentale Française* 1920:401–67.

Huete, A. R., and C. J. Tucker. 1991. Investigation of soil influences in AVHRR red and near-infrared vegetation index imagery. *International Journal of Remote Sensing* 12:1223–42.

Kellert, S. R., J. N. Mehta, S. A. Ebbin, and L. L. Lichtenfeld. 2000. Community natural resource management: Promise, rhetoric, and reality. *Society and Natural Resources* 13:705–15.

Lélé, S. 1991. Sustainable development: A critical review. *World Development* 19:607–21.

Leprieur, C., Y. H. Kerr, and J. M. Pichon. 1996. Critical assessment of vegetation indices from AVHRR in a semi-arid environment. *International Journal of Remote Sensing* 17:2549–63.

Marty, A. 1993. La gestion des terroirs et les éleveurs: Un outil d'exclusion ou de négociation? *Revue Tiers Monde* 34:329–44.

Middleton, N., and D. Thomas, eds. 1997. *World atlas of desertification.* New York: United Nations Environment Programme.

Niamir-Fuller, M., ed. 1999. *Managing mobility in African rangelands.* London: Intermediate Technology Publications.

Niemeijer, D. 1996. The dynamics of African agricultural history: Is it time for a new development paradigm? *Development and Change* 27:87–110.

Painter, T., J. Sumberg, and T. Price. 1994. Your terroir and my "action space": Implications of differentiation, mobility and diversification for the Approche Terroir in Sahelian West Africa. *Africa* 64:447–63.

Ribot, J. 1996. Participation without representation: Chiefs, councils and forestry law in the West African Sahel. *Cultural Survival Quarterly* (Fall): 40–44.

———. 1999. Decentralisation, participation and accountability in Sahelian forestry: Legal instruments of political-administrative control. *Africa* 69:23–65.

Scoones, I., ed. 1994. *Living with uncertainty: New directions in pastoral development in Africa.* London: Intermediate Technology Publications.

Scott, J. C. 1998. *Seeing like a state: How certain schemes to improve the human condition have failed.* New Haven, CT: Yale University Press.

Stamp, L. D. 1940. The southern margin of the Sahara: Comments on some recent studies on the question of dessication in West Africa. *The Geographical Review* 30:297–300.

Stebbing, E. P. 1935. The encroaching Sahara: The threat to the West Africa colonies. *Geographical Journal* 85:506–24.

Stoddart, L. A., A. D. Smith, and T. W. Box. 1975. *Range management.* New York: McGraw-Hill.

Swift, J. 1996. Desertification: Narratives, winners, and losers. In *The lie of the land,* ed. M. Leach and R. Mearns, 73–90. London: International African Institute.

Thomas, D. S. G. 1993. Sandstorm in a teacup? Understanding desertification. *Geographical Journal* 159:318–31.

Toulmin, C. 1993. Gestion de terroir: Principles, first lessons and implications for action. Discussion Paper. New York: United Nations Sahelian Office, March.

Tucker, C. J., H. E. Dregne, and W. W. Newcomb. 1991. Expansion and contraction of the Sahara Desert. *Science* 253:299–301.

Turner, M. D. 1993. Overstocking the range: A critical analysis of the environmental science of Sahelian pastoralism. *Economic Geography* 69:402–21.

———. 1999a. Conflict, environmental change, and social institutions in dryland Africa: Limitations of the community resource management approach. *Society and Natural Resources* 12:643–57.

———. 1999b. Merging local and regional analyses of land-use change: The case of livestock in the Sahel. *Annals of the Association of American Geographers* 89:191–219.

———. 2003. Methodologial reflections on the use of GIS and remote sensing in human ecological research. *Human Ecology* 31 (2): 255–79.

Vandergeest, P. 1996. Mapping nature: Territorialization of forest rights in Thailand. *Society and Natural Resources* 9:159–75.

Vedeld, T. 1992. Local institution-building and resource management in the West African Sahel. *Forum for Development Studies* 1:23–50.

Warren, A. 1995. Changing understandings of African pastoralism and the nature of environmental paradigms. *Transactions of the Institute of British Geographers* 20:193–205.

Western, D., and R. M. Wright, eds. 1994. *Natural connections: Perspectives in community-based conservation.* Washington, DC: Island.

Westoby, M., B. Walker, and I. Noy-Meir. 1989. Opportunistic management for rangelands not at equilibrium. *Journal of Range Management* 42:266–73.

Zerner, C. 1996. Telling stories about biological diversity. In *Valuing local knowledge,* ed. S. B. Brush and D. Stabinsky, 68–101. Washington, DC: Island.

PART III

Transnational and Border Issues in Global Conservation Management

Border issues are a salient feature of the global expansion of conservation and sustainability policies. These issues are not entirely new, of course, since the role of borders has always been a particularly challenging feature of the policies for guiding resource use and conservation management. Indeed, border issues involving conservation and resource management have been prevalent since the colonial periods in many places throughout the world. Similarly, the heated or delicate status of border concerns has long been a characteristic of indigenous and local systems of resource management worldwide, which nowadays are increasingly interfaced with conservation areas that are formalized as "official" or government designations. The importance of border issues globally is now multiplying more quickly than ever due to the large overall increase in the number of conservation territories, the involvement of global organizations, and the increasingly complex styles of spatial management with multiple areal units and the growing application of spatial technologies, particularly geographical information systems (GIS) and the global positioning system (GPS), to create and manage bounded units with an unprecedented extent of borders.

A number of the most important border issues facing current efforts at conservation and sustainability are transnational, where coordination of environmental management is called for across national borders. Transnational border issues have become characteristic of the globalization of conservation and sustainability policies. The best-known examples of transnational border issues are associated with transboundary protected areas (TPAs), also known as Peace Parks, and conservation and sustainability corridors, such as the Mesoamerican Biological Corridor (MBC), both discussed in chapter 1. Our volume contributes to these understandings by focusing on the transnational border issues that arise in conjunction with combined biodiversity conservation and river basin development (chap. 8), ecoregion concepts for transnational protected areas (chap. 9), and the transnational dimensions of conservation ideas and planning (chap. 10).

The combination of river basin management and biodiversity conservation is a confluence of goals for transnational resource management that appears logical, at least at first glance (chap. 8). The spatial and environmental properties of river basins, as well as biodiversity, suggest the need for territorial units of management that cross national boundaries and ensure transnational cooperation.[1] In addition, both river basin development and biodiversity conservation have been favorite projects for global organizations during recent decades. The globalization and coupling of these management objectives is perhaps more evident in the Mekong River countries of Southeast Asia than anywhere else in the world. Yet, as chapter 8 demonstrates, a powerful tension is created in this example of globalization. While globalization in general, and the Mekong project in particular, tends to promote combining management goals in the context of transnational sustainability, these aims may require policies that are contradictory to one another, even though they may share the similar context of seeking transnational sustainability. Chapter 8 evaluates how the proposals for transnational river basin development would induce a number of effects that run strictly counter to the policy aims of biodiversity conservation. Focusing on this tension is important for conservation and sustainable use policy, since transnational initiatives are more likely than ever found in conjunction with other globalization trends.

Biodiversity conservation, as a defining focal point for conservation and sustainable use, is frequently conceived of as requiring project initiatives that are transnational in scope. Indeed the spatial foundations of biodiversity conservation are inherently transnational (chap. 9). The ecoregion, which is a region-scale territory of distinct environments and biota, is also of considerable interest as potentially practical for planning and management purposes. Within the past decade the ecoregion has become a preferred unit, typically situated within a global geographical framework of ecoregions worldwide, for the grand design of global biodiversity conservation and sustainable use. Unsurprisingly, it is also widely seen as a potential template for the design of transnational initiatives. Indeed ecoregions are usually a foundation for the design rationales and reasoning for both TPAs and conservation corridors, which are described at some length in chapter 1. Yet, as chap-

---

1. Often more than one type of major objective is frequently contained in projects in the new transnational policies for conservation and sustainability. Global organizations and multigovernment planning agencies have strong motivations for the combination of goals in order to build a broader base of political, economic, and public opinion support and accrue broader management power (see the discussion of conservation corridors in chap. 1).

ter 9 argues, the possibilities for transnational biodiversity conservation are laden with tensions that arise as a result of national and local factors. It offers an analysis of the spatial design of protected-area coverage and its implications for ecoregion-based biodiversity conservation in the Andean countries of Colombia, Ecuador, Peru, and Bolivia. The analysis shows that the difficulties with transnational coordination are one of the main reasons for the disjuncture between protected-area coverage and ecoregion-based planning goals for biodiversity conservation. In the case of these tropical Andean countries, the prospect of transnational initiatives must be seen as necessarily negotiating and acquiring new spatialities as a result of these political and development concerns, rather than being viewed solely as imperfections or impediments to abstract spatial ideals.

A deeply transnational or global dimension is also readily apparent in the geographic foundation of scientists and scientific knowledges that support biodiversity conservation. One highly important dimension is the role of networks of international scientists and Western scientific ideas. The transnational or global aspect of these international scientific interactions typically offers a network style of spatial interaction, rather than the territorial forms that are most common to conservation. Such networks of scientific expertise that frequently extend into policy or project influence are extremely important in contemporary issues of biodiversity conservation and sustainable use. Chapter 10 examines the interaction of international and global scientific networks with national counterparts in the realm of conservation ideas that have informed the establishment of protected areas within the country of Peru. The situation of Peru is an important example since it receives a high level of international and global scientific activities and conservation interest. The current importance of Peru to global conservation and sustainability policies is driven in part by the disproportionate prioritization of biodiversity "hot spots" within the country. That present-day importance is combined with a considerable degree of national scientific and public awareness of conservation that is of longstanding interest within Peru.

Chapter 10 argues that the prominent role of international scientists and scientific ideas did not originate de novo with the recent prevalence of global conservation and sustainability issues in Peru. Rather the experience of the country shows the formative influence of several foreign scientists in the creation of conservation ideas and the establishment of protected areas in Peru dating back to the 1800s. The chapter focuses on how this legacy is a revealing comparison to the recent influences in Peru of the global environmental sciences, especially the global biodiversity and conservation sciences. It demonstrates that the recent emphasis of global scientists and scientific ideas on tropical rain forest areas offers a prioritization that does not correspond closely to the evolution of a mainstream Peruvian scientific perspective on protected areas that has sought to balance the representation of these humid tropical areas with semi-arid and arid coastal and high Andean habitats. At issue is whether and how the growing influence of the conservation ideas of global environmental science will be further melded with Peruvian national interests in the ongoing practice, policy, and planning of protected areas.

8 Conservation Initiatives and "Transnationalization" in the Mekong River Basin

CHRIS SNEDDON

The Mekong is a transnational river basin of considerable interest to development planners, international financial institutions, international environmental advocates, bilateral aid donors, and managers of global capital investment funds. Part of their interest relates to the significant development pressures facing the basin in the form of a long-dormant agenda of constructing a cascade of ten to twelve massive dams on the river's main channel. This cascade, originally conceived in the late 1950s and 1960s, has once again captured the imagination of the Thai, Lao, Cambodian, and Vietnamese national governments and assorted bilateral donors and international financial institutions.[1] At the same time, there is a tremendous amount of confusion, not to mention disagreement, over the ultimate costs and benefits of damming the Lower Mekong. Exploring the merits of these debates is not my intention here. Rather, my primary concern is to explicate the current institutional environment for transboundary conservation in the Mekong Basin from a perspective that problematizes the strategies of conservation actors and recognizes the impacts of extrabasin political-economic processes.

Emerging at least in part as reaction to the developmental pressures outlined above, there has been a growing interest in the conservation of the Mekong Basin—and Southeast Asia more generally—as a biodiversity "hot-

---

1. A variety of reports and studies focused on hydroelectric potential in the Lower Mekong Basin have identified more than 200 hydroelectric dam projects with an estimated total generating capacity of up to 60,000 megawatts. An indicative basin plan from the early 1990s, representing the perspective of the basin's riparian governments, identifies seventy-eight projects, of which fifty-eight are basinwide in scope. These include eleven large-scale, run-of-the-river hydroelectric dams (see Chooduangngern 1996). See also MRC 2001b for more recent hydroelectric development plans.

spot" and a region of "mega-biodiversity" (Dudgeon 2000). A diverse assortment of international environmental nongovernmental organizations (NGOs), research institutes, bilateral donors, development organizations, and environmental advocacy groups, in combination with the Mekong River Commission, have invested considerable resources in designing and debating conservation projects that will ostensibly sustain the basin's ecological systems and contribute in some manner to improving basin livelihoods. The result is a partially actualized network of protected areas, or "conservation territories" (see Zimmerer, chap. 1 in this volume), and a commensurate network of institutional actors who ostensibly govern these protected areas. Much of the rationale for this "conservationist" interest resides in recognition of the Mekong's unique ecological attributes and its status as arguably the least altered large river system in the world.

Yet conservation efforts in a transnational river basin raise both conceptual and practical questions. How do conservation efforts, asserted most forcefully by international environmental actors, intersect with current plans to implement a host of large-scale water resource development projects at the scale of the entire basin? Furthermore, is the sort of conservation being advocated within the international "community" of environmental organizations—a conservation that explicitly defines territories within sovereign states—synchronous with state-level and community-based conservation initiatives? What is being "conserved" and in whose interest?

To address these questions, I first explicate the scope of transboundary conservation initiatives in the Mekong Basin and describe their basic characteristics. As part of a more general process of environmental globalization (Zimmerer, chap. 1 in this volume), conservation advocates in the Mekong have astutely, yet problematically, inserted themselves into social conflicts over water in the basin during the past decade. A majority of conservation initiatives in the basin, I argue, have yet to come to grips with either the ways in which proposed development activities will fundamentally alter the basin's ecohydrological dynamics or the extrabasin political-economic processes that will determine when and how these development activities will be carried out. In other words, conservation of the Mekong is an exemplar of the contradictions of globalization, as well as the contradictions of conservation networks as they reconfigure our understandings of geographical scale. To elaborate on this argument, I then examine the Asian Development Bank's Greater Mekong Subregion initiative as a case in point of the scalar politics of transnational conservation and development efforts. I conclude with a plea for conservation initiatives to be mindful of socioecological processes that occur at scales that transcend the boundaries of geographically delimited conservation territories.

Transboundary Conservation in the Mekong

Conservation initiatives occurring in the Mekong basin, particularly those that fall under the "transboundary" category, must account for ecohydrologic and political-economic processes typically associated with broader geographical scales if they hope to promote ecological integrity and resilience and reduce social vulnerabilities.[2] This assertion underscores the importance of merging scales of analysis in order to explain the multiple factors contributing to (and detracting from) conservation initiatives associated with specific locales or regions, a point reinforced as one of the central themes of the present volume (see also Turner 1999). I thus do not want to fall into the analytical trap of simply identifying "contextual" factors and moving on to the nitty-gritty of individual conservation initiatives and projects, a move common in the literature associated with community-based conservation, people-park relationships, and the like.[3] My approach is roughly similar to Zimmerer and Young's (1998) "regional-national-global framework" for understanding conservation in developing countries, although even this categorization begins to break down in the face of the complexity of local-global political-ecological interactions. Those ecohydrologic and political-economic processes typically construed as "regional," "transnational," and "global" often constitute multifaceted networks of actors, in effect producing novel spatialities of human-environment relationships that transcend associations with single geographical scales.

A detailed ecological accounting of the Mekong's biological diversity and the ways in which it is being transformed by human activities is beyond the scope of the present work.[4] A primary point to highlight is the coevolution of the Mekong's myriad species of fish, aquatic invertebrates, aquatic bird life, riverine mammals, and aquatic and riparian vegetation with the highly seasonal flows of the Mekong's main channel and tributaries. Like the rest of Southeast Asia, the Mekong's ecohydrological functioning is highly influenced by a monsoonal climate, producing a dry season (roughly mid-October through mid-May) of low flows and relative water scarcity and a wet season

---

2. For fuller treatment of the concepts of integrity, resilience and vulnerability, as applied in both ecological and socio-economic senses, see Adger (2000).

3. This critique does not deny the validity and usefulness of more micro-scale analyses; in fact, such cases are absolutely crucial in defining the ways in which community dynamics related to, for example, gender and other kinds of social relations intersect with nationally and internationally defined conservation areas (see Neumann 1998; the cases collected in Zimmerer and Young 1998; Brown, chap. 4, WinklerPrins, chap. 5, and Sundberg, chap. 11 in this volume).

4. Dudgeon (2000) summarizes the most critical conservation concerns in the region, and the International Centre for Environmental Management (ICEM 2003) offers a detailed accounting of the Lower Mekong region's most pressing conservation concerns.

(roughly June through mid-October) of high discharge rates and region-specific inundated floodplains.[5] Any disruption of this pronounced seasonality, such as almost always occurs following large-scale flow regulation via dam construction, is likely to have significant and deleterious consequences for aquatic species (ICEM 2003, 63–65).

This assertion is underscored by the emergence in recent years of several paradigms that seek to explain the dynamics of large river basins. These include the river continuum concept (the longitudinal gradient from headwaters to estuary); the serial discontinuity concept (the "resetting" of biophysical characteristics downstream of impoundments; Stanford and Ward 2001); the riparian control concept (a comparison of organic matter budgets, particularly the importance of allochthonous debris); the nutrient spiraling concept (the transformation of nutrients from dissolved to particulate states as they move downstream); the flood pulse concept (the lateral dynamics between river channels and their associated floodplains); and the ecotone concept (the presence of interconnected patches within riverine landscapes; Poff et al. 1997). Investigations of large river systems based on these notions stress that impoundments and other alterations of flows disrupt the connectivity of the longitudinal (upstream-downstream), lateral (flood plains), vertical (riparian zones), and temporal aspects of a basin's ecohydrologic dynamics. The severance of connectivity along these different dimensions due to dam construction almost always results in adverse impacts on basin biota.

To cite one example, there is evidence that a significant percentage of Mekong fish species (as high as 90 percent) depend on flood plain habitats for reproduction and feeding and that the most economically important Mekong species undertake transboundary, seasonal migrations (some between 500 and 1,000 kilometers) attuned to well-timed changes in flow rates (MRC Secretariat 2001; Coates et al. 2003). By regulating river flow rates, large-scale development of the Mekong's main channel will irrevocably alter the relationship between fish species and the river's seasonal changes in water conditions. Given that total annual catch in the Mekong's inland fisheries is estimated at between 1.6 and 2.8 million metric tons—worth roughly US$1.4 billion—and that fish are by far the most important source of animal protein for basin residents (Bao et al. 2001), an altered Mekong would almost certainly undercut a crucial aspect of basin livelihoods (White 2000). Furthermore, the impacts of such alterations will be geographically differential, with the inland fishery-dependent communities of Laos and Cambodia, as

5. As an example of the tremendous differences involved, peak flows at a stream gauge station near Phnom Penh in September-October regularly exceed 45,000 cubic meters per second (cms), while dry-season flows in March and April at the same site average 1,500 cms. The river's average annual total discharge amounts to 450 billion cubic meters (Kite 2000).

well as the Mekong's delta region in Vietnam, being most affected (see Gum 2000; Chomchanta et al. 2000; and Sneddon and Nguyen 2001).

Such concerns explain in part the recent materialization of the Mekong as a region of considerable interest in terms of biological diversity and its subsequent transnationalization within the practices of international conservation organizations.[6] A regional workshop held in May 2002 brought together senior officials from the Mekong region's protected-area management agencies and environmental and economic development consultants to review the prospects for expanding the area officially designated as "protected" to 15 percent of the entire region (MRC 2002; see fig. 8.1), and some have projected that up to 22 percent of the region's territories will attain some kind of protected-area status by the late 2000s (ICEM 2003, 13). International conservation organizations, primarily based in the United States and Europe, have assumed a leading role in advocating Mekong conservation. For example, the World Wildlife Fund–Worldwide Fund for Nature (WWF) recently released a report titled *Towards a Vision for Biodiversity Conservation in the Forests of the Lower Mekong Ecoregion Complex* (Vietnam News Service 2001). The WWF's Ecoregion Action Programme focuses on the Greater Annamite mountain range along the border of Laos and Vietnam, which encompasses a diverse assembly of large mammal and bird species, many of which are globally endangered. Indeed, many of these initiatives advocate an "ecoregional" approach to conservation. Within this framework, an ecoregion is a "relatively large unit of land and water containing a geographically distinct assemblage of species, natural communities and environmental conditions" (World Commission on Protected Areas 2000). Yet it is far from clear how these territorially defined "ecoregions" are linked to the Mekong's broader ecohydrologic dynamics (e.g., seasonal flooding), which operate at the basin scale. Indeed, it is also unclear how the current patchwork approach to protected areas in the Lower Mekong region will mesh with plans fostered by the Asian Development Bank (discussed below) that envisage the region, or "subregion," as the most appropriate geographical scale for the coordination of economic development. There are several ongoing initiatives to identify and institutionalize transboundary protected areas (TBPAs) in the lower Mekong region, but formalization of these arrangements has been slow (ICEM 2003, 150–62).

In addition to the work of the WWF, Conservation International (CI)

6. International environmental organizations such as the IUCN (the World Conservation Union), the World Wildlife Fund (WWF) Worldwide Fund for Nature, and Conservation International (CI) are "transnational" in the sense that they comprise global networks of scientists, government representatives and environmental advocates (e.g., IUCN's membership includes 140 states, seventy-seven government agencies, and more than 800 NGOs), and their scope in terms of conservation projects spans the planet.

FIGURE 8.1 Map of Upper and Lower Mekong Basin, with protected areas indicated

launched an initiative in 2000 within Cambodia, which it identifies as a critical nation within the "Indo-Burma hotspot" region of biodiversity conservation. The ultimate goal of CI's initiative is to protect an area identified as the Central Cardamoms Protected Forest, an "almost pristine" region, amounting to 6 percent of the nation's territory, with relatively high concentrations of biological diversity in terms of mammal, bird, reptile, and amphibian species endemic to Cambodia (CI 2002). Reflecting similar goals (and employing similarly effusive language), The Nature Conservancy (TNC) initiated its Great Rivers Project in the northwestern portion of China's Yunnan province in the late 1990s. This region of "lush valleys, precipitous river gorges and rugged, ice-capped mountains" includes part of the Upper Mekong[7] and is the focus of TNC's work because it is "one of the most vital centers of plant diversity in the northern temperate hemisphere" (TNC 2002). The World Conservation Union, or IUCN, has several ongoing projects located throughout the basin, focusing in particular on wetland conservation in the delta region of Vietnam and the broader Mekong lowlands that extend into Cambodia's Tonle Sap region (IUCN 2002).

Reflecting the emergence of publicity and resources for conservation in the Mekong, several of the riparian states have seconded their international counterparts through an assortment of national programs to ostensibly sustain biodiversity. For example, the Cambodian government in 1993 authorized a novel protected-area system composed of national parks, wildlife sanctuaries, special reserves, and multiple-use zones that covers 18 percent of the nation's territory (CI 2002), and some of the Cambodian government's environmental agencies aim to implement a national biodiversity conservation program (Rao and Seila 2001). Concurrently, the government of Laos created a protected-area system in 1993 that now includes twenty National Biodiversity Conservation Areas (NBCAs) covering some 13 percent of the country (Chape 2001). In both the Cambodian and Lao cases, national-level recognition of biodiversity conservation was encouraged and, to some degree, orchestrated with overarching guidance from the IUCN, other prominent conservation groups, and bilateral aid donors. Most recently, the Australia-based International Centre for Environmental Management—in affiliation with IUCN's Asia office, assorted government agencies, and the Mekong River Commission—reviewed the conservation status of the four countries of the Lower Mekong Basin (ICEM 2003). One of the goals of this program is to

---

7. This chapter's central concern is the political ecology of the Lower Mekong, although the Chinese government's ongoing interventions in the Lancang (Upper Mekong) basin—two dams have been completed and seven more are projected—will undoubtedly produce downstream impacts in terms of biophysical, institutional, and political dynamics.

examine the role of protected areas in economic development through, for example, ecotourism programs and community-based conservation projects.

While the actual amount of financial resources being poured into Mekong conservation resources is unclear,[8] the widening scope of basin biodiversity initiatives highlights an important shift in the state's role in environmental management, and one related to the ongoing environmental globalization described by Zimmerer in chapter 1. In essence, the Mekong's riparian states have relinquished much of the power for formulating environmental governance agendas to international environmental groups and bilateral aid donors. This amounts to a further "internationalization" or "transnationalization" of the state (Glassman 2001; Robinson 2001), but in this case in terms of environmental policies and management rather than developmental trajectories. In a fashion akin to the evolution of protected-area networks in South America (see Young and Rodríguez, chap. 10, and Sierra, chap. 9 in this volume), international conservation groups have been more than willing to fill this environmental governance "gap" primarily through partnerships with government agencies and basin-level institutional actors such as the Mekong River Commission. (The Mekong River Commission, or MRC, is the formal institution charged with overseeing transnational development efforts in the Mekong basin; it is composed of representatives of the governments of the four member states, as well as advisors from donor countries and international agencies.) The growing importance of such partnerships suggests the need for analysts of conservation to examine networks of actors—in this case built around the activities of conservation-oriented, transnational environmental organizations—that extend far beyond the contours of traditionally defined conservation territories (e.g., parks and reserves) and to trace how these networks help define "conservation" at national and local scales. In the case of the Mekong, these novel spatialities of conservation are being constructed in contradictory ways.

Conservation of What and for Whom?
One of the most intractable dilemmas facing conservation organizations at work in the Mekong concerns how to negotiate the complex terrain defined by the national-level economic development objectives of the riparian states, those same states' conservation efforts, and local social movements stand-

---

8. According to ICEM, overseas development assistance (ODA) for the "environment sector" (with protected areas making up 80 percent of this total) in Vietnam alone increased from around US$10 million annually in the late 1980s to well over US$180 million per year in the late 1990s. Similarly, ODA to biodiversity conservation in Lao PDR increased from around US$6 million in 1996 to roughly US$35 million in 2000. However, ICEM (2003, 24–26) points out that environment-oriented ODA in Vietnam has declined in recent years in ODA to protected areas in Lao PDR has "all but shut down" after peaking in 2000.

ing in opposition to both. Despite careful attention to the human components of the Mekong basin, particularly the roughly 60 million primarily rural inhabitants who remain to a large extent resource dependent, the conservation organizations cited above have gone about the business of promoting biodiversity-oriented projects in ways that at times seem contradictory to norms of social justice and sustainable livelihoods. In other words, it is critical to ask to what extent conservation in the Mekong is advancing in a manner that risks becoming a kind of "green orientalism" (Lohmann 1993). There is now a diverse literature regarding the ample mistakes of past conservation efforts throughout the third world, which disregarded and further marginalized human communities who had the misfortune to live in or near "protected" areas (Hirsch 1997; Adams and McShane 1992; Neumann 1998; Oates 1999; Athanasiou 1996). International conservation organizations continue to initiate alliances with governmental and intergovernmental actors in an effort to extend their spheres of influence. This reflects a clear prioritization of specific biological conservation goals above objectives related to the social resilience of the Mekong's human communities and the ecological services upon which their livelihoods depend.

Throughout the 1990s, some conservation organizations at least tacitly supported government-sponsored dam projects on Mekong tributaries in Laos in spite of their projected impacts on local human communities and aquatic ecosystems. In the case of the Nam Theun 2 hydroelectric dam in Laos (fig. 8.1), both IUCN and the Wildlife Conservation Society (WCS) have supported the involvement of the World Bank in financing the dam in the hopes that this would provide leverage for conservation of the Nakay Plateau. This grassland-forest ecoregion, 470 square kilometers of which will be inundated by the dam's reservoir, consists of primary montane forests on lands rising from 600 to nearly 2,300 meters in the Annamite mountain range along the Laos-Vietnam border. This area of high biodiversity and unique species—it contains at least fourteen globally endangered large mammals (including the celebrated saola, or Vu Quang ox) and myriad threatened bird species— was officially recognized by the Lao government as the Nakay-Nam Theun NBCA in 1993. As a partial trade-off, IUCN and WCS agreed to support construction of Nam Theun 2, or at least those components of the project that will, they argue, result in biodiversity conservation. Although the agreement is not yet finalized, Laos agreed in principle, under the umbrella of the World Bank's Global Environment Facility, to devote 1 percent of the dam's annual hydropower revenues toward watershed management and conservation in the attendant NBCA (Australian Mekong Resource Centre 1999). In addition, some reports contend that IUCN will receive a direct payment of US$65 million from the World Bank, which has been involved in supporting the project, in exchange for support of Nam Theun 2 (Monbiot 1997).

Critics charge that the 680-megawatt, US$1.1 billion dam—which the Laos government sees as a vehicle for obtaining much-needed foreign exchange via the sale of hydroelectricity to Thailand—will result in further migration to the plateau, degradation of the region's forests, and the resettlement of existing communities of resource users (approximately 5,700 families). Furthermore, critics point out that there are few guarantees that conservation efforts as envisioned will actually be implemented upon completion of the dam (Monbiot 1997; International Rivers Network 2004). Despite such charges, Hans Friederich, the head of IUCN's Aquatic Ecosystem Programme for Asia, argues that there are situations where "dams may be good for biodiversity conservation and local livelihoods," but these "issues are complicated and have many angles" (Friederich 2000). Indeed, there are no easy answers. Representatives of conservation groups argue that an unyielding opposition to economic development projects—for example, the Nam Theun 2 dam project in Laos—fails to recognize the on-the-ground political-economic realities of resource-dependent nations with little interest in and even less resources for the conservation of biodiversity.[9] In their minds, hydroelectric development is a relatively small price to pay for at least some measure of protection of unique and threatened ecological systems such as those in the upland areas of the Mekong basin. Yet this stance places them at odds with other actor-networks—prominently international NGOs such as the International Rivers Network (IRN) and regional NGOs such as the Bangkok-based Towards Ecological Recovery and Regional Alliance (TERRA)—who are working in opposition to dam projects that threaten not only nonhumans but also the countless human communities who will be uprooted and otherwise have their livelihood options severely reduced by the construction of large dams.

Fish Conservation in the Mekong
Given the recent surge in the Mekong as a region of intense interest to conservation organizations and their funders, the general lack of even basic ecological knowledge of the basin's aquatic ecosystems is marked. Indeed, aquatic ecologist David Dudgeon (2000, 239) notes that tropical Asia, in general, "has a rich flora and fauna yet, compared to the Neotropics and Africa, does not seem to invoke the same concern over biodiversity conservation" and that "comparatively little ecological research" has been undertaken or reported in the mainstream ecological literature. Within this frame of general neglect, freshwater biodiversity, particularly that associated with river systems, has been especially overlooked (Dudgeon 2000). Ecological knowledge of the Mekong reflects these observations. Despite estimates of fish diver-

9. WCS staff member, personal communication, 15 April 2002.

sity in the basin's waters that run from 500 to 1,700 species—behind only the Congo and Amazon basins in terms of fish species richness—very little knowledge exists regarding fish migration timing and patterns, reproductive and feeding behavior, interspecific dynamics, and other fundamental ecological processes. This profound ignorance of fish biodiversity is mirrored in the cases of invertebrate biodiversity, mammalian species associated with riverine habitats, and a wide variety of aquatic flora in the Mekong Basin (Dudgeon 1992, 2000). As Turner (chap. 7 in this volume) highlights in a study of the science-conservation-politics nexus in the Sahel, scientific ideas and practices mobilized for conservation purposes rarely proceed in straightforward fashion, and often produce unintended consequences in terms of socioecological change. The inattention to fisheries in the Mekong may be due in part to scientific oversight, but the material impacts of such neglect on Mekong livelihoods are profound, in large part because this neglect helps maintain the relatively low status accorded fish and aquatic ecosystems within both conservation and development agendas.

Ironically, fish species of the Mekong (and perhaps fish of other large tropical river systems as well) have been doubly neglected. In the first instance, wild capture fisheries as a crucial component of the livelihood strategies of basin residents[10]—and thus ostensibly an important component of any water resource development program's agenda—have been largely ignored by the organizations (i.e., the Mekong Committee and its more recent version, the MRC) responsible for coordinating information collection and technical activities under the broad regime of cooperative development in the basin (see Sneddon and Nguyen 2001). The lack of knowledge concerning Mekong fisheries and the biology and ecology of fish species is not innocent but rather reflects the prioritization of other water resource development objectives (e.g., hydroelectric generation, irrigated agriculture, flood control) on the part of the basin's states. Wild capture fisheries, either as part of subsistence livelihoods or as critical supplemental food and income sources, have simply been beyond the radar screen of Mekong institutions over the past four decades. Virtually nothing is known regarding the massive migrations of fish species in the river system, which on occasion appear to migrate as multispecies assemblages from one main channel location to another, from mainstream to tributary, or from tributary to tributary (MRC Secretariat 2001). As iterated previously, an estimated 90 percent of Mekong fish species depend on inundated floodplains for primary breeding and juvenile feeding habitat, and the release of nutrients from flooded vegetation in these regions results in fish

10. For example, millions of basin inhabitants near the Tonle Sap ("Great Lake") of Cambodia and the Cambodia-Vietnam delta region obtain 80 percent of their total protein intake from fish (Gum 2000).

productivity measures that triple those of the main channel (Kite 2000; Coates et al. 2003). There is thus little question that upstream impoundments will cause major obstacles to migratory patterns and that subsequent regulation of the river to reduce seasonal flood pulses will prove disastrous for many species and for the basin residents who depend wholly or partially on fisheries.

In the second instance, freshwater fish and other aquatic species have, to my knowledge, never achieved the same significance within the hallways and reports of international environmental NGOs as terrestrial species, particularly large mammals and other "charismatic megafauna." Such an emphasis is, in part, a strategic calculation by these organizations to draw financial and political support, yet it may also reflect a bias that privileges land-based species (in part because they are more effective in building memberships and organizational resources) at the expense of fish. While exploring the foundations of such a bias in the present work is impractical, I do wish to note its presence in the ways in which conservation initiatives in the Mekong are being conceived and carried out. In the Nakay-Nam Theun 2 case mentioned previously, investigations of the fish species that comprised an important component of the livelihoods of the several thousand ethnic minorities displaced by the dam's reservoir never entered into the government's environmental impact assessments (Monbiot 1997). A representative of one of the conservation organizations involved in the case, the WCS, freely admits that at the time of negotiations over the dam they had little knowledge of aquatic biodiversity and were more concerned with terrestrial species and habitats.[11]

Given the historically low priority accorded biodiversity conservation within the riparian states of the Mekong, which has translated into relatively little concern over conservation within the intergovernmental institution charged with environmental governance of the basin as a whole, the MRC,[12] conservation ecologists and associated international environmental representatives are in a difficult position. The difficulty is compounded by conservation organizations' relative lack of interest in aquatic biodiversity, although this may be changing.[13] There are also signs emanating from the MRC

11. WCS staff member, personal communication, 15 April 2002.

12. A recent report by the MRC concerning hydropower development does make reference to the significant levels of biodiversity present in basin, particularly in the montane regions of Lao PDR and Vietnam (MRC 2001b, 31). The report also cites the number of fish species in the river system as "more than 1,300" (the largest estimate to date that I am aware of), and identifies certain "aquatic flagship species" of the basin (32). However, it is not clear from this designation whether this is a reference to these species value as popular symbols or as ecological keystone species, or both. The report also notes that few of the biodiversity-rich wetlands of the Mekong have any kind of protected status and "there is no regional wetlands policy or strategy prepared" (32).

13. For example, there are several ongoing projects focused on Cambodia's critical and highly threatened Tonle Sap ecosystem. Similarly, IUCN in coordination with the United Nations Develop-

that fisheries have assumed greater importance within their governance and research frameworks.[14] Yet conservation practices, of fish species or otherwise, also need to account for political-economic processes, associated with scales that transcend the basin, that are actively reconfiguring the pace and manner of water resource development in the Mekong region.

The Greater Mekong Subregion Initiative

There is little doubt that large-scale development as currently planned and promoted in the Mekong Basin will significantly alter the ecohydrological characteristics of existing and potential protected areas (ICEM 2003). It is thus worth investigating how another set of actors is attempting to recast the Mekong almost solely in economic terms to greatly facilitate large-scale hydroelectric development. The Greater Mekong Subregion (GMS) development plan, launched in June 1993 under the aegis of the Asian Development Bank (ADB), raises questions about both (a) the efficacy of Mekong-scale governance and conservation in the face of efforts to construct an economic region interdependent with global financial and trade networks and (b) the need for an independent and accountable institutional framework at the Mekong scale. These questions highlight the changing character of the formal institution charged with overseeing transnational development efforts, the MRC, and deficiencies in its capacities to promote resilience and integrity in the basin in the context of conservation initiatives or otherwise. While the GMS initiative engages with conservation issues in only cursory fashion (e.g., by highlighting the potential for ecotourism), I argue that conservation initiatives must come to grips with the way the GMS program seeks to insert the Mekong basin, its resources, and its people firmly within networks of economic globalization.

As conceived by the ADB, the GMS initiative focuses on those sectors— energy, telecommunications, transportation, trade, and tourism—deemed

---

ment Programme (UNDP) is initiating a comprehensive review of the Lower Basin's wetlands with a particular focus on conserving the river's highly threatened aquatic species, such as the giant catfish (*Pangasianodon gigas*), the giant Mekong barb (*Catlocarpio siamensis*), the freshwater dolphin (*Orcaella brevirostris*) and the highly endangered Siamese crocodile (*Crocodylus siamensis*). Avian-oriented groups such as Birdlife International and the International Crane Foundation have focused attention on Vietnam's delta region, where, among other reserves, the Tram Chim National Park is home to almost the entire global population of the eastern sarus crane (*Grus antigone sharpii*; Friederich 2000). For evidence of a global shift in perspective that gives freshwater biodiversity greater status, see Groombridge and Jenkins (1998).

14. Recent reports sponsored by the MRC stress the critical role that fish species play in the livelihoods of basin residents. In addition, a long-term research initiative carried out under the auspices of the MRC is studying the ways in which local knowledge—provided by basin residents directly engaged in data collection activities—can enhance ecological studies of Mekong fish species (MRC 2001a; Poulsen and Jorgensen 2001; Coates et al. 2003).

crucial to the overall economic growth and "modernization" of the region (Jerndal and Rigg 1999, 50–51). At the heart of the program is an agenda that promotes economic growth through the attraction of private capital to stimulate development and, in theory, alleviate poverty. In its own words, the ADB is working toward "creating the enabling environment for domestic and private foreign investors and shifting the role of government from owner-producer to facilitator-regulator" (from a March 2000 ADB report on "Private Sector Development" quoted in Adams 2001, 14). Since the early 1990s, a variety of transnational and national actors, under the guidance of the ADB, have initiated four development forums and six capital investment funds designed to attract private investment to the Mekong region (Stensholt 1996, 201). To cite just one example of the sums involved, investments in priority projects (dams, telecommunications infrastructure, transport systems) were expected to exceed US$1 billion annually by the year 2000 (Gill 1996, 5), a goal that has yet to be reached. Furthermore, the GMS program theorizes a "natural process of sub regional integration at work, spurred by the profit incentive of market forces, so that private sector developments often lead the process of co-operation" (ADB, quoted in Jerndal and Rigg 1999, 50).

One way to theorize this trend toward "regionalization" in the Mekong context is to locate it within the political-economic worldview associated with neoliberalism and the "Washington consensus," and analyze how such a view is being altered through regional cooperation. As Gore (2000) notes,

> Policies geared towards regional integration and cooperation are important elements of an overall approach to economic development that has emerged in response to the perceived shortcomings of the neoliberal "Washington consensus." The idea is to promote increased international competitiveness through the creation of, for example, regional production chains, and also nurture the development of regional markets in order to reduce demand-side constraints on growth. (798)

Rather than each nation-state "going it alone" in the global economy, as might be preferred within a neoliberal model, the idea is to reconfigure the Mekong as a transboundary economic region, one that will then be capable of competing more effectively within hostile global markets. However, the state, as under neoliberal prescriptions, should play a minimal role in directing economic growth (as a "facilitator-regulator" in ADB's term). The private sector, as primary engine of growth, becomes crucial in this regard. Thus, the regionalization of the Mekong within the GMS initiative is interpenetrated by a transnational network of capital investment, whose application toward development projects within the region will profoundly transform the basin's socioecological relationships. The "Mekong" subregion envisaged within the GMS is not the biophysically-defined river basin, but a complex series of economic networks that perceives the ecohydrological dynamics of the basin solely in terms of the resources it will provide to development initiatives.[15]

This rapid push toward economic growth through private investment raises significant doubts regarding the ability of civil society organizations to achieve environmental and social goals within the Mekong states. The increasingly strong linkages between the MRC and the ADB, under the umbrella of the GMS initiative, represent perhaps the greatest threat to the capability of Mekong-level institutions to promote conservation, ecological sustainability, and social resilience within the basin. The MRC is becoming increasingly dependent on the ADB for financial resources, and the MRC's role under the aegis of the GMS appears to be limited to providing "technical, logistical and information support to the ADB in its ambitious master planning of the Mekong region" (Nette 1997, 32). These institutions cooperate in the identification of priority projects and preliminary feasibility studies, particularly in the energy sector. As evidence of formalization of these linkages, the MRC and the ADB recently signed a partnership agreement "to enhance co-operation and co-ordination of their efforts in the region" (MRC 2001b, 50).

While the MRC and its previous incarnations have always had a problematic role vis-à-vis the promotion of environmental and social goals (see Bakker 1999; Sneddon and Nguyen 2001), its linkages with ADB are likely to contribute to further reductions in its capacity to advocate for basin-oriented socioecological goals such as those related to biodiversity conservation and protected areas. As noted above, the ADB has aggressively pursued a strategy of luring private capital to the Mekong region in the hopes of moving ahead with ambitious infrastructure development plans. The MRC's recent report on hydropower development in the basin underscores these observations. The MRC's mandate has undergone some rather profound shifts since the lower basin's riparian states (Thailand, Laos, Cambodia, and Vietnam) signed a new agreement in 1995. Rather than advocating, planning, and arranging the financing for specific dam projects, one of its principal roles over the past four decades, the MRC's new role reflects an emphasis on "providing information and policy advice on broader, basin-wide issues," which is a "comparative advantage" of the MRC as a river basin organization (MRC 2001b, 11–12). Without the capacity to link itself to specific projects, a function apparently

15. Global political-economic dynamics, however, conspired to slow down implementation of the GMS policies and goals. There is little doubt that the 1997 Asian financial crisis altered the GMS plan's ambitious slate of projects. The ADB's US$12 billion bailout package to Thailand certainly diverted some of the funds that might otherwise have gone toward GMS projects ("Mekong Plans Suffer Delay from Turmoil" 1998).

now the purview of the basin states, who might choose to individually or collectively work in tandem with private developers, the MRC has been effectively removed from a key stage of the dam planning process and, at least partially, from promoting socioecological concerns.

Some credence is given to this view by the prominence accorded the role of private capital in future Mekong development schemes, which mirrors the priorities of the GMS initiative. The MRC hydropower report clearly recognizes that the era of publicly financed water projects is over and perceives private capital as the primary, perhaps only, means of continuing along the path of large-scale hydroelectric development. Thus, the authors of the report focus attention on creating an institutional and regulatory environment in the Mekong countries that is conducive to influxes of private capital. Ironically, the delinking of the MRC from individual projects, while representing a net reduction in decision-making power, may provide it with a good deal more latitude to advocate for conservation and other development priorities, such as its recent concern over inland capture fisheries (see MRC 2001a), which account for basinwide hydrologic and ecological dynamics.

To sum up, conservation initiatives in the Mekong basin are ultimately at the mercy of extrabasin processes and institutions (as exemplified by the ADB's GMS initiative) that have been neglected in the strategic thinking and planning of international conservation advocates. The GMS initiative, mediated by ADB, represents a scheme to enroll the complex ecohydrogical agents of the Mekong River basin, and the human actors whose livelihoods at least in part depend on the continued integrity of the basin's biophysical network, into a transnational economic network. Such a network would have global aspirations and prioritize rapid energy development, more liberal trade regimes, greater levels of foreign investment, and cooperative tourism development. International conservation actors in the Mekong, meanwhile, have largely failed to engage at these politically relevant transnational and global scales.

Reflections on "Transnationalization" and Scale in the Mekong

As highlighted above, there are two concurrent and contradictory processes of "transnationalization" occurring in the Mekong basin. One concerns the efforts of transnational organizations engaging with riparian states to produce a network of conservation territories as a means of promoting biodiversity conservation. A second involves ongoing efforts by the ADB, with the partial collusion of the MRC, to insert Mekong socioecological dynamics into global economic circuits by, initially, attracting private capital to the region and, subsequently, securing Mekong resources firmly within the global economy. While both sets of initiatives are proceeding at a rapid pace, actual knowledge of the basin's ecological processes and structures, knowledge that ostensibly underpins the conservation programs and would presumably enhance the "sustainability" of the GMS initiative (a stated concern in ADB documents), is lagging behind.

Furthermore, international environmental organizations are inserting themselves into the political-ecological dynamics of Mekong development in contradictory ways that in some places privilege nonhumans over humans (e.g., the Nakay-Nam Thuen 2 case in Laos) and in a broader sense privilege certain types of nonhumans over others (e.g., the relative inattention to fish species of the basin). Conservation initiatives have also neglected the manner in which powerful "transnational" (or "global") actors such as the ADB are privileging transboundary economic scales (e.g., the GMS) in order to advance a development trajectory that may be quite at odds with basinwide conservation efforts. Thus, the relative successes or failures of "transnational conservation" hinge on the political construction of scales that are both interwoven throughout and transcendent of the basin scale. Analysts and proponents of Mekong conservation must therefore wrestle with "new scales of importance" (see Zimmerer, chap. 1 in this volume) in conservation ideas and practices.

One way of conceptualizing these novel constructions of geographical scale is to see the international conservation actors as a more or less connected set of human and nonhuman agents engaged in extending their "global" network; the most visible, fixed, and bounded expressions of this network are conservation territories (see Whatmore and Thorne 1997). From this perspective, conservation territories are not simply spaces that bound landscapes of high ecological value, determined by scientific criteria, but also are the product of political contestation at diverse scales, beyond whatever scale is seemingly the most "relevant" in ecological terms. Consequently, the meaning of a conservation territory derives not so much from its inherent characteristics but rather its position within a global network of similar territories, and the identities of the human and nonhuman agents within the territory are reworked and redefined through the constant action of the network. To develop this argument further, we could ask who is left out of such networks, through either design or neglect; what types of power effects such networks render; and how conservation alliances intersect with their developmental counterparts. Thus, perhaps the most difficult question confronting environmental regimes and conservation initiatives in the Mekong Basin is how to recalibrate unequal power relationships (e.g., between resource-dependent communities and development-minded state agencies) in a way that encourages social resilience and ecological integrity, and this is intimately connected to questions of scale.

It is fairly clear in the case of the Mekong, and probably for many transboundary river basins, that the processes driving basin changes—and the discourses that bolster and justify these processes—are associated with geographical scales, such as the transnational or the global, that rarely align with the basin in question. Similarly, there may be a dissonance between the working scale of political engagement for international conservationists and the scales that are the real locus of power in a globalized world. As we have seen, the strategies of international conservationists in the Mekong region have thus far focused primarily on collaborations with state agents, instead of engagement with the transnational or international institutions that effectively control the finances and establish the rules for development planning. The end result is a patchwork of conservation territories that may be inconsistent with basin-scale conservation needs. While the achievements of conservation initiatives in the Mekong have been impressive in terms of scope (e.g., the total coverage of protected areas), the long-term impacts of these endeavors are open to question. The capacities of the extended networks of actors associated with the ADB and developmental interests within the Mekong's riparian states to produce "power effects" (Murdoch and Marsden 1995)—in short, to manipulate the basin according to their priorities—are much greater than the extended networks of international conservation initiatives. Through efforts to enlist the ecohydrologic entities of the Mekong into their respective actor networks, both development interests and international conservation interests have actively constructed a version of the "Mekong" that fits their respective needs: in the case of the former, to promote greater control over resources and economic development, and in the case of the latter, to promote biodiversity conservation. Yet all of these diverse interests may ultimately conflict with those of the majority of the basin's resource-dependent residents.

To conclude on this point, the tension between conservation aims and economic globalization discussed above suggests novel strategies for international conservation groups. Such new spatial strategies transcend a traditional focus on rigidly delineated parks and protected areas, whether designated for strict protection or sustainable use, toward spatialities such as "conservation networks" (see Zimmerer, chap. 1, Mutersbaugh, chap. 2, and Brown, chap. 4 in this volume), "off-park conservation" (Thackway and Olsson 1999; WinklerPrins, chap. 5 in this volume), and the broad notion of "ecoregions" (see Olson 1998; and Sierra, chap. 9 in this volume). No one is (yet) forwarding the radical proposition that the entire Mekong basin be construed as the most appropriate scale for a designated conservation territory. At one level, however, taking fish seriously as both a transboundary resource crucial to basin livelihoods *and* as a focus of transboundary conservation efforts in many ways directs attention to this politically uncomfortable proposition. The raising up of fish and fishing-based livelihoods as a conservation priority would, inter alia, demand that conservation efforts confront directly

the likely decimation of basin fisheries if proposed hydroelectric development schemes under the GMS banner proceed as planned. It is doubtful that
conservation organizations are yet willing to risk the ire of governmental allies by assuming such a stance, but off-park conservation considerations in
the Mekong basin will remain a salient and politically troublesome issue in
the years to come.

At another level, the ecological and hydrological dynamics of the Mekong
River and its tributaries, spatially defined as *basin* processes, encompass the
protected-area network of the region. It will be difficult to maintain the integrity of conservation "islands" (i.e., protected areas) if large-scale alteration
of the basin occurs, and there are crucial considerations at the intersection
of conservation goals and livelihoods (e.g., fisheries) that transcend officially
designated conservation territories. As these examples demonstrate, and
given their growing importance as actors in the governance of the Mekong,[16]
international environmental organizations must undertake a careful read-
ing of the very fluid scales and networks of conservation that now characterize the basin if they hope to advance their aims.

References

Adams, C. 2001. The Asian Development Bank, capital flows and the privatization of infrastructure
    projects in the South. In *Profiting from poverty: The ADB, private sector and development in Asia,* ed. Fo-
    cus on the Global South, 11–22. Bangkok: Focus on the Global South.

Adams, J. S., and T. O. McShane. 1992. *The myth of wild Africa: Conservation without illusion.* New York:
    W. W. Norton.

Adger, N. 2000. Social and ecological resilience: Are they related? *Progress in Human Geography* 24 (3):
    347–64.

Athanasiou, T. 1996. *Divided planet: The ecology of rich and poor.* Boston: Little, Brown.

Australian Mekong Resource Centre (AMRC). 1999. The contested landscapes of the Nam Theun, Lao
    PDR. Sydney: AMRC. Available at http://www.mekong.es.usyd.edu.au/case_studies/nam_theun/
    index.htm.

Bakker, K. 1999. The politics of hydropower: Developing the Mekong. *Political Geography* 18:209–32.

Bao, T. Q., K. Bouakhamvongsa, S. Chan, K. C. Chhuon, T. Phommavong, A. F. Poulsen, P. Rukawoma,
    et al. 2001. Local knowledge in the study of river fish biology: Experiences from the Mekong.
    Mekong Development Series no. 1. Phnom Penh, Cambodia: Mekong River Commission, July.

Chape, S. 2001. An overview of integrated approaches to conservation and community development
    in the Lao People's Democratic Republic. *Parks* 11 (2): 24–32.

Chomchanta, P., P. Vongphasouk, S. Chanrya, C. Soulignavong, B. Saadsy, and T. Warren. 2000. A pre-
    liminary assessment of Mekong fishery conservation zones in southern Lao PDR, and recom-
    mendations for further evaluation and monitoring. Living Aquatic Resources and Research Cen-
    ter (LARReC) Technical Report no. 0001. Vientiane, Lao PDR: LARReC, April.

Chooduangngern, S. 1996. The Mekong Basin development plan. In *Development dilemmas in the Mekong
    Subregion,* ed. B. Stensholt, 185–93. Clayton, Australia: Monash Asia Institute, Monash University.

16. Both IUCN and WWF have signed memoranda of agreement with the Mekong River Commission and in doing so have achieved semi-official status as "partner" NGOs within the framework of
Mekong governance.

CI (Conservation International). 2002. Asia-Pacific: Cambodia. Available at http://www.conservation
.org/xp/CIWEB/regions/asia_pacific/cambodia/cambodia.xml.

Coates, D., O. Poeu, U. Suntornratnana, T. T. Nguyen, and S. Viravong. 2003. Biodiversity and fisheries
in the Mekong River Basin. Mekong Development Series no. 2. Phnom Penh, Cambodia: Mekong
River Commission, June.

Dudgeon, D. 1992. Endangered ecosystems: A review of the conservation status of tropical Asian
rivers. *Hydrobiologia* 248:167–91.

———. 2000. Riverine biodiversity in Asia: A challenge for conservation biology. *Hydrobiologia* 418:1–13.

Friederich, H. 2000. The biodiversity of the wetlands in the Lower Mekong Basin. Paper prepared for
the Fourth Regional Consultation of the World Commission on Dams. 26–27 February, Hanoi,
Vietnam.

Gill, I. 1996. Mekong: Dismantling the barriers. *ADB Review* (July–August): 3–8.

Glassman, J. 2001. State power beyond the "territorial trap": The internationalization of the state. *Political Geography* 18 (6): 669–96.

Gore, C. 2000. The rise and fall of the Washington Consensus as a paradigm for developing countries.
World Development 28 (5): 789–804.

Groombridge, B., and M. Jenkins. 1998. *Freshwater biodiversity: A preliminary global assessment.* Cambridge,
UK: World Conservation Monitoring Centre (IUCN, WWF, UNEP), World Conservation Press.

Gum, W. 2000. Inland aquatic resources and livelihoods in Cambodia: A guide to the literature, legislation, institutional framework and recommendations. Consultancy report, Oxfam GB (Great
Britain) and NGO Forum on Cambodia. Phnom Penh, Cambodia: Oxfam, November.

Hirsch, P. 1997. Introduction: Seeing forests for trees. In *Seeing forests for trees: Environment and environmentalism in Thailand,* ed. P. Hirsch, 1–36. Chiang Mai, Thailand: Silkworm Books.

ICEM (International Centre for Environmental Management). 2003. Regional report on protected
areas and development. Review of protected areas and development in the Lower Mekong River
region. Indooroopilly, Queensland, Australia: ICEM, November.

International Rivers Network (IRN). 2004. Nam Theun 2, Laos: Another World Bank disaster in the
making. Berkeley, CA: IRN, January.

IUCN (World Conservation Union). 2002. In the spotlight: The Mekong Wetlands biodiversity project.
Gland, Switzerland: IUCN. Available at http://www.iucn.org/themes/wani/mekong.htm.

Jerndal, R., and J. Rigg. 1999. From buffer state to crossroads state: Spaces of human activity and integration in the Lao PDR. In *Laos: Culture and society,* ed. Grant Evans, 35–60. Chiang Mai, Thailand:
Silkworm Books.

Kite, G. 2000. Developing a hydrological model for the Mekong Basin: Impacts of basin development
on fisheries productivity. Colombo, Sri Lanka: International Water Management Institute (IWMI).

Lohmann, L. 1993. Green orientalism. *The Ecologist* 23 (6): 202–4.

Mekong plans suffer delay from turmoil. 1998. *The Nation* (Bangkok), 10 March.

Monbiot, G. 1997. Conservationists who are enemies of the earth. *The Guardian,* 6 August.

MRC (Mekong River Commission). 2001a. Local knowledge in the study of river fish biology: Experiences from the Mekong. Mekong Development Series no. 1, Phnom Penh, Cambodia: MRC, July.

———. 2001b. *MRC hydropower development strategy.* Phnom Penh, Cambodia: MRC Water Resources and
Hydrology Programme, Mekong River Commission, October.

———. 2002. Protected areas in region set to expand. MRC Press Release no. 10/02. Phnom Penh, Cambodia: MRC, 3 May.

MRC Secretariat. 2001. The MRC programme for fisheries management and development cooperation:
Annual report 2000/2001. Bangkok: MRC, May.

Murdoch, J., and T. Marsden. 1995. The spatialization of politics: Local and national actor-spaces in environmental conflict. *Transactions of the Institute of British Geographers* 20:368–80.

Nette, A. 1997. Mekong River Commission: Going with the money. *Watershed* 2 (3): 31–34.

Neumann, R. 1998. *Imposing wilderness: Struggles over livelihood and nature preservation in Africa.* Berkeley: University of California Press.

Oates, J. 1999. *Myth and reality in the rain forest: How conservation strategies are failing in West Africa.* Berkeley: University of California Press.

Olson, D. M. 1998. The Global 200: A representation approach to conserving the earth's most biologically valuable ecoregions. *Conservation Biology* 12 (3): 502–15.

Poff, N. L., J. D. Allan, M. B. Bain, J. R. Karr, K. L. Prestagaard, B. D. Richter, R. E. Sparks, and J. C. Stromberg. 1997. The natural flow regime. *BioScience* 47 (11): 769–84.

Poulsen, A. F., and J. V. Jorgensen. 2001. Mekong fisheries: The cornerstone in fisheries research. *Catch & Culture* 6 (4): 4–7.

Rao, S., and D. Seila. 2001. Strategy for biodiversity conservation in Cambodia. Paper presented at the SEAG-Symposium on Resource Management: Private-Public Partnership and Knowledge Sharing, Los Baños, Philippines, 27–31 August.

Robinson, W. I. 2001. Social theory and globalization: The rise of a transnational state. *Theory and Society* 30:157–200.

Sneddon, C., and B. T. Nguyen. 2001. Politics, ecology and water: The Mekong Delta and development of the Lower Mekong Basin. In *Living with environmental change: Social vulnerability, adaptation and resilience in Vietnam,* ed. W. N. Adger, P. M. Kelly, and N. H. Ninh, 234–62. London: Routledge.

Stanford, J., and J. V. Ward. 2001. Revisiting the serial discontinuity concept. *Regulated Rivers: Research and Management* 17:303–10.

Stensholt, B. 1996. The many faces of Mekong co-operation. In *Development dilemmas in Mekong Region,* ed. B. Stensholt, 1–23. Clayton, Australia: Monash Asia Institute, Monash University.

Thackway, R., and K. Olsson. 1999. Public/private partnerships and protected areas: Selected Australian case studies. *Landscape and Urban Planning* 44:87–97.

TNC (The Nature Conservancy). 2002. Yunnan province. Available at http://nature.org/wherewework/asiapacific/china/work/art5098.html.

Turner, M. D. 1999. Merging local and regional analyses of land-use change: The case of livestock in the Sahel. *Annals of the Association of American Geographers* 89 (2): 191–219.

Vietnam News Service. 2001. Environmentalists tot up Mekong Delta biodiversity. Hanoi: Vietnam News Service, 25 October.

Whatmore, S., and L. Thorne. 1997. Nourishing networks: Alternative geographies of food. In *Globalising food: Agrarian questions and global restructuring,* ed. D. Goodman and M. Watts, 287–304. London: Routledge.

White, W. 2000. Infrastructure development in the Mekong Basin: Risks and responses. Report prepared for Oxfam America. Acton, MA: Foresight Associates, July.

World Commission on Protected Areas (WCPA). 2000. Regional approaches to biodiversity management. Biodiversity Brief no. 5. Brussels: European Commission.

Zimmerer, K., and K. Young. 1998. Introduction: The geographical nature of landscape change. In *Nature's geography: New lessons for conservation in developing countries,* ed. K. Zimmerer and K. Young, 3–34. Madison: University of Wisconsin Press.

9 A Transnational Perspective on National Protected Areas and Ecoregions in the Tropical Andean Countries

RODRIGO SIERRA

From Local to Global Biodiversity Conservation in the New Millennium
This chapter examines the regional outcomes of almost a century of national-scale biodiversity conservation efforts in a set of countries that have similar ecological characteristics and political economic history: the tropical Andean countries of Venezuela, Colombia, Ecuador, Peru, and Bolivia.[1] In this region, as in most of the world, the rapid expansion of protected areas has been a key aspect of the globalization of environmental conservation. While the designation of protected areas has been often used as an indicator of the success of global environmentalism (Zimmerer, chap. 1 in this volume), the efficiency of the global system has been called into question repeatedly for a variety of reasons, from failure to include a representative sample of the world's biodiversity, to their often doubtful long-term viability (e.g., Olson and Dinerstein 1998; Margules and Pressey 2000), to the social and economic costs arising from restrictions imposed on the use of the resources available in these areas (Agrawal and Gibson 1999; Brandon, Redford, and Sanderson 1998).

Despite paradigm shifts over the last century (Brandon, Redford, and Sanderson 1998; Schwartzman, Moreira, and Nepstad 2000) and the diversification of the conservation debate to address the social costs and the sometimes uncertain environmental benefits of protected areas, biodiversity conservation worldwide continues to rely primarily on the protection of specific areas where biodiversity and ecological processes are out of reach of most people (Margules and Pressey 2000; Terborgh 1999). Those who advocate "fortress conservation" argue that even the simple designation of an area as

1. Biodiversity includes all the diversity of life, from wildlife to domesticated crops and animals and natural and, in some cases, human-made ecosystems. In this chapter, however, the definition of biodiversity is, for convenience, narrower. It refers primarily to the diversity of genes, species, and ecosystems occurring in natural or wilderness areas.

"protected" is enough to improve its conservation potential (Bruner et al. 2001; Rodríguez and Rodríguez-Clark 2001; Weiler and Seidl 2004). However, for even the best-protected areas to contribute to the overall objective of bio-diversity conservation, they must contribute to including a representative sample of the world's biological diversity and maintain the physical processes that make them viable over the long run (Margules and Pressey 2000). These processes include the necessary conditions that facilitate biodiversity persist-ence, such as ecosystem complementarity (e.g., ecological gradients that fa-cilitate genetic flow and seasonal migration) and population characteristics that are associated with evolutionary potential (e.g., genetic variability).

The international dynamics of biodiversity conservation policy reveal im-portant paradoxes or contradictions of globalization. On the one hand, in-ternational conservation organizations, such as Conservation International (CI), the World Wildlife Fund (WWF), and The Nature Conservancy (TNC), and donors, such as the United Nations (UN) and the World Bank's Global Envi-ronmental Facility (GEF), are increasingly playing key roles in shaping na-tional protected-area networks (see, for example, Young and Rodríguez, chap. 10, and Sundberg, chap. 11 in this volume). These actors, however, often do not respond to national biodiversity conservation visions but to global and regional frameworks for optimizing their efforts and targeting investment (Sierra, Campos, and Chamberlin 2002). On the other hand, because of glob-alization, policy and economic decisions—including those that determine land management arrangements—are becoming increasingly local, espe-cially in the developing world. This results from the economic and political restructuring of developing countries, through neoliberal reforms that call for rapid decentralization, which transfers decision making to local agents and bolsters the importance of local conditions. In the arena of biodiversity conservation, this means passing on decision-making and management re-sponsibilities to subnational conservation agencies and organizations. Mex-ico, Costa Rica, and Ecuador, for example, have established subnational ad-ministrative units to manage existing protected areas and to create new ones where needed. These regions have relative autonomy but also have increasing responsibilities, including generating the funds needed to manage existing protected areas and create new ones.

Biodiversity conservation, then, has effectively become polarized between the global and local scales: environmental-monitoring and conservation-funding decisions are dominated by global institutions, while management of conservation areas and execution of conservation strategies are increas-ingly performed by local-scale governmental and nongovernmental organi-zations (NGOs). This means that it is possible that certain biodiversity ele-ments of global importance remain unprotected by any network, or that, unknown to conservation officials in one country or area, some are protected to levels that could be considered globally excessive, especially if other biodiversity elements are facing greater risk of destruction. Under ideal circumstances, if every country located its protected areas according to biogeographical representativeness and persistence objectives, the sum of all protected areas would generate a regional, and ultimately a global, reserve network protecting a lasting sample of all ecosystems and species in the planet. Of course, this idealized location process is not occurring. National-level studies show that the protected-area networks within countries often fail to incorporate significant parts of their geographic biodiversity units (e.g., Rodríguez and Young 2000; Young and Rodríguez, chap. 10 in this volume; Sierra, Campos, and Chamberlin 2002). A key question remains, however: are national gaps in conservation covered by protection measures in other countries?

Within this context, this study assesses if and how traditional conservation frameworks, which usually operate at national and subnational levels, result in an efficient regional network—that is, one that satisfies the representativeness objective of a new global conservation paradigm. Specifically, this study examines the efficiency of the tropical Andean region's protected-area network based on three criteria: (a) the uniqueness of the biodiversity of the region, (b) the level of protection afforded to this biodiversity, and (c) the level at which the reserve network targets biodiversity at risk. The underlying assumption in this analysis is that a globally efficient conservation network is built from regionally efficient conservation networks and that these, in turn, are directly dependent on the efficiency of national conservation actions. From a regional point of view, and for these countries' networks to be efficient, it would be necessary that they complement each other. This analysis also shines light on what remains to be done and on the potential pitfalls of a fragmented management system that relies on an increasingly local vision of conservation needs and priorities. However, this study does not attempt to identify specific areas for new reserves. The decision of locating a reserve in one place or another must be worked out at a scale different from the one employed in this analysis. At that scale, social and economic factors must also be considered and factored into decisions that maximize the conservation value and minimize the social cost of reserves (Sierra, Campos, and Chamberlin 2002).

A Whole Made of Parts: The Evolving Global View of Biodiversity Conservation

These are challenging but also promising times for biodiversity conservation. A key challenge arises from the need for a conservation framework that articulates objectives that often transcend multiple scales (e.g., global versus local) and spheres of action (e.g., social versus biological). Over the last thirty years, it has become increasingly clear that factors affecting biodiversity globally, such as climate change and mass extinction, are linked to local environmental and social processes, such as habitat loss and land degradation. Research has also shown that environmental processes in general, and ecosystem functions in particular, are likely to be affected by changes in species diversity (Loreau et al. 2001). Local and regional environmental change also modifies the existing context of evolutionary processes, potentially affecting the characteristics of global biodiversity in the future. Large terrestrial and marine predators, for example, may become even scarcer since it will probably become increasingly harder to find wilderness areas large enough to sustain viable populations of these species. In the next 100 years, it is expected that land use change, followed by climate change and nitrogen deposition, will have the greatest impact on terrestrial biodiversity. For freshwater ecosystems biotic exchange is more important (Sala et al. 2000).

A new framework is needed not only because we understand better the relationships among the parts of the global environment but also because, as Polanyi (2001) shows, every component of human culture is in state of constant evolution. The economic, technological, and even value systems that constitute the new global culture continuously redefine the ways in which people interface with the environment in general, and with biodiversity in particular. The challenge lies in that while "old" cultures had time to adapt to slowly changing environmental conditions, the new one has to do it on the run. Global changes in biodiversity are now observable over the span of decades, sometimes even shorter periods, rather than the centuries or millennia of the past (Daly and Farley 2004), and massive extinction events are expected in the near future (Birdlife International 2000; Jenkins 2003). In response to this evolving global view of biodiversity conservation, international treaties and funding have become major drivers of conservation efforts, often overshadowing local priorities and resources that had been dominant a few decades ago (e.g., Chapman 2003; Ramutsindela 2004; Young and Rodríguez, chap. 10 in this volume). Regionally, conservation goals have been analyzed in the context of development objectives (e.g., Gonzales et al. 2003), and local conservation efforts increasingly recognize the role that social and economic conditions, such as community participation, tenure security, and power structures, have in defining conservation outcomes (e.g., Agrawal and Gibson 1999; Sundberg, chap. 11 in this volume).

The development of spatial analysis and spatially based conservation theories promise new tools for meeting these challenges or at least for shedding some light on the path to global biodiversity conservation. The availability of local, national, regional, and even global spatial information on biodiversity, land cover (habitat) change, and protected-area networks, while still imperfect, permits preliminary research on how biodiversity conservation is effected at these hitherto disjointed scales of analysis and decision making. Until recently, studies of regional-level conservation gaps and priorities were difficult because the information on regional biodiversity and protected areas was scarce or in formats that did not lend themselves to reliable comparative analysis. Maps delineating biodiversity units, and especially units with conservation significance (i.e., unique biotic assemblages), were only applicable in coarse studies, mostly at global levels (e.g., Takhtajan 1986; Udvardy 1975) and were not compatible with the information on protected areas, often developed to be used within specific countries.

In the last decade, however, new, finer-scale geographic databases have been developed from these coarse-scale sources for specific regions (e.g., Dinerstein et al. 1995; Omernik 1995) or for the world overall (e.g., Olson et al. 2001). While the validity of the biological units they identify is still being debated, primarily because of their lack of spatial correspondence with major vegetation types (Wright, Murray, and Merrill 1998; O. Huber, personal communication) and levels of detail that still simplify local variations (Sierra, Campos, and Chamberlin 2002), these databases are becoming popular as a means of identifying geographic priorities for conservation in large regions. While these databases cannot replace national- or local-level information in the analysis of specific conservation needs, or identify specific areas to be declared protected, they do offer insight about the gaps in the protection of biodiversity at regional to global scales. Regional-level analyses are the building blocks of global conservation efforts and the basis for specific reserve design efforts, which must take place at national to local scales (Sierra, Campos, and Chamberlin 2002). At these finer scales, reserve viability is often defined by interest groups, power structures, commodity prices, and many other factors (Illsley and Richardson 2004) that are not evident at large geographic scales.

Scale, Data, and Methods

In this study, the regional biodiversity of the five Andean countries, which comprise an area of 4.8 million km², was defined using the map of ecoregions of the world created by Olson et al. (2001). The coherence of the biogeographic origins of the region's diversity is also demonstrated by the small number of floristic complexes identified by Takhtajan (1986): the Caribbean region, Guyana Highlands, Amazon region, and Andean region, the latter being by far the most extensive of the four.

The unit of analysis, an ecoregion or ecosystem region, is a homogeneous geographic area that shares similar environmental resources and ecosystems (Bailey 1998; Olson et al. 2001; Omernik 1995). From a conservation point of view, an ecoregion is a management and analysis unit because a random selection of elements can be expected to be representative and proportional to the whole universe of elements for that ecoregion. The ecoregion concept is one of the latest attempts to classify geographic areas into homogeneous zones based on their biological characteristics. This idea has the advantages of a relatively high level of cartographic detail and the support of a large number of expert contributors (Olson et al. 2001). Earlier regional categorization efforts—such as those by Cabrera and Willink (1973) and Udvardy (1975), based on biogeographic criteria; Hueck and Siebert (1972) and Takhtajan (1986), using fitogeographic criteria; and Bailey (1998), using physical variables—offer levels of detail that are too coarse for regional analysis. To be sure, ecoregions are still a simplification of ecological and biological variability and may contain smaller, biogeographically distinct units. For example, Forero (1982) and Hernandez Camacho et al. (1992) recognize, respectively, eight and twenty biogeographic districts within the Colombian Chocó, which is considered one ecoregional unit by Olson et al. (2001) and, hence, by this study. On the other hand, one of the advantages of the ecoregion concept is that it can be expanded into more detailed classification schemes. Indeed, any biogeographical scheme needs to be hierarchical, so that it can be applied at different levels of detail and regional coverage (Bailey 1998; Sierra 1999).

The baseline data for protected areas were derived from UN statistics (United Nations 2002), modified and updated based on current national databases (e.g., INRENA for Peru, SNAP for Ecuador). Only reserves corresponding to the World Conservation Union's (IUCN's) first three categories (Strict Nature Reserve or Wilderness Area, National Parks, and Natural Monument) were included in the analysis. These reserves significantly restrict human use and thus serve as a vehicle for biodiversity conservation and the maintenance of ecological processes on their own right. Cartographic coverage was available for 131 of the 136 protected areas in categories I to III of IUCN established as of mid-2001. The area of the five reserves not included in this study corresponds to only 0.3 percent of the total area protected in the region in these categories as of 2001.

For the representativeness criteria, an ideal, but arbitrary, level of representation of 10 percent of the area of was used. This level of representativeness has been considered as appropriate and feasible by some decision makers in the study region (Sierra, Campos, and Chamberlin 2002). Other studies have proposed similar objectives. Hummel (1989), for example, proposed an objective of 12 percent of the area of a geographic biodiversity unit, following guidelines set by the UN Environment Programme. Other, more ambitious, conservation targets have also been proposed. Strittholt and Boerner (1995), for example, used a conservation goal of 25 percent of the original area of each ecosystem in their study, and the work by other researchers suggests that, in some cases, even more thorough coverage (e.g., between 50 and 97 percent) may be needed to ensure long-term species and ecosystem function conservation (e.g., Noss 1995; Ryti 1992).

Information on risk was derived from recent land cover data produced by the Joint Research Centre of the European Union (Eva et al. 2004). This 1 km resolution data set identifies the vegetation cover type of South America as of 2000. Vegetation types that corresponded to nonnatural cover were used to mask out and calculate the remnant area of each ecoregion in the Andean region. Risk is assumed to be inversely proportional to the level of remnancy of each ecoregion: that is, the lower the remnant area the higher the risk.

Protecting Regional-Level Biodiversity Based on National Protected-Area Networks in the Tropical Andes

Seventy-five of the ecoregions of the world fall in varying degrees in the Andean region. This means that approximately one of every twelve ecoregions in the world is represented in an area equivalent to 3.6 percent of the land surface of the planet, making this one of the world's most biologically diverse regions. Furthermore, forty-five of the seventy-five ecoregions have 95 percent or more of their total area within the region, and consequently can be considered unique to this region (fig. 9.1). Overall, over two-thirds of the ecoregions in the tropical Andean region have restricted distributions, with 75 percent or more of their area occurring within the region. Many represent highland biodiversity, such as Eastern cordillera montane forests in the northern Andes and the wet and dry puna ecoregions in the southern Andes. Others represent the biodiversity of lowland environments, both west and east of the Andes, such as the dry forest in the Tumbez-Piura region along the coasts of northern Peru and southern Ecuador and the savannas of the Beni in eastern Bolivia. Yet others correspond to non-Andean biodiversity, such as Tepuis and Guyanan savanna in Venezuela.

As of 2001, there were still seventeen ecoregions that were not represented in the regional protected-area network. Of these, nine are exclusively found in the region, and their global representation can be achieved only through regional action (groups 1A and 1B in table 9.1).[2] Group 1A also identifies ecoregions that have a very high level of risk of destruction. In these ecoregions, more than half of the original area has already been converted to some type of human-made land cover. This is a measure of the need of immediate protection action. For example, only 30 percent of the original area of the Magdalena Valley dry forests remain under their natural state. Group 1B identifies those ecoregions that while not currently protected retain at least half of their area under natural cover and therefore face lower levels of risk. Interestingly, the ecoregions that show the highest deficiencies in protection are of two types: mangroves and dry environment.

---

2. Due to measurement uncertainties arising from the scale of analysis it is assumed here that any value below 1 percent indicates almost null coverage.

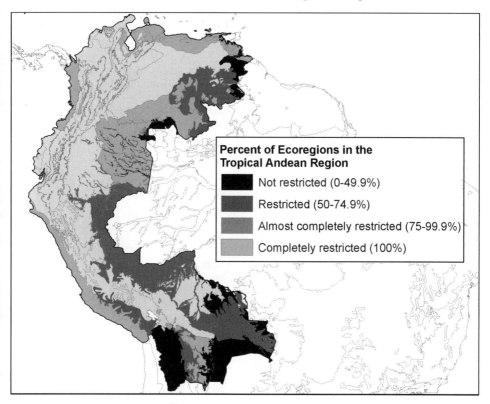

Percent of Ecoregions in the
Tropical Andean Region

▮ Not restricted (0-49.9%)

▮ Restricted (50-74.9%)

▮ Almost completely restricted (75-99.9%)

▮ Completely restricted (100%)

FIGURE 9.1. Ecoregions of the tropical Andes by level of geographic restriction

Groups 2A and 2B in table 9.1 identify those ecoregions that are completely restricted to the Andean region and enjoy some level of protection. The level of protection is, however, less than 10 percent of its total area and is, therefore, considered by this study to be deficient. Overall, of the fifty-two ecoregions with 75 percent or more of their area in the tropical Andean region, thirty-three have less than 10 percent of their respective areas protected. Of the ecoregions in group 2, those in group 2A face the most significant risks. In fact, the Sinu dry valley forests, where almost 90 percent of the original cover has been replaced by agricultural cover, is, according to this analysis, the most endangered ecoregion of all. Interestingly, each ecoregion with a distribution that is considered completely restricted and that does not occur within the region's protected-area network is found within one country. But twelve ecoregions with restricted distribution (75–99 percent) that are not covered up to the 10 percent representative target are shared by one or more countries (fig. 9.2).

Ecoregions in group 3B in table 9.1 are also completely restricted to the Andean region but are represented in the region's reserve network by over 10

TABLE 9.1. Restricted ecoregions of the tropical Andean region and their level of coverage by the network of protected areas and their level of anthropogenic transformation

Group	Ecoregion name	Area (km²)	% area in region	% protected total	Area protected in region	% protected in region	% of area transformed
1A	Magdalena Valley dry forests	19,748.00	100.00	0.00	0.0	0.0	69.0
	Cauca Valley dry forests	7,369.50	99.90	0.00	0.0	0.0	63.4
	Manabi mangroves	1,105.50	100.00	0.00	0.0	0.0	63.4
	Patia Valley dry forests	2,268.50	100.00	0.00	0.0	0.0	61.9
1B	Maracaibo dry forests	30,829.00	99.70	14.80	0.0	0.0	56.8
	Marañón dry forests	11,340.00	100.00	0.00	0.0	0.0	14.9
	Piura mangroves	117.75	100.00	0.00	0.0	0.0	4.2
	Sechura desert	183,760.00	98.60	2.00	1,129.8	0.6	3.4
	Beni savanna	134,704.75	100.00	16.30	1,027.8	0.8	0.3
2A	Sin Valley dry forests	25,216.00	100.00	3.30	626.3	2.5	89.6
	Catatumbo moist forests	23,260.00	100.00	28.00	1,884.0	8.1	72.0
	Magdalena-Urabá moist forests	77,244.25	100.00	1.50	1,041.3	1.3	70.6
	Western Ecuador moist forests	33,733.50	100.00	4.50	1,229.5	3.6	70.1
	Guayaquil flooded grasslands	2,883.50	100.00	2.00	56.8	2.0	67.1
	Lara-Falcón dry forests	17,563.25	99.90	16.00	1,427.3	8.1	64.6
	Ecuadorian dry forests	21,020.00	100.00	2.00	428.0	2.0	64.2
	Apure-Villavicencio dry forests	70,342.75	100.00	14.70	3,398.8	4.8	61.8
	Magdalena Valley montane forests	105,894.00	100.00	1.60	1,703.5	1.6	53.4
	Cauca Valley montane forests	32,136.50	100.00	2.60	847.3	2.6	52.6
2B	Guajira-Barranquilla xeric scrub	31,804.25	100.00	31.60	515.8	1.6	48.2
	Araya and Paria xeric scrub	5,467.50	95.00	5.30	292.3	5.3	44.7
	La Costa xeric shrublands	72,793.50	99.90	14.20	1,951.5	2.7	35.8
	Venezuelan Andes montane forests	30,172.00	100.00	50.00	2,278.0	7.6	29.3
	Tumbes-Piura dry forests	41,069.00	100.00	4.70	948.0	2.3	20.5
	Cordillera Central paramo	12,141.50	100.00	9.20	1,003.5	8.3	15.6
	Central Andean wet puna	119,599.25	100.00	14.10	2,937.5	2.5	14.4
	Llanos	408,527.00	100.00	10.70	23,146.5	5.7	13.4
	Napo moist forests	252,565.25	100.00	15.20	20,431.3	8.1	10.4
	Peruvian Yungas	188,221.25	100.00	13.00	11,049.8	5.9	10.0
	Paraguana xeric scrub	16,305.50	98.90	11.70	846.8	5.2	6.9
	Bolivian montane dry forests	85,514.25	99.80	9.80	883.5	1.0	3.0
3B	Northwestern Andean montane forests	81,061.25	100.00	12.50	9,283.5	11.5	43.9
	Cordillera Oriental montane forests	68,736.00	100.00	26.10	14,116.0	20.5	43.0
	Santa Marta montane forests	4,840.75	100.00	73.00	2,362.5	48.8	33.4
	Magdalena-Santa Marta mangroves	2,412.75	100.00	16.00	442.0	18.3	32.9
	Coastal Venezuelan mangroves	4,866.00	100.00	33.40	992.8	20.4	29.9
	Eastern Cordillera real montane forests	102,188.50	100.00	18.10	12,180.3	11.9	24.1
	Northern Andean paramo	30,071.25	100.00	33.60	9,632.3	32.0	23.3
	Cordillera La Costa montane forests	15,221.00	99.90	58.30	3,472.5	22.8	18.1
	Gulf of Guayaquil-Tumbes mangroves	2,854.00	100.00	10.60	319.3	11.2	16.5
	Esmeraldes/Chocó mangroves	4,546.75	100.00	8.90	501.0	11.0	15.5

TABLE 9.1. *(continued)*

Group	Ecoregion name	Area (km²)	% area in region	% protected total	Area protected in region	% protected in region	% of area transformed
	Ucayali moist forests	115,270.75	100.00	21.50	14,510.5	12.6	9.5
	Orinoco wetlands	6,569.00	98.80	30.80	1,050.0	16.0	8.9
	Bolivian Yungas	95,349.00	100.00	54.80	19,619.3	20.6	8.1
	Cordillera de Merida paramo	2,881.00	100.00	72.20	1,221.3	42.4	1.6
	Santa Marta paramo	1,257.75	100.00	99.90	1,239.8	98.6	1.2
4B	Chocó-Darién moist forests	59,449.25	80.80	12.30	1,492.5	2.5	19.1
	Solimoes-Japurá moist forest	134,025.25	78.00	7.40	6,009.3	4.5	0.1
5B	Negro-Branco moist forests	171,797.00	76.60	35.20	20,722.8	12.1	1.2
	Caqueta moist forests	174,736.25	92.90	36.50	27,918.3	16.0	7.3
	Orinoco Delta swamp forests	25,925.50	82.10	59.20	5,952.5	23.0	2.3
	Tepuis	49,861.75	94.80	65.40	16,138.8	32.4	0.7

Note: Only ecoregions with 75 percent or more of their total area within the region are included.

percent of their original area; in the case of the Santa Marta paramo ecoregion, almost all of its original area is in protected-area status. Maybe because of this protection, none of these is also at the level of risk seen in the ecoregions in groups 1 and 2. Both groups 4 and 5 are ecoregions with significantly restricted distributions (i.e., more than 75 percent of their area in the region) but are protected at various levels. Group 4 identifies ecoregions with deficient protection while group 5 identifies those with acceptable levels of representativeness. In all cases, the risk levels are low when compared to ecoregions in groups 1 and 2.

Figure 9.3 shows the periods in which the different protected areas were added to the individual reserve networks of the tropical Andean countries. By far the largest growth in recent times has been that of the reserves of Venezuela, with almost 70 percent of their area added in the 1990s. Indeed, Venezuela has an interesting history of park establishment. It was the first country in the region to establish a strictly protected area—the Henri Pittier National Park near Caracas, established in 1939—but afterward very little activity took place until the 1970s. Up to the 1950s, almost 90 percent of all the area protected in the region was in Colombia, mainly through the large Sierra de la Macarena National Park, created in 1948. Until then, Ecuador and Peru had no protected areas at all. Reserve creation picked up speed in the 1950s in Ecuador and in the 1960s in Bolivia, but it was in the 1970s that seem to have been the most important period of conservation efforts in the region. Between 20 (Venezuela) and 58 percent (Ecuador) of the each country's protected-area coverage was established in the 1970s. Based on the area in reserves, the reserve networks of Bolivia, Colombia, and, to a lesser extent, Ecuador, had almost stabilized by the 1980s. Peru, on the other hand, has

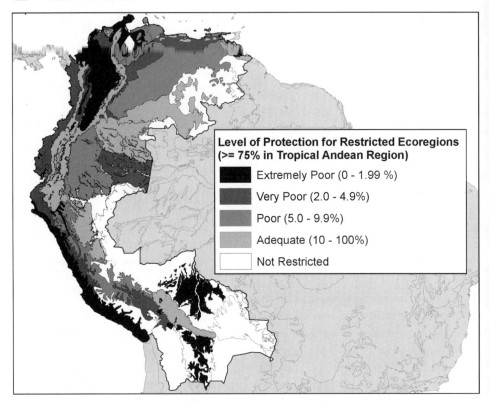

Level of Protection for Restricted Ecoregions (>= 75% in Tropical Andean Region)

■ Extremely Poor (0 - 1.99 %)

■ Very Poor (2.0 - 4.9%)

■ Poor (5.0 - 9.9%)

▨ Adequate (10 - 100%)

□ Not Restricted

FIGURE 9.2. Level of protection (representativeness) of restricted distribution ecoregions

increased protected-area coverage drastically since 2000, almost doubling the previous area (this includes the change of category of one of Peru's largest reserves, the Tambopata National Reserve).

By 2001, a network of 136 reserves protected samples of fifty-nine of the seventy-five ecoregions in the region. Table 9.2 shows that earlier reserves expanded the ecoregional coverage rapidly but that the latest round of reserves has not been equally efficient for improving the protection of the region's biodiversity. For example, at the beginning of the 1950s the four existing national parks included samples of seven ecoregions. By the end of the 1970s, fifty-two ecoregions were represented in eighty-two reserves, with twenty-three of these added in fifty-eight new reserves in that decade alone, a ratio close to 0.9 new ecoregions for each new reserve. However, during the last twenty years of the twentieth century and the beginning of this century, fifty-four additional reserves added samples of only four new ecoregions, a ratio of 0.07 new ecoregions for each new reserve.

To some extent this should be expected, as early on almost any new reserve would have protected hitherto unprotected ecoregions. Furthermore, these

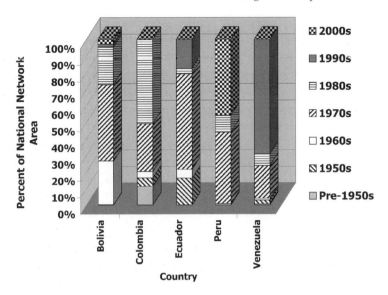

FIGURE 9.3. Evolution (by area) of the protected-area network tropical Andes, by country

TABLE 9.2.   Ecoregional coverage of the reserve network of the tropical Andean region by time

	Pre-1950s	1950s	1960s	1970s	1980s	1990s	2000s
Number new	4	6	14	58	33	19	2
Number total	4	11	24	82	115	134	136
New ecoregions covered	7	9	13	23	3	4	0
Total ecoregions covered	7	16	29	52	55	59	59
Area added	11,408.3	8,258.8	18,994.3	121,544.8	74,567.3	114,587.8	24,985.0
Area total	11,408.3	19,667.1	38,661.4	160,206.2	234,773.5	349,361.3	374,346.3
Average reserve size, all	2,852.1	1,787.9	1,610.9	1,953.7	2,041.5	2,607.2	2,752.5
Average reserve size, new	2,852.1	1,376.5	1,356.7	2,095.6	2,259.6	6,030.9	12,492.5

statistics simplify, maybe excessively, historical improvements in coverage as some really important conservation initiatives are not visible in this analysis. For instance, the Ucayali moist forests had been first protected in the 1980s, but most of its protected area really came as a result of the creation of the Cordillera Azul National Park in 2001. Still, the overall result seems to be that, at an ecoregional level, new rounds of reserve designation have not been as effective as early ones. Indeed, if the great increase in regional protected-area coverage noted by Green and Paine (1997, cited in Zimmerer and Carter 2002) is examined using ecoregion representativeness criteria, then the conservation efforts of the last twenty-five years have not been as effective as those before 1980. To be sure, many of the new reserves increased the level of protection of these ecoregions, but considering that the average size of the new reserves in the 1980s was more than twice the average size of reserves in the previous decades, and that in the 1990s new reserves were even bigger, almost five times larger on the average that the older reserves, the efficiency of older reserves seems to be far greater.

Interestingly, the 1990s also brought to the Andean region new international initiatives attempting to take a regional view to the conservation problem in the region (see also Young and Rodríguez, chap. 10 in this volume). Large conservation NGOs, such as TNC, CI, and WWF, began regional programs that targeted threatened ecosystems rather than countries. This shift in focus reflects the transborder nature of biodiversity, the globalization of conservation objectives, and the international reach of large, nonstate conservation institutions. For example, CI has three major transnational conservation areas of interest: the Chocó-Manabí Corridor, in Colombia and Ecuador; the Condor-Cutucú Corridor, in Ecuador and Peru; and the Vilcabamba-Amboró Corridor in Peru-Bolivia. One WWF project targets the Pastaza watershed shared by Ecuador and Peru. With funds from the GEF and in coordination with local partners, TNC is searching for new reserve sites in five ecoregions, ranging from Panama to Bolivia: Chocó-Darién, Cordillera Real Oriental, Yungas Peruanas, Yungas Bolivianas, and the Chaco. In some instances, international organizations have cooperated with local governments to create new reserves or even created their own protected areas. For example, CI brokered the protection of approximately 350,000 hectares of forests next to Bolivia's Madidi National Park by purchasing logging concession rights from a private company.

Discussion and Conclusions

Today there are calls to reevaluate the conservation paradigms of the last decades of the twentieth century, from the idealistic people-oriented approaches of the 1980s and 1990s, to more pragmatic efforts based on protected natural reserves and a global perspective of biodiversity (Brandon, Redford, and Sanderson 1998; Terborgh 1999). While these calls do not go unchallenged (e.g., Vacklay et al. 2001; Wilshusen et al. 2002), the creation of reserves, even paper ones, has been noted as a positive step for conservation. A recent study by Bruner et al. (2001) found, for example, that the designation of a protected area effectively reduces the threats to biodiversity in that area, even with modest investment and limited management resources. The efficiency of reserves increased with resources, but even relatively poor reserves performed better than expected. My study offers striking support for this claim, if not a clear-cut cause-and-effect relationship between reserve establishment and conservation quality. Table 9.1 shows clearly that areas in the five Andean countries that have been better protected historically are today in better conservation condition. For protected-area programs, globalization has had a healthy impact, by facilitating national and international conservation efforts to expand the Andean region network of protected areas. However, as this study has shown, much work remains to be done. While protected-area coverage has expanded significantly in recent years, new parks and reserves are usually not sited in ecoregions that are in critical need of protection.

Because the financial and political resources available for conservation are limited, especially in developing countries, individual countries could maximize their investment in conservation by emphasizing biodiversity areas that face high risk of destruction at global scales, leaving lower-risk areas for later work. A more selective approach to establishing protected areas would also be more economically efficient and politically expedient, by limiting the opportunity costs incurred by biodiversity conservation, that is, the sum of sustainable and unsustainable benefits that are forgone by society due to restrictions placed on resource users. However, the realization of socioeconomic objectives in conservation-with-development or "sustainable use" projects will be ephemeral successes if the main objective of biodiversity conservation is abandoned. Whether one takes an activist or technical position, it should be clear to all that social welfare requires some level of ecological integrity. After all, biodiversity is not only a romantic, biocentrist concept of genes, species, and ecosystems, but also a key to maintaining the goods and services that the environment generates. A new global framework of biodiversity conservation must recognize not only the dependence of life on the system itself but also its underlying structure and functions. Of critical importance for such a global perspective is the recognition that biodiversity and evolutionary processes in one part of the environment depend on the conditions in other parts of the environment, not only organisms but also biophysical processes.

While we should expect that conservation theories and objectives change again in the future, most countries will continue to depend on their network of protected areas as their primary tool to promote and ensure biological conservation. Yet the evidence is growing that we are far from our objective of the systematic, representative coverages that are held desirable in global conservation. The results of this study indicate that after a century of conservation efforts, and seventy years of protected-area development, the protected-area network of the tropical Andean region still has significant gaps. This is an illustration of the dislocation of global and local conservation objectives. Part of the problem is that there is an information gap between these two scales of operations (see also Turner, chap. 7, and Sneddon, chap. 8 in this volume). Global and regional analysis of conservation needs requires levels of generalization that are not efficient for local analysis. Addressing local conservation problems and establishing protected areas require detailed, site-specific information about, for example, the biological and ecological value of alternative sites, the socio-economic determinants of reserve viability, and local administrative conditions. These typically cannot be examined at regional levels (Sierra, Campos, and Chamberlin 2002). Generating data at such level of detail would require considerable time and financial resources, none of which are available. Country designs that do not take into account larger, regional conditions are unlikely to protect biodiversity and ecosystem services at a global level. Hence, it is critical that national networks are assessed in relation to other countries' networks, and especially in relation to the networks of ecologically complementary countries. It was also shown here that the category of ecoregions that is at highest risk consists of the type of ecoregions that is found in just one country and should hence be a priority for conservation if regional objectives are to be met, even if that means that other unprotected but shared ecoregions have to wait.

This new geography of conservation management runs counter to the global tendencies of decentralization. It requires thinking not in country or administrative units but in ecological units that cross boundaries and require the concerted efforts of conservation agents across countries and sometimes even regions. Articulating these two increasingly polarized arenas is one of the great new challenges of biodiversity conservation. Even in an increasingly globalized world, the organization of global biodiversity is the aggregate of each country designing and building a representative protected-area network while taking into account the larger cross-national ecological matrix. Yet, it is still not clear how transnational coordination will work in or for conservation, at least in the case of biodiversity conservation and protected areas in the Andean countries. On the other hand, these same changes create the opportunity of a fluid interface with outside organizations, both national and international, that seek to support such local efforts with funds and expertise. Otherwise, without these linkages, local conservation efforts would be likely to encounter significant difficulties in meeting their conservation objectives, let alone in contributing to objectives defined at other geographical scales. This situation would be similar to that of their national-level predecessors. Indeed, because in the past national-level conservation actions have generally been subordinated to economic and political priorities, it can be expected that local non-biological criteria will also dominate local efforts to conserve biodiversity.

From a regional and global biodiversity point of view, a country could use its resources more efficiently by focusing its efforts on globally-important biodiversity elements that are not protected elsewhere. In an era of global biodiversity concerns, priority setting and new reserves need to take into account not only what is the threat and representativeness within a given country, but also how new reserves add to the conservation of regional, cross-national biodiversity and ecological functions. One thing is certain, however: global biodiversity conservation will probably continue to depend on the national networks of protected areas as the primary tool to promote and ensure biological conservation.

References

Agrawal, A., and C. Gibson. 1999. Enchantment and disenchantment: The role of community in natural resource conservation. *World Development* 27 (4): 629–49.

Bailey, R. 1998. *Ecoregions: The ecosystem geography of oceans and continents.* New York: Springer.

Birdlife International. 2000. *Threatened birds of the world.* Cambridge, UK: Lynx Editions and Birdlife International.

Brandon, K., K. Redford, and S. Sanderson, eds. 1998. *Parks in peril: People, politics, and protected areas.* Washington, DC: Island.

Bruner, A., R. Gullison, R. Rice, and G. da Fonseca. 2001. Effectiveness of parks in protecting tropical biodiversity. *Science* 291:125–28.

Cabrera, A., and A. Willink. 1973. *Biogeografía de America Latina.* Departamento de Asuntos Científicos de la Organizacion de Estados Americanos. Serie de Biología, Monografía no. 13. Washington, DC: Organizacion de Estados Americanos.

Chapman, D. 2003. Management of national parks in developing countries: A proposal for an international park service. *Ecological Economics* 46 (1): 1–7.

Daly, H., and J. Farley. 2004. *Ecological economics: Principles and applications.* Washington, DC: Island.

Dinerstein, E., D. Olson, D. Graham, A. Webster, S. Primm, M. Bookbinder, and G. Ledec. 1995. *A conservation assessment of the terrestrial ecoregions of Latin America and the Caribbean.* Washington, DC: World Bank.

Eva, H., A. Belward, E. de Miranda, C. di Bella, V. Gond, O. Huber, S. Jones, M. Sgrenzaroli, and S. Fritz. 2004. A land cover map of South America. *Global Change Biology* 10:731–44.

Forero, E. 1982. La flora y la vegetación del Choco y sus relaciones fitogeográficas. *Colombia Geográfica* 10 (1): 77–90.

Gonzales, E. K., P. Arcese, R. Schulz, and F. L. Bunnell. 2003. Strategic reserve design in the central coast of British Columbia: Integrating ecological and industrial goals. *Canadian Journal of Forest Research* 33 (11): 2129–40.

Green, M., and J. Paine. 1997. State of the world's protected areas at the end of the twentieth century. Paper presented at IUCN World Commission of Protected Areas Symposium on Protected Areas in the 21st Century: From Islands to Networks, Albany, Australia, 24–29 November.

Hernandez Camacho, J., A. Hurtado, R. Ortiz, and T. Walshburger. 1992. Unidades biogeográficas de Colombia. In *La diversidad biológica de Iberoamérica,* ed. G. Haffer, 105–51. Veracruz, Mexico: Instituto de Ecología. Hueck, K, and P. Seibert. 1972. Vegetationskarte von Südamerika. Stuttgart, Germany: Fischer Verlag.

Hummel, M., ed. 1989. Endangered spaces: The future of Canada's wilderness. Ontario: Key Porter Books.

Illsley, D., and T. Richardson. 2004. New national parks for Scotland: Coalitions in conflict over the allocation of planning powers in the Cairngorms. *Journal of Environmental Planning and Management* 47 (2): 219–42.

IUCN (World Conservation Union). 1994. Guidelines for protected area management categories. Gland, Switzerland: IUCN.

Jenkins, M. 2003. Prospects for biodiversity. *Science* 302:1175–77.

Loreau, M., S. Naeem, P. Inchausti, J. Bengtsson, J. Grime, A. Hector, D. Hooper, M. Huston, D. Raffaelli, B. Schmid, D. Tilman, and D. Wardle. 2001. Biodiversity and ecosystem functioning: Current knowledge and future challenges. *Science* 294:804–8.

Margules, C., and L. Pressey. 2000. Systematic conservation planning. *Nature* 405:243–53.

Noss, R. 1995. Maintaining ecological integrity in representative reserve networks. Toronto and Washington, DC: World Wildlife Fund–Canada and World Wildlife Fund U.S.

Olson, D., and E. Dinerstein. 1998. The Global 200: A representation approach to conserving the earth's most biologically valuable ecoregions. *Conservation Biology* 12 (3): 502–15.

Olson, M., E. Dinerstein, E. Wikramanayake, N. Burgess, G. Powell, E. Underwood, E. D'amico, et al. 2001. Terrestrial ecoregions of the world: A new map of life on earth. *BioScience* 51:933–38.

Omernik, J. 1995. Ecoregion: A framework for managing ecosystems. *George Wright Society Forum* 12:35–50.

Polanyi, K. 2001. *The great transformation: The political and economic origins of our time*. Boston: Beacon.

Ramutsindela, M. 2004. Globalization and nature conservation strategies in 21st-century Southern Africa. *Tijdschrift voor Economische en Sociale Geografie* 95 (1): 61–72.

Rodríguez, J., and K. Rodríguez-Clark. 2001. Even "paper parks" are important. *Trends in Ecology & Evolution* 16 (1): 17.

Rodríguez, L., and K. Young. 2000. Biological diversity of Peru: Determining priority areas for conservation. *Ambio* 29:329–37.

Ryti, R. 1992. Effect of the focal tax on on the selection of nature reserves. *Ecological Applications* 2 (4): 404–10.

Sala, O., F. Chapin, J. Armesto, E. Berlow, J. Bloomfield, R. Dirzo, E. Huber-Sanwald, et al. 2000. Global biodiversity scenarios for the year 2100. *Science* 287:1770–74.

Schwartzman, S., A. Moreira, and D. Nepstad. 2000. Rethinking tropical forest conservation: Perils in parks. *Conservation Biology* 14 (4): 1351–57.

Sierra, R., ed. 1999. *Propuesta preliminar de un sistema de clasificación de vegetación para el Ecuador continental*. Quito, Ecuador: Proyecto INEFAN/GEF and EcoCiencia.

Sierra, R., F. Campos, and J. Chamberlin. 2002. Conservation priorities in continental Ecuador: A study based on landscape and species level biodiversity patterns. *Landscape and Urban Planning* 59:95–110.

Strittholt, J., and R. Boerner. 1995. Applying biodiversity gap analysis in a regional nature reserve design for the edge of Appalachia, Ohio (USA). *Conservation Biology* 9:1492–1505.

Takhtajan, A. 1986. *Floristic regions of the world*. Los Angeles: University of California Press.

Terborgh, J. 1999. *Requiem for nature*. Washington, DC: Island.

Udvardy, M. 1975. World biogeographical provinces. IUCN Occasional Paper no. 18. Gland, Switzerland: IUCN.

United Nations. 2002. Prototype nationally designated protected areas database. Available at http://www.unep-wcmc.org/. Accessed 12 May 2002.

Vacklay, K., A. Bruner, R. Gullison, R. Rice, and G. da Fonseca. 2001. The effectiveness of parks. *Science* 293:1007.

Weiler, S., and A. Seidl. 2004. What's in a name? Extracting econometric drivers to assess the impact of national park designation. *Journal of Regional Science* 44 (2): 245–62.

Wilshusen, P., S. Brechin, C. Fortwangler, and P. West. 2002. Reinventing the square wheel: Critique of a resurgent "protection paradigm" in international biodiversity conservation. *Society and Natural Resources* 15:17–40.

Wright, R., M. Murray, and T. Merrill. 1998. Ecoregions as a level of ecological analysis. *Biological Conservation* 86:207–13.

Zimmerer, K., and E. Carter. 2002. Conservation and sustainability in Latin America and the Caribbean. In *Latin America in the 21st century: Challenges and solutions*, ed. G. Knapp, 207–50. Austin: University of Texas Press.

10 Development of Peru's Protected-Area System: Historical Continuity of Conservation Goals

KENNETH R. YOUNG AND LILY O. RODRÍGUEZ

One of the ways that globalization is expressed in biodiversity conservation is that global policies and transnational institutions begin to intervene in national and local conservation issues, creating tensions and contradictions. This intervention may at times be considered a positive method to redress inequalities in power, wealth, social justice, or access to information and natural resources, by involving international mediation or participation in the resolution of those issues. In some cases, the opposite may be true wherein those inequalities are exacerbated or new inequalities are produced (e.g., Neumann 1995; IADB 1997; Laurie and Bonnett 2002; Pendras 2002; Coombes 2003; Speth 2003; Vanden 2003; Gwynne and Kay 2004; Sundberg, chap. 11 in this volume). Potentially negative aspects of globalization include consequences of more direct economic linkages between small-scale producers and the global prices offered for natural resources, which can transform environmentally sustainable agricultural and extractive practices into economically unsustainable relics (e.g., Place 1988; Salafsky et al. 2001; Marshall and Newton 2003).

Social institutions that have functioned in the past to mediate disputes and moderate ecological pressures may not be able to adapt to changing economic and political conditions. In addition, the rapid fluxes in capital that accompany neoliberal economic reforms can and will cause dramatic changes in national and local contexts. International investments in resource extraction may be shifted toward the countries and places where oversight and mitigation of environmental damage are minimal. All of these factors can make conservation efforts illusory on indigenous lands and in protected areas (Stevens 1997; Zerner 2000; Terborgh et al. 2002; Young 2004; Zimmerer, chap. 1 in this volume). In fact, there are also potent criticisms of certain kinds of transnational aid efforts (Hancock 1989; Maren 1997; Scott 1998).

However, globalization and international policies also potentially link together social actors with well-meaning intentions and positive actions. By understanding the unintended negative consequences of globalization and decentralization, and by fomenting positive outcomes, it might be possible to help conservation efforts be more successful in particular cases and places. Historical perspectives can provide useful comparisons and contrasts (e.g., Frank 1997; Anderson 1999; Evans 1999; Campbell 2002; Kingsland 2002; Mendel and Kirkpatrick 2002). Biodiversity conservation also requires the identification of the special places and species that warrant concern (Margules and Pressey 2000; Scott et al. 2002). These can then be managed and conserved in an appropriate manner, whether that be by protected areas, extractive systems, endangered species programs, reintroductions of locally extinct species, ecological restoration, or some combination of these strategies (e.g., Young 1997; Young et al. 2002).

Protected natural areas provide habitat for native plant and animal species. They can also be important sources of income, of controversy, or in some cases both. Peru now has about 13 percent of its territory within a national system of protected areas (fig. 10.1). This represents a considerable investment of land—roughly twice as much as in the United States in terms of percentage area (Scott et al. 2001)—and also of human and economic resources. In this chapter, we evaluate the history of the development of this system. While doing this, we attempt to elucidate the relative importance of external (transnational and global) and internal (national and local) influences acting upon this history. This can be viewed as an example of how nature conservation and globalization processes have interacted over the past several decades. A historically oriented evaluation of conservation territories is crucial to disentangling the relative roles and power of the social actors and institutions involved. Different points in time will be characterized in reference to the identification of priority areas for biodiversity and the establishment of protected areas. The information comes from an evaluation of an environmental bulletin produced in the 1940s to 1960s in Peru, plus other sources documenting the development of the protected-area system.

Exploration and Early Conservation Period

Europeans arrived in South America in the sixteenth century in order to extract natural resources. At that time the Inca rulers controlled large herds of camelids in the Andean highlands (Lumbreras 1974). Although the evidence is sparse and open to various interpretations (Bauer 1992; Stern 1993), it is generally believed that the pre-Hispanic utilization of natural resources was more sustainable that that practiced in the centuries that followed. This perception exists because the Spaniards arrived with the explicit goal of extracting natural resources, especially precious minerals, from their new colony and shipping raw materials and commodities overseas. The infrastructure

	National Park
1	Cutervo
2	Tingo María
3	Manu
4	Huascarán
5	Cerros de Amotape
6	Río Abiseo
7	Yanachaga-Chemillen
8	Bahuaja-Sonene
9	Cordillera Azul
	National Reserve
10	Pampa Galeras
11	Junín
12	Paracas
13	Lachay
14	Titicaca
15	Salinas y Aguada Blanca
16	Calipuy
17	Pacaya Samiria
18	Tambopata
	National Sanctuary
19	Huayllay
20	Calipuy
21	Lagunas de Mejía
22	Ampay
23	Manglares de Tumbes
24	Tabacones Namballe
	Historical Sanctuary
25	Chacamarca
26	Pampa de Ayacucho
27	Machupicchu
28	Bosque de Pomac
	Protected Forest
29	A.B. Canal Nuevo Imperial
30	Puquio Santa Rosa
31	Pui Pui
32	San Matias - San Carlos
33	Pagaibamba
34	Alto Mayo
	Communal Reserve
35	Yanessha
36	El Sira
37	Amarakaeri
	Hunting Area
38	El Angolo
39	Sunchubamba
	Scenic Reserve
40	Nor Yauyos-Cochas
	Reserved Zone
41	Laquipampa
42	Apurímac
43	Pantamos de Villa
44	Tumbes
45	Algarrobal El Moro
46	Chancaybaños
47	Aymara Lupaga
48	Gueppi
49	Rio Rimac
50	Allpahuayo-Mishana
51	Santiago Comaina
52	Alto Purus
53	Cordillera de Colan

FIGURE 10.1. Peru, showing political departments and the protected areas

and urban settlements of the colony were at least partially designed or re-adapted for this purpose by the 1570s (Lockhart 1968; Davies 1984; Stern 1993). Among the many European explorers and colonists were naturalists in search of useful plant and animal species (Lamas 1980a; Bonfiglio 2001). For example, there were repeated attempts in the 1700s and 1800s to localize stands of cinchona trees (*Cinchona* spp.), whose bark would form the raw material for industrial production of quinine, a medicine used to treat and prevent malaria (Ruiz 1940; Condamine 1993; Honigsbaum 2001). Renowned naturalists, such as Alexander von Humboldt, were motivated not only by intellectual curiosity but also the quest for natural resources of economic utility (von Humboldt 1821; Núñez and Petersen 2002).

Explorers continued to arrive and document the flora and fauna of Peru during the early Republic years (1821 to 1910). A. Mathews, R. Spruce, and H. A. Weddell were among the explorers searching for cinchona in eastern Peru (Weberbauer 1945; Honigsbaum 2001). Most expeditions were of relatively short duration, such as the 1851–1852 trip of William Lewis Herndon (Kinder 2000) and also the 1852–1853 trip of Clements Markham (Blanchard 1991). The research career of Antonio Raimondi marked the first time that such exploration switched from short-term visits from Europeans to long-term sustained efforts. Raimondi arrived in Peru in 1851 and spent two decades walking over much of the country (Santillana Cantella 1989). He provided the first detailed maps and accounts of the country's geology and vegetation. This information was of general interest to the international scientific community, but more important was vital to planning by the Peruvian government (e.g., Raimondi 1940); in fact, his plant and rock collections were declared national patrimony and housed in San Marcos University in Lima. Lamas (1980b) lists more than twenty other scientists and scientific collectors that spent extended periods during those years in Peru or became permanent residents.

Eventually, European scientists in Peru built the foundations for locally based science, by passing their insights and methods on to Peruvian students and colleagues. August Weberbauer exemplified this cohort of European scientists whose influence continued at least through the first half of the twentieth century. His description of flora observed during his extensive travels (Weberbauer 1945) is still used as a basic reference today (León 1995, 2002). Indeed, there are still botanists in Peru that identify themselves as "disciples of Weberbauer." On the zoological side, Maria and Hans Koepcke did pioneering studies on marine and terrestrial ecological systems in Peru, focusing on animal adaptations and distributions (Koepcke and Koepcke 1952; H.-W. Koepcke 1961; M. Koepcke 1961).

In June 1940, scientists and educators formed a national committee (Comité Nacional de Protección a la Naturaleza) to help establish national parks and reserves in Peru, with members in Lima, Iquitos, and Cusco. One of

TABLE 10.1.   Selected topics covered in the *Boletín del Comité Nacional de Protección a la Naturaleza*

Year	Topics
1948	Bird migration, "yareta" (*Azorella*) used for fuel
1951	Conservation efforts elsewhere in the world, possibilities for cork production on the coast
1952	Guano islands
1953	Environmental conservation, botanical gardens, vicuña
1954	Marine fisheries, orchids
1955	Marine resources, whale hunting, soil conservation, seabirds
1956	Forests of Cutervo, vicuña, Parinacochas lake
1957	Environmental education, Junín frog
1958	West slope forests, camelids, marine plankton
1959	Guano islands and conservation, coastal plants
1960	Marine resources, eucalypts for the mining industry, Amazonian fauna, urban environmental problems, coastal climate

the sources of inspiration for this group was a 1940 visit to Peru by Gilbert Pearson, then president of the Audubon Society in the United States, which exemplified the success of private organizations in promoting scientific study and conservation of natural resources (Graham 1990). In 1944, the committee began publishing a technical and news bulletin (*Boletín del Comité Nacional de Protección a la Naturaleza*). Analyzing this publication provides insights into the organization's ideology, goals, and achievements. Conservation concerns in the early years included the protection of large natural areas and the sustainable harvest of terrestrial and marine resources, from fuelwood to fishes (table 10.1). There was a strong emphasis on coastal resources and locations, although Andean and Amazonian places and biota were also mentioned. Despite important changes in conservation science and policymaking over the years, this list of conservation priorities is quite similar to those found in much more recent Peruvian documents (e.g., Oficina Nacional de Evaluación de Recursos Naturales 1986; CDC-UNALM 1991; Dirección General de Forestal y Fauna 1991; Rodríguez 1996; Consejo Nacional del Ambiente 2001).

In early 1942, the government of Peru ratified a treaty with other New World countries that required the signatories to begin study of how national parks, reserves, and monuments were to be established and implemented (Barreda 1951, 131). Attempts to establish protected areas in Machu Picchu, Vilcabamba, and other sites in the early 1940s were sporadic, disorganized, and fruitless (see Ruiz Alarco and Giesecke 1953 for a complete list). Natural scientists interested in conservation lobbied for protection of their favored places. Barreda (1951, 131–33) recommended protecting the seasonal desert fog ecosystem of Lomas de Atiquipa; the mesquite forests of Piura and Lambayeque; the high Andean forests in Ancash and Puno; the tropical alpine puna between Arequipa and Puno, and also in Ayacucho; and finally the Amazon, where immense areas were available for conservation. Weberbauer (1951) suggested the following priorities: (a) seasonal fog vegetation, or *lomas;* (b) mesquite forests; (c) moist forests of the western slopes of the Peruvian Andes; and (d) inter-Andean valleys. Meanwhile, the Koepckes, coming from a zoological perspective, passionately argued for the recognition of the conservation value of small forest relicts on the western slopes of the Andes (Koepcke and Koepcke 1958).

The formation of a national organization for conservation, the international parks treaty, and debate among conservation scientists in Peru resulted in the formal establishment of a protected-area system and the creation of the first national park in the early 1960s. These events can also be seen as the culmination of a social transformation that was centuries in the making. Rampant extraction of natural resources for colonial or national economic development was now tempered by the work of national and resident scientists who not only documented the biological diversity and natural history of Peru but also advocated for conservation of its natural wealth. Within the scientific community itself, there had been a shift in focus from identifying extractable resources to evaluating the conservation value of plant and animal species and tracts of land. By the end of the 1950s, Peru's scientists were conversant with the major biogeographical patterns of the country and had identified a number of priorities that could be included within protected areas.

The Conservation Period: 1960s to 1990s

The nucleus of Peru's protected-area system was put in place in the 1960s, principally under the elected government of Fernando Belaúnde (1963–68), and then expanded by military governments in the late 1960s and the 1970s (table 10.2). This protected-area system was implemented and administered by a new bureaucracy in the national Agriculture Ministry, originally known as Servicio Forestal y de Fauna, later converted into the Dirección General Forestal y de Fauna, and now called INRENA. Because of their economic importance, marine resources were managed by other agencies that employed different, less conservation-minded rationales. Schweigger's (1947) pioneering synthesis on the Peruvian littoral zone was scientific in orientation, but the Peruvian Fishery Ministry (or its current equivalent in the Ministry of Production) has always focused policy on extraction rather than on conservation (e.g., Roemer 1970). This tendency has been tempered by the use of seasonal restrictions placed on harvested species that are reproducing, and by long-standing restrictions on who harvests bird guano from the islands offshore (Majluf 1996).

At the start of this period, two national parks were established by the

TABLE 10.2. Chronology of major events in the establishment of Peru's national system of protected areas

Year	Event(s)
1961	Cutervo, first national park in northern Andes
1963	first two national forests designated (Biabo-Cordillera Azul, Mariscal Cáceres) in Amazon
1965	Tingo María, second national park in Amazon; two national forests (Alexander von Humboldt, Pastaza-Morona-Maranon) in Amazon
1967	Pampas Galeras, first national reserve, in high Andes
1973	Manu, third national park, includes highlands and Amazon environments
1974	Junín, second national reserve; Huayllay, first national sanctuary; Chacamarca, historical sanctuary; all in highlands
1975	Paracas, third national reserve, on coast; Huascaran and Cerros de Amotape, fourth and fifth national parks, in highlands
1977	Lachay, fourth national reserve, on coast
1978	Titicaca, fifth national reserve, in highlands of southern Peru
1979	Salinas and Aguada Blanca, sixth national reserve, in southern highlands
1980	Pampas de Ayacucho, second historical sanctuary, in highlands
1981	Machu Picchu, third historical sanctuary, on Amazon drainages; Calipuy, seventh national reserve, in northern highlands
1982	Pacaya Samiria, eighth national reserve, in Amazon
1983	Río Abiseo, sixth national park, includes highland and Amazon environments; Pampas de Heath, third national sanctuary, with lowland tropical savanna
1984	Lagunas de Mejia, fourth national sanctuary, on southern coast
1986	Yanachaga-Chemillén, seventh national park, includes Amazonian environments
1987	Ampay, fifth national sanctuary, in highlands
1988	Manglares de Tumbes, sixth national sanctuary, on coast; Tabaconas-Namballe, seventh national sanctuary, in highlands of northernmost Peru
1999	Bahauja-Sonene, eighth national park, in Amazon
2001	Cordillera Azul, ninth national park, in Amazon
2003	Otishi, tenth national park, in Amazon
2003	Alto Purús, eleventh national park, in Amazon

Belaúnde administration, one in tropical montane forest in northern Peru (Cutervo) and one in tropical premontane and lowland forest in central Peru (Tingo María). In addition, large tropical rain forest areas were designated as national forests and the first national reserve, Pampas Galeras, was created to protect high Andean plants and animals. Comparing tables 10.1 and 10.2 demonstrates that these protected areas were a product of the biodiversity conservation concerns identified in Peru beginning in the 1940s. For example, the montane forests included in Cutervo National Park had been identified in the 1950s as being of importance due to their biological and ecological uniqueness and also due to the rapidity with which they were being cut and degraded. Cutervo's forests had been used during World War II for extraction of *Cinchona* bark for quinine; it was also known to be near an active deforestation front, with such interesting animals as the cave-dwelling oilbird (*Steatornis caripensis*) threatened as a result. Tingo María National Park presented a similar situation, with an oilbird colony threatened by long-term colonists of the upper Amazon.

In 1975, a forestry and wildlife law established four categories of conservation units: national parks, national reserves, national sanctuaries, and historical sanctuaries (table 10.3). And by 1980 most of the protected-area system was in place (table 10.2; see also Barker 1980). One of the world's largest tropical national parks, Manu, was established in 1973, and the world's longest protected tropical mountain chain was created with Huascarán National Park in 1975. Other superlatives include important high Andean reserves (Junín, Titicaca, and Salinas and Aguada Blanca), a large coastal reserve (Paracas), and a significant area of tropical dry forest (Cerros de Amotape). This was complemented by the designation of several small historical sanctuaries (Chacamarca, Pampa de Ayacucho), and an inland reserve near the Pacific (Lachay). At this time, most of the major ecological regions of Peru had been included.

During the 1980s this process continued, now under the elected governments of Belaúnde's second term and Alan Garcia (table 10.2). By 1988 two new national parks protecting complete altitudinal gradients from the high Andes down to lowland Amazonia (Río Abiseo, Yanachaga-Chemillén) had been established. In addition, the ecologically and archaeologically important Machu Picchu area was made a historical sanctuary, a large tropical swamp was made into a national reserve (Pacaya Samiria), and the small tracts of mangroves and tropical savannas in the country were almost completely protected by inclusion in, respectively, the national sanctuaries of Manglares de Tumbes and Pampas de Heath. These achievements were made despite concurrent economic crises and increasing violence associated with narcotics and guerrilla groups (Poole and Rénique 1992; Stern 1998; Young 2004). Smaller designated areas included high Andean (Calipuy), coastal (Lagunas de Mejía), and tropical montane forests (Ampay, Tabaconas-Namballe).

The case of Río Abiseo National Park, created in 1983, serves to show linkages between past events, including the discoveries of the early naturalists, and more recent conservation efforts. This park is a center of endemism for cloud forest vertebrates (Leo 1995). One such species is a large primate, the yellow-tailed woolly monkey (*Lagothrix flavicauda*), only known from the wet montane forests of Peru's northeastern Andes (Eisenberg and Redford 1999). Alexander von Humboldt first discovered this species in 1812. It was "rediscovered" and studied in the 1970s and 1980s (Macedo and Mittermeier 1979; Leo 1980, 1982, 1984). Its presence in Río Abiseo was used to help justify the park's creation. It is protected nowhere else and has not been successfully maintained in captivity. Thus, it is possible to link a series of events from the 1800s to the present, interconnecting scientific explorations and biodiversity conservation.

TABLE 10.3. Current national protected areas of Peru

Name and category	Year established	Area (hectares)	Location (department[s])	Geographical region
National parks				
Cutervo	1961	2,500	Cajamarca	Highland
Tingo María	1965	18,000	Huánuco	Amazon
Manu	1973	1,692,137	Cusco, Madre de Dios	Amazon, Highland
Huascarán	1975	340,000	Ancash	Highland
Cerros de Amotape	1975	91,300	Piura, Tumbes	Western
Rio Abiseo	1983	274,520	San Martín	Amazon, Highland
Yanachaga-Chemillen	1986	122,000	Pasco	Amazon, Highland
Bahuaja-Sonene	2000	1,091,416	Madre de Dios, Puno	Amazon, Highland
Cordillera Azul	2001	1,353,190	Huánuco, Loreto, San Martín, Ucayali	Amazon
Otishi	2003	305,973	Cusco, Junín	Amazon
Alto Purús	2003	2,510,694	Madre de Dios, Ucayali	Amazon
National reserves				
Pampa Galeras	1967	6,500	Ayacucho	Highland
Junin	1974	53,000	Junin, Pasco	Highland
Paracas	1975	335,000	Ica	Western
Lachay	1977	5,070	Lima	Western
Titicaca	1978	36,180	Puno	Highland
Salinas y Aguada Blanca	1979	366,963	Arequipa, Moquegua	Highland
Calipuy	1981	64,000	La Libertad	Highland
Pacaya Samiria	1982	2,080,000	Loreto	Amazon
Tambopata	2000	274,690	Madre de Dios	Amazon
Allpahuayo-Mishana	2004	58,070	Loreto	Amazon
National sanctuaries				
Huayllay	1974	6,815	Pasco	Highland
Calipuy	1981	4,500	La Libertad	Highland
Lagunas de Mejia	1984	690	Arequipa	Western
Ampay	1987	3,635	Apurimac	Highland
Manglares de Tumbes	1988	2,972	Tumbes	Western
Tabaconas-Namballe	1988	29,500	Cajamarca	Highland
Megatoni	2004	215,869	Cusco	Amazon
Historical sanctuaries				
Chacamarca	1974	2,500	Junín	Highland
Pampa de Ayacucho	1980	300	Ayacucho	Highland
Machu Picchu	1981	32,592	Cusco	Amazon, Highland
Bosque de Pomac	2001	5,887	Lambayeque	Western
Landscape reserves				
Nor Yauyos-Cochas	2001	221,268	Junin, Lima	Highland
Reserved zones				
Laquipampa	1982	11,347	Lambayaque	Western
Pantanos de Villa	1989	263	Lima	Western
Tumbes	1994	75,102	Tumbes	Western

(continued)

TABLE 10.3.   *(continued)*

Name and category	Year established	Area (hectares)	Location (department[s])	Geographical region
Reserved zones (continued)				
Algarrobal el Moro	1995	321	La Libertad	Western
Chancaybanos	1996	2,628	Cajamarca	Highland
Aymara Lupaca	1996	300,000	Puno	Highland
Gueppi	1997	625,971	Loreto	Amazon
Rio Rimac	1998	~14	Lima	Western
Santiago-Comaina	2000	1,642,567	Amazonas, Loreto	Amazon
Codillera de Colan	2002	64,115	Amazonas	Amazon
Cordillera Huayhuash	2002	67,590	Ancash, Huánuco, Lima	Highland
Protected forests				
A. B. Canal Nuevo Imperial	1980	18	Lima	Western
Puquio Santa Rosa	1982	73	La Libertad	Western
Pui Pui	1985	60,000	Junin	Amazon
San Matias-San Carlos	1987	145,818	Pasco	Amazon
Pagaibamba	1987	2,078	Cajamarca	Highland
Alto Mayo	1987	182,000	San Martín	Amazon
Communal reserves				
Yanesha	1988	34,744	Pasco	Amazon
El Sira	2001	616,413	Huánuco, Pasco, Ucayali	Amazon
Amarakaeri	2002	402,336	Cusco, Madre de Dios	Amazon
Ashaninka	2003	184,469	Junin	Amazon
Matsiguenga	2003	218,946	Cusco	Amazon
Game reserves				
El Angolo	1975	65,000	Piura	Western
Sunchubamba	1977	59,735	Cajamarca	Highland

Note: The reserved zones are awaiting definitive categorization and delimitation. Updated lists are available on the INRENA web page (http://www.inrena.gob.pe).

In 1990, Peru's national system was restructured with additional categories, and given the name of Sistema Nacional de Areas naturales Protegidas por el Estado (SINANPE). The 1990s were the years of Alberto Fujimori's government. As documented in table 10.2, this was the least significant period in terms of protected-area conservation in the past forty years. This is especially the case given that the establishment of Bahauja-Sonene as a national park was simply a case of redesignating lands that already had been under other protected status. The irony is that this was the decade in which Peru received the most international financial support for the protected-area system. Most of this money was distributed through international environmental nongovernmental organizations (NGOs) to their Peruvian counterparts, as was true elsewhere (Bramble and Porter 1992; Price 1994; Fisher 1997; Keese 1998; Bryant 2002; Redford et al. 2003; Zimmerer, chap. 1 in this volume). Presumably those monies went for programs meant to encourage local and national support for protected areas. By the mid-1990s, international NGOs, through their national partners, were often as important as the national government in local management decisions made in some parks and reserves.

Another characteristic of the system, starting in the late 1980s, was the proliferation of "reserved zones." These areas are temporally designated for conservation protection by the government, but are not categorized and typically not staffed. This status was supposedly temporary, but it has continued for many years for several important areas (table 10.3). Unfortunately, some of these reserved zones have been spontaneously colonized due to land speculation. Manu Reserved Zone was maintained in "reserved zone" status for two decades before finally being incorporated into the tourist use zone of Manu National Park. In at least one case, we have heard of the official designation of reserved-zone status being delayed to allow loggers to clear forests of large mahogany trees. The often-cited "official" statistics for protected-area coverage in Peru, sanctioned by international monitors such as the World Conservation Union (IUCN), the United Nations Environment Program (UNEP), and the World Conservation Monitoring Centre (WCMC; see Zimmerer, chap. 1 in this volume, for more information), need to be critically evaluated in light of the minimal protection offered by these reserved zones. Conservation coverage is less complete in reality than it is on paper (e.g., fig. 10.1).

Recently, conservationists in Peru have focused less attention on pristine, remote environments in favor of natural areas proximate to large populations. The Pantanos de Villa Reserved Zone (Cano and Young 1998), awaiting categorization as a wildlife refuge since 1989, is a wetland near Lima. Protecting this wetland has received support because of international and national interest in the migratory birds that use it seasonally; it is globally recognized by inclusion on the "Ramsar" List of Wetlands of International Importance. It has received substantial local support because of the Municipality of Lima's interest in having a natural environment available for recreation and environmental education for the greater metropolitan area. Thousands of schoolchildren come each year to walk the nature trails and see the exhibits in the visitors' center. This area has high local support: L. M. Young (1998) interviewed 494 households neighboring this protected area and found that 85 percent of the interviewees considered the area to be important and valuable for conservation. As a result, the potential exists for a successful blending of local, national, and international biodiversity conservation goals.

In conclusion, the active decades of protected-area establishment in Peru built upon incipient efforts in the past, particularly upon knowledge of the distribution of ecological systems and species of concern across the country. This can be seen in the case of one of Peru's most recent national parks,

Cordillera Azul, which was established in 2001, under the interim government of Valentín Paniagua. It is symbolic of both the continuity of internal and long-term efforts within Peru and the resources and interest that can occasionally be leveraged with external social actors. The report that first summarized the biodiversity importance of the area and suggested how a national park might be delimited was published by the Field Museum of Chicago, but it also included participation from several Peruvian institutions (Alverson, Rodríguez, and Moskovits 2001). The area they identified for conservation had, in fact, first been considered important for formal conservation in 1963 when it was included in the Biabo-Cordillera Azul National Forest, and even earlier, when in the 1940s it was suggested that both it and the area placed eventually in the 1965 Tingo María National Park be considered important for conservation (*Boletín del Comité Nacional de Protección a la Naturaleza*). In turn, those priorities were based on natural history observations accumulated over the previous two centuries.

Current Trends

In 1994, after Peru endorsed the Biodiversity Convention, a group of scientists and conservationists was convened to critically evaluate Peru's protected-area system and to rank priority areas for biodiversity in the country. The results of the workshop and follow-up activities were detailed in Rodríguez (1996) and summarized by Rodríguez and Young (2000); a similar effort for Ecuador was described by Sierra, Campos, and Chamberlin (2002). The complete system and the "strictly protected" part, consisting only of national parks, sanctuaries, and reserves, were both compared cartographically to ecological regions as mapped by Zamora Jimeno (1996) and to priority areas for diversity and endemism of plants and animals as mapped and identified by the workshop participants. Ecological regions with the least conservation coverage at that time included hot desert, warm desert, cold high Andean desert, desert scrub, steppe, humid steppe, and "puna" (a seasonally dry tropical alpine zone). All of these had total protected areas that were equivalent of 0 to 2 percent of the area of the respective ecological regions in Peru, which collectively cover 429,000 km^2. As seen in tables 10.1 and 10.2, in fact the coastal and Andean portions of Peru have always been considered important to Peruvian scientists and were included in the earliest conservation efforts. Despite this internal consistency in priority setting within Peru, it is likely that the external funding (and associated research interests) that came to Peru designated for Amazon sites has overridden national considerations.

That this need not have been the case can be appreciated by considering Peru's deserts. The drylands of Peru have been used for natural resources for millennia. Rostworowski de Diez Canseco (1981) reported evidence of careful and perhaps sustainable use of natural resources on the central coast of Peru just before and after European contact. For example, there was replanting of sedges extracted for basketry and other handicrafts, and also a type of aquaculture for mullets in coastal inlets and lagoons. Excavation and irrigation allowed planting of domesticated plants and also changed land cover patterns. Other coastal habitats differ in their seasonality of precipitation, much of which arrives in the form of a fine mist in this otherwise hyperarid climatic zone. These sites, known in Peru as "lomas," were used seasonally for grazing by camelids. Motivated by conservation concerns, León et al. (1997) documented the presence of rare and unusual plants in the Nazca area, which is visited by thousands of tourists each year to see the archaeological sites, and then proposed a possible way that a conservation corridor could be established. Similarly, Polk (2003) utilized change detection analyses with Landsat satellite images, in conjunction with interview data, to propose possible conservation corridor scenarios in southwestern Peru that would connect the Andean highlands to the coast, passing near the city of Arequipa. It would have seemed logical to target arid areas for more conservation efforts, given that they are nationally and internationally recognized for high local endemism of plant and animal species (Ferreyra 1961, 1977, 1985; Rundel et al. 1991; León, Cano, and Young 1995; Léon, Young, and Cano 1996; Dillon 1997). Since most of these lands cannot be farmed and by default belong to the central government, they would have been available at little or no upfront cost for biodiversity conservation purposes and with little risk of local disputes about the declaration of areas for conservation.

Despite all of these biological, historical, cultural, and practical reasons, however, there has been little recent conservation attention directed toward the drier or higher natural environments in Peru. The reasons for such neglect can only be appreciated by considering Peru's place within a globalized network of conservation science and policy. The bias toward protecting Amazonian environments in Peru originates with the activities of international environmental NGOs, which prioritize rain forest conservation on a global scale. Identified with the protection of biodiversity and carbon sinks, rain forests "sell" to members and donors, while coastal deserts do not. As the preceding discussion demonstrates, this biome has recognized conservation value. But in the competition for scarce and increasingly privatized financial resources for establishing protected areas, "marketing" matters, and rain forest advocates are powerful and well organized.

The 1994 workshop participants identified thirty-eight priority areas for coverage of species diversity and their uniqueness (endemism). Surprisingly, only nine of those areas had already been included completely within Peru's protected-area system, while ten were partially covered and nineteen had no coverage at all. The unprotected priority areas included ten found in the Amazon, in border areas where geopolitical concerns had minimized conservation efforts. Geopolitics effectively blocked efforts in the 1970s to give protected-area status to Gueppi, located in the triangle formed by the border of Peru with Colombia and Ecuador, but this ideology appears to be receding. Gueppi was made a reserved zone in 1997, and coordinated protected-area efforts are now in progress between Peru and Ecuador (Cordillera del Condor area), Peru and Brazil (Sierra Divisor), and Peru and Bolivia (Vilcabamba-Amboró conservation corridor; Schulenberg and Awbry 1997; Ponce 2001). In fact, the Foreign Relations Ministry now welcomes protected areas as one strategy for stabilizing international borders. Coordination among neighboring countries for biodiversity protection has many obvious benefits (Duffy 2001; Rodrigues and Gaston 2002; Sierra, chap. 9 in this volume). The remaining unprotected priority areas, in coastal or Andean environments, continue to be overlooked.

The priority areas identified in the 1994 workshop were officially recognized as important conservation goals by the Peruvian government in March 1999. The year 2001 marked important changes in the management of natural protected areas. New regulations for park management were introduced, and several new protected areas were created (table 10.3). The surface covered by protected areas increased to represent about 12.9 percent of the nation. Meanwhile, three reserved zones acquired definitive categories: Cordillera Azul National Park, Amarakaeri Communal Reserve, and Bosque de Pomac Historic National Monument. At the time of this writing, the national system of protected areas is composed of fifty-nine units, including eleven national parks (table 10.3). Only minimum-impact activities, such as tourism and scientific investigation, are permitted in these parks. All national parks have some level of investment by either the government (by direct income) or private organizations, yet most of them have barely enough to support conservation goals, and long-term support is not guaranteed.

There are seven national sanctuaries (table 10.3) where the habitat of a plant or animal species, as well as natural formations of scientific or landscape interest, are considered of value. In addition, there are four historical sanctuaries that represent places with important natural values that constitute the environment of historical actions of special national significance, containing places of archaeological or historical importance. The ten national reserves are established for biodiversity conservation and the sustainable use of wild flora and fauna resources, in benefit of the neighboring rural populations. In truth, at times there is little appropriate conservation by locals or any advantage to them (e.g., the Titicaca case described by Orlove 2002). There are four protected areas (Alto Mayo, Pagaibamba, Pui-Pui, San Matías-San Carlos) in cloud forests and two (Cañete, Puquio Santa Rosa) on the Pacific drainages that were established to guarantee, in theory, the protection of upper or lower watersheds. Here, resource use that will not modify the vegetation structure is permitted. Game reserves are areas of fauna management dedicated mainly to sport hunting activities; both are partly or completely managed through private contracts. There is only one landscape reserve to date, and no wildlife refuges have been implemented. Finally, communal reserves are established nationally for the conservation of wildlife through direct management by neighboring communities. The use and merchandising of these resources is to be done under management plans (e.g., Newing and Bodmer 2003). Only five reserves have been created (table 10.3). They are located in the Amazon side of Peru; no communal reserves exist for Andean or coastal ethnic groups.

Most recently created areas have little or no financial support, and protection forests and reserved zones often have few or no resources. This is reflected in the lack of infrastructure and personnel. Higher investment is badly needed in order to implement the system. In addition, land ownership issues in protected areas are chronic problems. This is a consequence of both the history of Peruvian land tenure and a lack of law enforcement. While some national conservation policies allow the presence of people inside protected areas, there is no clear mandate for dealing with the resulting land cover transformations and the implications of the growth of human populations. Land rights in Peru often have been inherited from generation to generation, without "legal" documents. Some land rights were assigned during the Spanish colony and still persist. Current civil laws recognize a possession right to those people who are working the land; thus, having a small farm produces rights that are transferable to the farmer's children. None of the protected areas have management plans or programs that effectively deal with these difficult issues.

Park-people programs that attempt to integrate protected areas with the needs and desires of local people are common in Peru and present in the buffer zones of all well-funded parks and reserves. These programs include ecotourism for both national and international visitors, with a mixture of goals and tactics, successes and failures. There is some shared responsibility among formal and informal institutions for these protected areas, in addition to involvement of local communities. The Natural Protected Areas office of INRENA is in charge of the administration of all protected areas. However, Pampa Galeras National Reserve, an area created with the specific purpose of managing wild vicuña populations, is currently managed by a group, Consejo Nacional de Camélidos Sudamericanos (CONACS), with the participation of local communities, and El Angolo Game Reserve is managed by a hunting club. In fact, under the new regulations, protected areas can be officially managed by private organizations, and INRENA is preparing the paperwork to call for private contracts to be granted for several national protected areas, to be funded with international monies arriving through PROFONANPE (Fondo de Areas Naturales Protegidos por el Estado).

In summary, current trends for protected areas include a mixture of transnational, national, and local efforts. There are numerous NGO-based efforts, some of strictly local origin, others that are essentially local mechanisms for implementing goals of international NGOs. In some cases, the nationally recognized biodiversity conservation goals have not been implemented, apparently because they do not coincide with objectives of the international conservation institutions. In addition, there are recurrent difficulties in connecting those national goals to successful local implementation.

Conclusion

By our estimation, only thirty-six of the existing protected areas in Peru have some kind of administration, and of those, six have only a park chief assisted by a single park guard. Our impression is that similar staffing shortages plague the protected areas of neighboring countries. In Peru, the category "protection forest" is particularly poorly implemented, a fact that has seriously affected the conservation status of these areas. Covering almost 400,000 ha, most of them located in cloud forests in eastern Peru, these sites are highly important for biodiversity conservation and watershed protection. The game reserves and communal reserves have not been developed adequately as management categories and have similar problems.

All developing countries will need to budget for the continued management of their protected areas. In Peru, the $2 million needed each year for the protected-area system comes from directly collected revenue and international cooperation. The amount gathered in this fashion is not sufficient to cover the system's operating expenses. The protected areas that generate most of the income for the whole SINANPE are two coastal units (Lachay, Paracas), one in the Andes (Huascarán), and two in the Amazon (Tingo María and Manu). The money collected by these areas is redistributed to all protected areas, through the SINANPE, excepting those areas that have their own funding sources. Internationally, private institutions and other governments provide funds and resources for protected areas, either directly, through local NGOs that invest in management and development projects in the areas or their buffer zones, or from PROFONANPE, a mixed public-private institution created to manage the National Fund for Protected Areas (FONANPE). Beginning in December 1992, its purpose has been to provide permanent financial support for conservation, protection, and management of Peru's protected areas. It can receive contributions and grants from individuals or legal entities, as well as manage financial resources from international cooperation. One of its main goals is to establish an endowment that could cover recurrent costs for the whole system. PROFONANPE started with a seed contribution of $5.2 million dollars from the Global Environmental Fund (GEF) and with the financial support of German cooperation GTZ-Proyecto FANPE. The fundraising is from investment portfolios, technical and financial assistance, and grants.

According to national legislation in Peru, with many parallels worldwide, the extraction of mineral resources and hydrocarbons is deemed incompatible with protected areas such as national parks, national sanctuaries, and historical sanctuaries, unless rights were acquired prior to the creation of the particular unit. However, current laws and investment policies, based on neoliberal economic goals, give priority to activities considered of "national interest." Legally, a protected area exists only on the surface of the landscape. Subsurface rights may be assigned to other users. When that use is incompatible, there is a legal vacuum as to how this should be resolved. There are also disputes among sectors of society and the national government over the type of environmental impact assessment done concerning these projects. Permits for a mining project, for example, come from the Ministry of Mining and decisions are made according to energy and mining requirements. Thus, national-level problems in biodiversity conservation management often originate with incompatible goals among governmental agencies, an issue of general concern (e.g., Thomas 2003). For example, despite theoretical (Sala et al. 2002) and practical (Friedlander et al. 2003; Russ et al. 2004) reasons for establishing marine reserves, Peru has not been able to establish a network of protected marine reserves because of the continued emphasis on species-based extraction of marine resources, typically by foreign fishing fleets, rather than area- or ecosystem-based approaches (e.g., Pikitch et al. 2004).

Globally, international programs and entities now intervene in affairs once thought to be national prerogatives. An example is in the arena of human rights violations, where an evolving set of legal and political mechanisms can provide leverage for and protection to victimized social groups and individuals (Foweraker, Landman, and Harvey 2003). In the case of biodiversity conservation, international groups of scientists and environmentalists can be crucial for identifying priorities for conservation (e.g., Stattersfield et al. 1998; Myers et al. 2000), whether that be for places of concern or particular species or ecosystem types; thus, to some extent it is the international community that is setting priorities and advocating in favor of the needs of other species. The Biodiversity Convention requires countries of the world to assess their own biodiversity conservation needs. It is crucial that the transnational and international NGOs and political consortia do not overwhelm these national and local efforts to set priorities. One way to do this would be to further explore ways that conservation networks can focus on synergistic rather than antagonistic or competitive planning and implementation.

Needless to say, successful conservation also requires support from the local communities most affected by any restrictions on land use (Brandon and Wells 1992; Western and Wright 1994; Berkes 2004). Much recent revisionism by policy analysts and social scientists has helped clarify some of the difficulties inherent in local and/or community-based approaches (e.g., Ravnborg and del Pilar Guerrero 1999; Schroeder 1999; Gbadegesin and Ayileka 2000; Kellert et al. 2000; Agrawal and Gibson 2001), including the degree of protectionism utilized or desirable (Kull 2002; Wilshusen et al. 2002), the way that local knowledge or tradition is involved in management (Goldman 2003; McDaniel 2003), the effect of recent economic (Dzingirai 2003) or demographic shifts (Gray 2002), and what methods can make these approaches more successful (Young 1993; Begossi et al. 1999; Bray et al. 2003; Trawick 2003; Pretty and Smith 2004). All of these processes and concerns can be illustrated with examples from Peru and appear likely to be important considerations in most developing countries.

At least in the case of Peru, the overlooked level has often been at the national level, where grandiose goals and sophisticated priorizations may be established, but from which few linkages extend that permit successful implementation in local situations. That this might be a more general omission is suggested by the recent update by Brooks, da Fonseca, and Rodrigues (2004) on protected areas that only mentions local and global concerns. In addition, as we have shown in this essay, the places and species found to be important during the era of natural history description in Peru have not been systematically included in more recent efforts, despite available summaries (e.g., Young and Valencia 1992; Rodriguez, Rodriguez, and Vasquez 1995). This can be seen by glancing at the reports and websites created by international environmental NGOs about Peru that do not cite the Peruvian scientific literature but do recycle as new ideas that were discussed within Peru for more than sixty years. Until the 1994 workshop documented by Rodríguez (1996), virtually all biodiversity priority setting in Peru was done behind closed doors, in government offices in Lima and in NGO suites in Washington, DC, or elsewhere.

While transnational influences can be blunt forces for change in conservation, especially when they are designed with little respect for past achievements or traditions, they can also be indispensable for forcing proposals through top-heavy governmental bureaucracies. This is especially the case when those proposals come linked to promises of financing. Peru's national government dedicates a large percentage of its income to the payment of international debt. Any proposal not tied to an outside funding source will likely be tabled; this would be the case of suggestions considered a national priority for conservationists but not viewed as globally significant. We suggest that a debt burden weighs heavily on all the biodiverse tropical countries. As a result, new conservation responsibilities and mandates need to be connected to short- and long-term funding (Bruner, Gullison, and Balmford 2004). In Peru, PROFONANPE will attempt to fill this void, as its endowment becomes a continuing source of within-country funds. Consejo Nacional del Ambiente (CONAM) is a relatively new governmental entity designed to oversee compliance with biodiversity treaties. These efforts might help return elements of sovereignty to national goal setting for biodiversity concerns. If successful, other countries could consider operationalizing these kinds of financial arrangements.

In Peru, as elsewhere, many sustainable development schemes based on extraction of natural products appear to be economic failures. For example, Assies (1997) and Vosti et al. (2003) pointed out difficulties inherent to restricting forest conversion while augmenting household incomes in the Brazilian Amazon. It could be argued for Peru that development projects promoting extraction or agriculture would be best implemented outside of the SINANPE system where biodiversity concerns by law and design take priority. The government of Fujimori opened extraction to foreign investment and most products sold or produced were subjected to competition from other countries. As a result, local projects in Peru (and in other countries subjected to neoliberal experiments) might well require subsidies and constant monitoring of market forces if they are to continue. Peru has a rich legacy of indigenous and other local organizations that have mediated resource use disputes in the past and present (Guillet 1992; Gose 1994; Zimmerer 1996; Bebbington 2001). Relatively little research has examined ways to connect national and regional level oversight and assistance to those local models based on decisions made by consensus. This lapse might be due to an overemphasis on transnational institutions as sources for inspiration and resources. The current protected-area system provides ample flexibility for searching for ways to interface local needs with those at national and international levels. In addition, we suggest that the needs of voluntarily isolated indigenous groups and low-intensity forest extractive systems are best met outside of the SINANPE system, at least as currently constituted in Peru.

In some ways, the 1990s in Peru were most similar to colonial days in terms of the degree of external controls and influences acting upon decision making in reference to living natural resources. This situation contrasts with the priorities and concerns of the 1940s, which rose organically from projects documenting natural history in the previous 100 years. These efforts then led to efforts in the 1950s to 1980s that were inspired by foreign examples, but for which many priorities were nationalistic in origin and stressed natural environments and species not found elsewhere (e.g., lomas, vicuñas, oilbirds; see also Cueto 1989). This history makes us reflect that national priorities should be a level of concern in biodiversity conservation, not just international concerns and local access and decision making. Environmental historians writing about other developing countries could also document how natural resource use intersects chronologically with the rise of protected-area conservation efforts. There are signs that the protected-area system is maturing in Peru and perhaps will become better able to negotiate a balance between external and internal forces. We base this opinion on the existence of increased freedom to try innovative methods. But we also feel that these efforts should be premised on a sensitivity and respect for the legacy of Peruvian research and priority setting.

Peru joins other biodiverse countries that are attempting to achieve conservation goals given both internal and external demands and constraints. Without the increased communication and transportation that characterize globalization, it is likely that many important natural environments would not be recognized and protected. Approaches that simultaneously link national aspirations with conservation projects on the ground and also with global biodiversity agendas are ways to build on the positive aspects of living in a more interconnected world.

Acknowledgments

We are very grateful for the information and comments provided by Blanca León, Jennifer Lipton, and Rodrigo Sierra. We also thank Karl Zimmerer for his organization of the event that resulted in this publication.

References

Agrawal, A., and C. C. Gibson, eds. 2001. *Communities and the environment: Ethnicity, gender, and the state in community-based conservation.* New Brunswick, NJ: Rutgers University Press.

Alverson, W. S., L. O. Rodríguez, and D. K. Moskovits, eds. 2001. *Perú: Biabo Cordillera Azul.* Rapid Biological Inventories Report no. 2. Chicago: The Field Museum.

Anderson, R. L. 1999. *Colonization as exploitation in the Amazon rain forest, 1758–1911.* Gainesville: University of Florida Press.

Assies, W. 1997. The extraction of non-timber forest products as a conservation strategy in Amazonia. *European Review of Latin American and Caribbean Studies* 62:33–53.

Barker, M. L. 1980. National parks, conservation, and agrarian reform in Peru. *Geographical Review* 70:1–18.

Barreda, C. A. 1951. Los recursos naturales del suelo peruano. Segunda parte. *Boletín del Comité Nacional de Protección a la Naturaleza* (Lima) 7:89–148.

Bauer, B. S. 1992. *The development of the Inca state.* Austin: University of Texas Press.

Bebbington, A. 2001. Globalized Andes? Livelihoods, landscapes and development. *Ecumene* 8 (4): 414–36.

Begossi, A., R. A. M. Silvano, B. D. do Amaral, and O. T. Oyakawa. 1999. Uses of fish and game by inhabitants of an extractive reserve (upper Juruá, Acre, Brazil). *Environment, Development and Sustainability* 1:73–93.

Berkes, F. 2004. Rethinking community-based conservation. *Conservation Biology* 18:621–30.

Blanchard, P., ed. 1991. *Markham in Peru: The travels of Clements R. Markham, 1852–1853.* Austin: University of Texas Press.

Bonfiglio, G., ed. 2001. *La Presencia Europea en el Perú.* Lima: Fondo Editorial del Perú.

Bramble, B. J., and G. Porter. 1992. Non-governmental organizations and the making of US international environmental policy. In *The international politics of the environment: Actors, interests, and institutions,* ed. A. Hurrell and B. Kingsbury, 313–53. Oxford: Clarendon Press.

Brandon, K., and M. Wells. 1992. Planning for people and parks: Design dilemmas. *World Development* 20:557–70.

Bray, D. B., L. Merino-Peréz, P. Negreros-Castillo, G. Segura-Warnholtz, J. M. Torres-Rojo, and H. F. M. Vester. 2003. Mexico's community-managed forests as a global model for sustainable landscapes. *Conservation Biology* 17:672–77.

Brooks, T. M., G. A. B. da Fonseca, and A. S. L. Rodrigues. 2004. Protected areas and species. *Conservation Biology* 18:616–18.

Bruner, A. G., R. E. Gullison, and A. Balmford. 2004. Financial costs and shorfalls of managing and expanding protected-area systems in developing countries. *Bioscience* 54:1119–26.

Bryant, R. L. 2002. False prophets? Mutant NGOs and Philippine environmentalism. *Society and Natural Resources* 15:629–39.

Campbell, L. M. 2002. Conservation narratives in Costa Rica: Conflict and co-existence. *Development and Change* 33:29–56.

Cano, A., and K. R. Young, eds. 1998. *Los Pantanos de Villa: Biología y conservación.* Serie de Divulgación no. 11, 1–238. Lima, Peru: Universidad Nacional Mayor de San Marcos, Museo de Historia Natural.

CDC-UNALM. 1991. Plan Director del Sistema Nacional de Unidades de Conservación (SINUC), una aproximación desde la diversidad biológica. Lima, Peru: Centro de Datos para la Conservación, Universidad Nacional Agraria la Molina.

Condamine, C. M. de la. 1993. Viaje a la America meridional por el rio de las Amazonas: Estudio sobre la Quina. 2nd ed. Quito, Ecuador: Abya-Yala. (Orig. pub. 1745.)

Consejo Nacional del Ambiente (CONAM). 2001. Convenio sobre Diversidad Biológica: Informe de su aplicación en el Perú. Lima, Peru: CONAM.

Coombes, B. L. 2003. Ecospatial outcomes of neoliberal planning: Habitat management in Auckland Region, New Zealand. *Environment and Planning B* 30:201–18.

Cueto, M. 1989. Andean biology in Peru: Scientific styles on the periphery. *Isis* 80:640–58.

Davies, K. A. 1984. *Landowners in colonial Peru.* Austin: University of Texas Press.

Dillon, M. O. 1997. Lomas formations, Peru. In *Centres of plant diversity: A guide and strategy for their conservation,* Vol. 3, *The Americas,* ed. S. D. Davis, V. H. Heywood, O. Herrera MacBryde, and A. C. Hamilton, 519–27. London: WWF and IUCN.

Dirección General de Forestal y Fauna. 1991. Plan nacional de acción forestal del Perú. Lima, Peru: Ministerio de Agricultura.

Duffy, R. 2001. Peace parks: The paradox of globalization. *Geopolitics* 6 (2): 1–26.

Dzingirai, V. 2003. The new scramble for the African countryside. *Development and Change* 34:243–63.

Eisenberg, J. F., and K. H. Redford. 1999. *Mammals of the neotropics: The central neotropics,* Vol. 3, *Ecuador, Peru, Bolivia, Brazil.* Chicago: University of Chicago Press.

Evans, S. 1999. *The green republic: A conservation history of Costa Rica.* Austin: University of Texas Press.

Ferreyra, R. 1961. Las lomas costaneras del extremo sur del Perú. *Boletín de la Sociedad Argentina de Botánica* 9:85–120.

———. 1977. Endangered species and plant communities in Andean and coastal Peru. In *Extinction is forever: The status of threatened and endangered plants of the Americas,* ed. G. T. Prance and T. S. Elias, 150–57. Bronx, New York: New York Botanical Garden.

———. 1985. Los tipos de vegetación de la costa peruana. *Anales del Jardín Botánico de Madrid* 40:241–56.

Fisher, W. F. 1997. Doing good? The politics and antipolitics of NGO practices. *Annual Review of Anthropology* 26:439–64.

Foweraker, J., T. Landman, and N. Harvey. 2003. *Governing Latin America.* Cambridge, UK: Polity.

Frank, D. J. 1997. Science, nature, and the globalization of the environment, 1870–1990. *Social Forces* 76:409–36.

Friedlander, A., J. S. Nowlis, J. A. Sanchez, R. Appeldoorn, P. Usseglio, C. McCormick, S. Bejarano, and A. Mitchell-Chui. 2003. Designing effective marine protected areas in Seaflower Biosphere Reserve, Colombia, based on biological and sociological information. *Conservation Biology* 17:1769–84.

Gbadegesin, A., and O. Ayileka. 2000. Avoiding the mistakes of the past: Towards a community oriented management strategy for the proposed national park in Abuja-Nigeria. *Land Use Policy* 17:89–100.

Goldman, M. 2003. Partitioned nature, privileged knowledge: Community-based conservation in Tanzania. *Development and Change* 34:833–62.

Gose, P. 1994. *Deathly waters and hungry mountains: Agrarian ritual and class formation in an Andean town.* Toronto: University of Toronto Press.

Graham, F., Jr. 1990. *The Audubon ark: A history of the National Audubon Society.* Austin: University of Texas Press.

Gray, L. C. 2002. Environmental policy, land rights, and conflict: Rethinking community natural resource management programs in Burkina Faso. *Environment and Planning* 20:167–82.

Guillet, D. W. 1992. *Covering ground: Communal water management and the state in the Peruvian highlands.* Ann Arbor: University of Michigan Press.

Gwynne, R. N., and C. Kay, eds. 2004. *Latin America transformed: Globalization and modernity.* 2nd ed. New York: Oxford University Press.

Hancock, G. 1989. *Lords of poverty: The power, prestige, and corruption of the international aid business.* New York: Atlantic Monthly Press.

Honigsbaum, M. 2001. *The fever trail: The hunt for the cure for Malaria.* London: Macmillan.

Humboldt, A. von. 1821. *An illustration of the genus cinchona: Comprising descriptions of all the officinal Peruvian barks, including several new species.* London: John Searle.

IADB (Inter-American Development Bank). 1997. *Latin America after a decade of reforms: Economic and social progress; 1997 report.* Washington, DC: Johns Hopkins University Press.

Keese, J. R. 1998. International NGOs and land use change in a southern highland region of Ecuador. *Human Ecology* 26:451–68.

Kellert, S. R., J. N. Mehta, S. A. Ebbin, and L. L. Lichtenfeld. 2000. Community natural resource management: Promise, rhetoric, and reality. *Society and Natural Resources* 13:705–15.

Kinder, G., ed. 2000. *Exploration of the valley of the Amazon, 1851–1852.* By William Lewis Herndon. New York: Grove.

Kingsland, S. E. 2002. Creating a science of nature reserve design: Perspectives from history. *Environmental Modeling and Assessment* 7:61–69.

Koepcke, H.-W. 1961. Synökologische Studien an der Westseite der peruanischen Andean. *Bonner Zoologischer Beitraege* 29:1–320.

Koepcke, H.-W., and M. Koepcke. 1952. División ecológica de la costa peruana. *Pesca y Caza* (Ministerio de Agricultura, Lima) 3:3–23.

Koepcke, H.-W., and M. Koepcke. 1958. Los restos de bosques en las vertientes occidentales de los Andes peruanos. *Boletín del Comité Nacional de Protección a la Naturaleza* (Lima) 16:22–30.

Koepcke, M. 1961. Corte ecológico transversal en los Andes del Perú central con especial consideración de las aves. Parte I: Costa, Vertientes occidentales y Región Altoandina. *Memorias del Museo de Historia Natural "Javier Prado"* 3:1–119.

Kull, C. A. 2002. Empowering pyromaniacs in Madagascar: Ideology and legitimacy in community-based natural resource management. *Development and Change* 33:57–78.

Lamas, G. 1980a. Introducción a la historia de la entomología en el Perú: I. Inicios y periodo exploratorio pre-Darwiniano. *Revista peruana de Entomología* 23 (1): 17–25.

———. 1980b. Introducción a la historia de la entomología en el Perú: II. Periodo de los viajeros, colectores y estudiosos especializados. *Revista peruana de Entomología* 23 (1): 25–31.

Laurie, N., and A. Bonnett. 2002. Adjusting to equity: the contradictions of neoliberalism and the search for racial equity in Peru. *Antipode* 34 (1): 28–53.

Leo, M. 1980. First field study of yellow-tailed woolly monkey. *Oryx* 15:386–89.

———. 1982. Conservation of the yellow-tailed woolly monkey (*Lagothrix flavicauda*). *International Zoological Yearbook* 22:47–52.

———. 1984. The effect of hunting, selective logging and clear-cutting on the conservation of the yellow-tailed woolly monkey (*Lagothrix flavicauda*). MA thesis, University of Florida, Gainesville.

———. 1995. The importance of tropical montane cloud forest for preserving vertebrate endemism in Peru: The Río Abiseo National Park as a case study. In *Tropical montane cloud forests*, ed. L. S. Hamilton, J. O. Juvik, and F. N. Scatena, 198–211. New York: Springer-Verlag.

León, B. 1995. Actualización de los nombres de pteridófitos en la obra de Weberbauer (1945). *Candollea* 50:173–93.

———. 2002. Significance of August Weberbauer's plant collecting for today's Río Abiseo National Park, northern Peru. *Taxon* 51:167–70.

León, B., A. Cano, and K. R. Young. 1995. La flora vascular de los Pantanos de Villa, Lima, Peru: Adiciones y guía a las especies comunes. *Publicaciones del Museo de Historia Natural UNMSM* (B) 38:1–39.

León, B., K. R. Young, and A. Cano. 1996. Observaciones sobre la flora vascular de la costa central del Perú. *Arnaldoa* 4:67–85.

León, B., K. R. Young, A. Cano, M. I. La Torre, M. Arakaki, and J. Roque. 1997. Botanical exploration and conservation in Peru: The plants of Cerro Blanco, Nazca. *BioLlania* (Venezuela) 6 (edición especial): 431–48.

Lockhart, J. 1968. *Spanish Peru, 1532–1560: A colonial society.* Madison: University of Wisconsin Press.

Lumbreras, L. G. 1974. *The peoples and cultures of ancient Peru.* Washington, DC: Smithsonian Institution Press.

Macedo, H. de, and R. Mittermeier. 1979. Redescubrimiento del primate peruano *Lagothrix flavicauda* (Humboldt 1812) y primeras observaciones sobre su biología. *Revista Ciencias* (Lima) 71:78–92.

Majluf, P. 1996. Protección de las puntas e islas guaneras en la costa peruana: Primer paso hacia un sistema de reservas marino-costeras. In *Diversidad biológica del Perú: Zonas prioritarias para su conservación*, ed. L. O. Rodríguez, 105–7. Lima, Peru: Proyecto FANPE, GTZ-INRENA.

Maren, M. 1997. *The road to hell: The ravaging effects of foreign aid and international charity.* New York: Free Press.

Margules, C. R., and R. L. Pressey. 2000. Systematic conservation planning. *Nature* 405:243–53.

Marshall, E., and A. C. Newton. 2003. Non-timber forest products in the community of El Terrero, Sierra de Manantlán Biosphere Reserve, Mexico: Is their use sustainable? *Economic Botany* 57: 262–78.

McDaniel, J. M. 2003. History and the duality of power in community-based forestry in southeast Bolivia. *Development and Change* 34:339–56.

Mendel, L. C., and J. B. Kirkpatrick. 2002. Historical progress of biodiversity conservation in the protected-area system of Tasmania, Australia. *Conservation Biology* 16:1520–29.

Myers, N., R. A. Mittermeier, C. G. Mittermeier, G. A. B. Da Fonseca, and J. Kent. 2000. Biodiversity hotspots for conservation priorities. *Nature* 403:853–58.

Neumann, R. P. 1995. Local challenges to global agendas: Conservation, economic liberalization, and the pastoralists' rights movement in Tanzania. *Antipode* 27:363–82.

Newing, H., and R. Bodmer. 2003. Collaborative wildlife management and adaptation to change: The Tamshiyacu Tahuayo Communal Reserve, Peru. *Nomadic People*, n.s., 7:110–22.

Núñez, E., and G. Petersen, eds. 2002. *Alexander von Humboldt en el Perú.* Lima: Banco Central de Reserva del Peru.

Oficina Nacional de Evaluación de Recursos Naturales (ONERN). 1986. *Perfil Ambiental.* Lima, Peru: ONERN.

Orlove, B. 2002. *Lines in the water: Nature and culture at Lake Titicaca.* Berkeley: University of California Press.

Pendras, M. 2002. From local consciousness to global change: Asserting power at the local scale. *International Journal of Urban and Regional Research* 26:823–33.

Pikitch, E. K., C. Santora, E. A. Babcock, A. Bakun, R. Bonfil, D. O. Conover, P. Dayton, P. Doukakis, D. Fluharty, B. Heneman, E. D. Houde, J. Link, P. A. Livingston, M. Mangel, M. K. McAllister, J. Pope, and K. J. Sainsbury. 2004. Ecosystem-based fishery management. *Science* 305:346–47.

Place, S. E. 1988. The impact of national park development on Tortuguero, Costa Rica. *Journal of Cultural Geography* 9:37–52.

Polk, M. H. 2003. An assessment of conservation approaches for the natural protected areas in arid southern Peru. MA thesis, University of Texas, Austin.

Ponce, C. F. 2001. Referencia de algunas nuevas iniciativas de reservas de biosfera en America latina, con especial énfasis en reservas transfronterizas (transboundary biosphere reserves). In *El Manu y otras experiencias de investigación y manejo de bosques neotropicales,* ed. L. O. Rodríguez, 251–64. Cusco, Peru: Pro-Manu, Convenio Perú-Unión Europea.

Poole, D., and G. Rénique. 1992. *Peru: Time of fear.* London: Latin America Bureau.

Pretty, J., and D. Smith. 2004. Social capital in biodiversity conservation and management. *Conservation Biology* 18:631–38.

Price, M. 1994. Ecopolitics and environmental nongovernmental organizations in Latin America. *Geographical Review* 84:42–58.

Raimondi, A. 1940. *El Perú: Tomo I, parte preliminar.* Lima, Peru: Escuela Tipográfica Salesiana.

Ravnborg, H. M., and M. del Pilar Guerrero. 1999. Collective action in watershed management: Experience from the Andean hillsides. *Agriculture and Human Values* 16:257–66.

Redford, K. H., P. Coppolillo, E. W. Sanderson, G. A. B. da Fonseca, E. Dinerstein, C. Groves, G. Mace, et al. 2003. Mapping the conservation landscape. *Conservation Biology* 17:116–31.

Rodrigues, A. S. L., and K. J. Gaston. 2002. Rarity and conservation planning across geopolitical units. *Conservation Biology* 16:674–82.

Rodriguez, A. F., A. M. Rodriguez, and R. P. Vasquez. 1995. *Realidad y perspectivas: La Reserva Nacional Pacaya Samiria.* Lima, Peru: Pro Naturaleza.

Rodríguez, L. O., ed. 1996. *Diversidad biológica del Perú: Zonas prioritarias para su conservación.* Lima, Peru: Proyecto FANPE, GTZ-INRENA.

Rodríguez, L. O., and K. R. Young. 2000. Biological diversity of Peru: Determining priority areas for conservation. *Ambio* 29:329–37.

Roemer, M. 1970. *Fishing for growth: Export-led development in Peru, 1950–1967.* Cambridge, MA: Harvard University Press.

Rostworowski de Diez Canseco, M. 1981. *Recursos naturales renovables y pesca: Siglos XVI y XVII.* Lima, Peru: Instituto de Estudios Peruanos.

Ruiz, H. 1940. Travels of Ruiz, Pavón, and Dombey in Peru and Chile (1777–1788). *Field Museum of Natural History Botanical Series* 21:1–372.

Ruiz Alarco, F., and A. A. Giesecke. 1953. Informe: Sobre gestiones que el Comité de Protección a la Naturaleza la Lecho ante el Supremo Gobierno, con el fin de proteger ciertas zonas geográficas, flora y fauna del pais, declarándolos Parques Nacionales. *Boletín del Comité Nacional de Protección a la Natrualeza* (Lima) 9:23–25.

Rundel, P. W., M. O. Dillon, B. Palma, H. A. Mooney, and S. L. Gulman. 1991. Phytogeography and ecology of the coastal Atacama and Peruvian deserts. *Aliso* 13:1–49.

Russ, G. R., A. C. Alcala, A. P. Maypa, H. P. Calumpong, and A. T. White. 2004. Marine reserve benefits local fisheries. *Ecological Applications* 14:597–606.

Sala, E., O. Aburto-Oropeza, G. Paredes, I. Parra, J. C. Barrera, and P. K. Dayton. 2002. A general model for designing networks of marine reserves. *Science* 298:1991–93.

Salafsky, N., H. Cauley, G. Balachander, B. Cordes, J. Parks, C. Margoluis, S. Bhatt, C. Encarnacion,

D. Russell, and R. Marcoluis. 2001. A systematic test of an enterprise strategy for community-based biodiversity conservation. *Conservation Biology* 15:1585–95.

Santillana Cantella, T. G. 1989. *Los viajes de Raimondi.* Lima, Peru: Occidental Petroleum Corporation of Peru.

Schroeder, R. A. 1999. Geographies of environmental intervention in Africa. *Progress in Human Geography* 23:359–78.

Schulenberg, T. S., and K. Awbrey, eds. 1997. The Cordillera del Cóndor region of Ecuador and Peru: A biological assessment. Conservation International Rapid Assessment Program Working Papers no. 7. Washington, DC: Conservation International.

Schweigger, E. 1947. *El litoral Peruano.* Lima, Peru.

Scott, J. C. 1998. *Seeing like a state: How certain schemes to improve the human condition have failed.* New Haven, CT: Yale University Press.

Scott, J. M., F. W. Davis, R. G. McGhie, R. G. Wright, C. Groves, and J. Estes. 2001. Nature reserves: Do they capture the full range of America's biological diversity? *Ecological Applications* 11:999–1007.

Scott, J. M., P. J. Heglund, M. L. Morrison, J. B. Haufler, M. G. Raphael, W. A. Wall, and F. B. Samson, eds. 2002. *Predicting species occurrences: Issues of accuracy and scale.* Washington, DC: Island.

Sierra, R., F. Campos, and J. Chamberlin. 2002. Assessing biodiversity conservation priorities: Ecosystem risk and representativeness in continental Ecuador. *Landscape and Urban Planning* 59:95–110.

Speth, J. G., ed. 2003. *Worlds apart: Globalization and the environment.* Washington, DC: Island.

Stattersfield, A. J., M. J. Crosby, A. J. Long, and D. C. Wege. 1998. Endemic bird areas of the world: Priorities for biodiversity conservation. BirdLife International Conservation Series no. 7. Cambridge, UK: BirdLife International.

Stern, S. J. 1993. *Peru's Indian peoples and the challenge of Spanish conquest.* 2nd ed. Madison: University of Wisconsin Press.

———, ed. 1998. *Shining and other paths: War and society in Peru, 1980–1995.* Durham, NC: Duke University Press.

Stevens, S., ed. 1997. *Conservation through survival: Indigenous peoples and protected areas.* Washington, DC: Island.

Terborgh, J., C. Van Schaik, L. Davenport, and M. Rao, eds. 2002. *Making parks work: Strategies for preserving tropical nature.* Washington, DC: Island.

Thomas, C. W. 2003. *Bureaucratic landscapes: Interagency cooperation and the preservation of biodiversity.* Cambridge, MA: MIT Press.

Trawick, P. 2003. Against the privatization of water: An indigenous model for improving existing laws and successfully governing the commons. *World Development* 31:977–96.

Vanden, H. E. 2003. Globalization in a time of neoliberalism: Politicized social movements and the Latin American response. *Journal of Developing Societies* 19 (2–3): 308–33.

Vosti, S. A., E. Muñoz Braz, C. L. Carpentier, M. V. N. D'Oliveira, and J. Witcover. 2003. Rights to forest products, deforestation and smallholder income: Evidence from the western Brazilian Amazon. *World Development* 31:1889–1901.

Weberbauer, A. 1945. *El mundo vegetal de los Andes Peruanos.* Lima, Peru: Ministerio de Agricultura.

———. 1951. La protección de la vegetación y la flora del Perú. *Boletín del Comité Nacional de Protección a la Naturaleza* (Lima) 7:51–54.

Western, D., and R. M. Wright, eds. 1994. *Natural connections: Perspectives in community-based conservation.* Washington, DC: Island.

Wilshusen, P. R., S. R. Brechin, C. L. Fortwangler, and P. C. West. 2002. Reinventing a square wheel: Critique of a resurgent "protection paradigm" in international biodiversity conservation. *Society and Natural Resources* 15:17–40.

Young, K. R. 1993. National park protection in relation to the ecological zonation of a neighboring human community: An example from northern Peru. *Mountain Research and Development* 13:267–80.

———. 1997. Wildlife conservation in the cultural landscapes of the central Andes. *Landscape and Urban Planning* 38:137–47.

———. 2004. Environmental and social consequences of coca/cocaine in Peru: Policy alternatives and a research agenda. In *Dangerous harvest: Drug plants and the transformation of indigenous landscapes,* ed. M. K. Steinberg, J. J. Hobbs, and K. Mathewson, 249–73. New York: Oxford University Press.

Young, K. R., C. Ulloa Ulloa, J. L. Luteyn, and S. Knapp. 2002. Plant evolution and endemism in Andean South America: An introduction. *Botanical Review* 68:4–21.

Young, K. R., and N. Valencia, eds. 1992. Biogeografía, ecología y conservación del Bosque Montano en el Perú. *Memorias del Museo de Historia Natural, Universidad Nacional Mayor de San Marcos* 21:1–227.

Young, L. M. 1998. Características e actitudes de los vecinos de los Pantanos de Villa. In *Los Pantanos de Villa: Biología y conservación,* ed. A. Cano and K. R. Young, 117–30. Serie de Divulgación no. 11. Lima, Peru: Universidad Nacional Mayor de San Marcos, Museo de Historia Natural.

Zamora Jimeno, C. 1996. Las regiones ecológicas del Perú. In *Diversidad biológica del Perú,* ed. L. O. Rodríguez, 137–41. Lima, Peru: Proyecto FANPE, GTZ-INRENA.

Zerner, C., ed. 2000. *People, plants, and justice: The politics of nature conservation.* New York: Columbia University Press.

Zimmerer, K. S. 1996. *Changing fortunes: Biodiversity and peasant livelihood in the Peruvian Andes.* Berkeley: University of California Press.

PART IV

Decentralization and Environmental Governance in Globalization

The powerful trend toward decentralization—broadly defined as the emphasis on local-scale initiatives that involve governance as well as the more systematic spatial dispersal and political restructuring of governance to local levels such as villages and communities—offers one of the most sharply etched of the Janus-faced prospects of environmental globalization. On one side of this dual image, the trend contains a number of laudable goals and potential merits. Decentralization for environmental management offers the powerful promise of a greater level of political representation, as plans call for the increasing participation of and control by local people in governance of resources and conservation territories. The planned empowerment of local people in these circles is resonant with the progressive goals of various political agendas. Indeed, local empowerment echoes a variety of the deeply seated tenets of populism, regionalism, and localism, and the efforts toward the autonomy or semiautonomy of indigenous or ethnic groups. In developing countries in particular the promise of decentralization is especially elevated in comparison to undemocratic forms of government, including autocratic or authoritarian political regimes. Yet, the other side of this dual image of decentralization offers many disadvantages. Perhaps most notably, the majority of impressions generated through fine-grained social and political analysis suggest that many "on-the-ground" attempts at decentralization are deficient in meeting both the social and environmental goals that they purport to serve. Since many decentralization efforts are now ten to fifteen years old, it has recently become possible to unpack this trend into several defining dimensions. The main dimensions that are addressed in this section are the relationship between decentralization of environmental governance and national political considerations (chap. 11); local conflicts over management plans, particularly zoning-based initiatives (chap. 12); and the socioenvironmental effects of decentralized land tenure and resource management (chap. 13).

National political considerations may seem highly removed from the concerns of biodiversity conservation, even in the context of globalization. The realm of biodiversity tends to evoke the image of unique flora and fauna that are particular to certain local places, which decentralization-based conservation initiatives would be thought to reinforce as predominantly local issues. Yet, as chapter 11 demonstrates, global biodiversity conservation may often adopt an emphasis on local projects and initiatives and attempt to ignore the legacies of national politics, but end up reinforcing the class and ethnic inequalities of these state-level political structures. The country of Guatemala and the case of the Maya Biosphere Reserve (MBR), one of the cornerstones of the ambitious Mesoamerican Biological Corridor (MBC), put into high relief the tension between the promise of local, decentralized initiatives for biodiversity conservation and the powerful interplay with national politics. Clearly, the pluralities of political discourse that have been created as a result of global biodiversity conservation in the buffer zone communities of the MBR are an improvement over the authoritarian past, when open discussions and political dissent were plainly impossible. Still, chapter 11 shows that the on-the-ground politics of local projects are fraught with power inequalities and influential misrepresentations that continually undermine the potential for real local democratization and environmental governance. These difficulties are not merely implementation problems; rather, they point to how conservation and sustainability policies, even at the local level, must become an active part of national political debates that may well entail such issues as race and ethnicity, which are seemingly tangential yet ultimately central to the chance for success.

Decentralization in environmental governance and democracy in practice are broadly similar in seeming to be messy by nature. This similarity does not suggest, however, that democratization is necessarily evident in the social and political conflicts that are associated with decentralization initiatives. Chapter 12 examines the genesis of social and political conflicts in the case of decentralization-based plans for land use sustainability in the country of Burkina Faso. As this chapter shows, the scope and frequency of conflicts have multiplied as a result of the countrywide attempts at the devolution of governance to village-level land use planning, a project that has been propelled by mandates from global lenders, namely the World Bank and the International Monetary Fund. Evaluating these conflicts requires understanding the relation of the social and political processes of environmental decentralization to the new spatial configurations, in the form of both territories and networks, that are also integral to the recent changes. The zonation of land use is one of the spatial backbones of sustainability efforts and governance plans in decentralization policies at the local level in Burkina Faso and elsewhere in developing countries. Likewise, zonation-based spatial

designs are highly common in the proliferating plans for the local governance of explicitly conservation-related activities. Zonation certainly brings
to mind the image of creating or reinforcing the stability, fine-scale sensibil-
ities, and social organization of local environmental management. Yet, as
chapter 12 points out, the on-the-ground zonation efforts in Burkina Faso
frequently exacerbate local social and political inequality, producing violent
tensions that cast doubt on the appropriateness of decentralization programs that are typical forms of environmental globalization.

The relation of property regimes to environmental management is one of
the key tensions in the global process of decentralization. The worldwide establishment of private property regimes is a major outcome of the neoliberal
agenda that has dominated economic globalization since the 1980s and
which has been referred to on occasion as the "privatization of everything."
This form of globalization, the uppercase Globalization that is referred to in
chapter 1, is closely tied to many facets of environmental globalization and
the expansion of land and resource markets. Privatization of property is represented as a form of decentralization in which resource control is devolved
to the most local level possible, to the individual or the household. Furthermore, the transfer to private parties of land and other resources formerly
held by states, collectives, or communities is frequently equated with sustainable management of the environment through policy, rhetoric, or implicit
reasoning. In this school of thought, which intertwines environmentalist and
neoliberal economic logics, it is assumed that combating environmental
degradation, such as deforestation, biodiversity loss, and desertification, requires establishing rights to private property. This widespread assumption,
global in scope, seems to eclipse, almost without comment, the many noteworthy accounts of sound, existing forms of environmental management
that are dependent on common-property forms of ownership. Chapter 13 examines the case of land and resource privatization, carried out through the
programs for decollectivization, that occurred from the 1980s onward in Inner Mongolia in northern China as part of the national government's effort
to combat desertification. This Chinese example of the establishment of private property can be seen as precursor of the decollectivization and privatization of land and resources that has occurred worldwide since the 1980s. As
chapter 13 demonstrates, the somewhat greater lapse of time since decollectivization in China enables a cross-temporal analysis of the type of property
control—namely, privatization—and environmental management (the use of
water and vegetation resources in this case). It shows that privatization may
improve environmental quality at the micro level of individually held properties, but the sustainability of resources over larger areas is reduced and social inequalities are exacerbated.

11 Conservation, Globalization, and Democratization: Exploring the Contradictions in the Maya Biosphere Reserve, Guatemala

JUANITA SUNDBERG

The global expansion of conservation policies and protected areas is an important political process currently transforming cultural landscapes in Latin America. This process brings new agendas and new actors to select places, while boundaries are redrawn to create new territories to protect biodiversity. As Karl Zimmerer notes in chapter 1, the globalization of conservation intersects with other global processes, such as the expansion of neoliberalism, a model premised upon a shrinking of the state and devolution of power to the private sector. In the context of conservation, neoliberalism implies the devolution of funding, policy-making, and management to international environmental nongovernmental organizations (NGOs). In Latin American countries, the trend empowering international NGOs to make decisions with profoundly local effects is occurring alongside processes of democratization that entail expanding rights and responsibilities for citizens.

How do conservation, globalization, and democratization articulate to shape the daily practices of conservation in Latin America? How do individuals and collectives—differently situated socially, politically, and geographically—conceptualize and negotiate these articulations? To explore the complexity of environmental globalization, this chapter presents a case study analysis of how these three processes articulate in one site: the Maya Biosphere Reserve, a large protected area in northern Guatemala. In particular, I draw from my ethnographic research on the daily practices of conservation, what Donald Moore (1998) calls the "micro-politics" of conservation. This approach allows me to move beyond the global or national scale to analyze the implications of conservation, globalization, and democratization in the daily lives of social groups who are most directly affected by shifting governance models. As I illustrate, the effects of globalization are contradictory, as it fortifies, disrupts, and reconfigures existing power relations.

I begin this chapter with a brief analysis of globalization, conservation, and democratic reform as they pertain to Guatemala. Although globalization and democratization created space for the emergence of a transnational alliance of national and international environmental activists with enough political prestige to transform conservation policy, the initial relationship between conservation and democratization was uneasy. What this meant for the implementation of the Maya Biosphere Reserve is the subject of the next section. In the final section of the chapter, I draw upon my ethnographic research with two very different groups of people: a community of agricultural migrants and a group of indigenous women. As I illustrate, the daily practices of conservation projects in the context of globalization and democratization have contradictory implications in the lives of these two local groups. I conclude this chapter with some thoughts about how this case contributes to broader discussions about globalization and conservation.

This chapter draws upon qualitative research undertaken between 1996 and 1997 on the politics of conservation in Guatemala and the Maya Biosphere Reserve. I conducted additional fieldwork in August 2000, focusing specifically on the relationship between environmental protection and processes of democratization.[1] Unless otherwise noted, all quotations are taken from my taped interviews and field notes, which I have translated from the Spanish. With the exception of some public officials, all names have been omitted or changed to protect the identity of the men and women working both to protect Guatemala's biophysical landscapes and to create a more just society.

Writing as a feminist political ecologist, I wish to note that the analysis presented here is necessarily selective and partial (after Haraway 1991). I did not speak to everyone, nor am I speaking for anyone in particular. Rather, I have constructed this narrative to make a particular argument about globalization, conservation, and democratization in Guatemala in the hopes that future generations of conservationists will consider the ways in which environmental protection intersects with other political processes at multiple social and geographical scales.

Conservation, Globalization, and Democratization in Guatemala

As Zimmerer points out in chapter 1, the late twentieth century has witnessed a dramatic increase in conservation territories. In Latin America, the expansion of protected areas has been particularly impressive. Zimmerer and Carter (2002, 208) note that, as of 1997, "approximately 45% of the Caribbean, 16% of Central America, and 10% of South America were under IUCN [World

1. I am very grateful to the University of British Columbia for funding my research in August 2000 with a Humanities and Social Science grant. My research in 1996–97 was funded by a Fulbright scholarship.

FIGURE 11.1. The Maya Biosphere Reserve accounts for one-third of Guatemala's territory (cartography by Paul Jance)

Conservation Union/International Union for the Conservation of Nature] protected area status" (see also Sierra, chap. 9, and Young and Rodríguez, chap. 10 in this volume). Guatemala is no exception to this trend. Indeed, legislation in 1989 established the country's system of protected areas as well as an administrative agency, the National Council of Protected Areas (Consejo Nacional de Áreas Protegidas, or CONAP). This legislation created the Maya Biosphere Reserve in the northern department of Petén to protect 1.6 million hectares of lowland tropical forest, rich with diverse species of flora and fauna (CONAP 1996; fig. 11.1). The Maya Biosphere Reserve alone accounts for 30 percent of the national territory.

In Guatemala, as in other parts of Latin America, driving the increasing numbers of protected areas is the global expansion of conservation-oriented

NGOs. Individual activists from international organizations were key to the creation of new protected-area legislation in Guatemala, and were instrumental in choosing the biosphere reserve as the most appropriate conservation model for the Petén (Berger 1997; Sundberg 1999). Moreover, although CONAP had the legal authority to implement the Maya Biosphere Reserve, the Guatemalan government signed an agreement in 1990 with the United States Agency for International Development (USAID) to participate in the management of the reserve (USAID 1989). The USAID contracted three North American NGOs to carry out the project: The Nature Conservancy, CARE International, and Conservation International. As noted, this devolution of responsibility to NGOs is tied to the restructuring of the economy along neoliberal models of state-society relations (Bulmer-Thomas 1996). In the case of Guatemala, the contraction of the state gave international actors a disproportionate say over the direction of conservation priorities and agendas (see Christen et al. 1998 for similar examples in other Latin American countries).

The globalization of conservation intersects with democratic reform in Guatemala, which was initiated in 1985 when the military permitted national elections to appoint the first civilian regime since 1966, thus ending two decades of military rule. With democratization came the resurgence of civil society, including environmental activism. Indeed, Mumme and Korzetz (1997, 46) argue, "liberalization and democratization create a host of new opportunities for environmental mobilization and policy development in the region by altering the formal, political, policy, and societal conditions for the insertion of environmental values into the agendas of Latin American governments." The director of a national environmental NGO in Guatemala echoes this claim: "Democracy in this country means rescuing from the hands of the state the authority and participation that it has usurped in the last thirty years." Out of the political opening created by the return to civilian rule in the 1980s emerged an environmental movement made up of individuals from the professional and elite classes (Berger 1997). These environmental activists played critical roles in creating conservation policies and institutions as well as implementing the newly established protected areas (Sundberg 1999).

In sum, the articulation between conservation, globalization, and democratization has infused the political and environmental scene in Latin American countries with new actors and governance models, thereby disrupting established political systems and methods of operation and allowing new ideas to blossom. Moreover, the presence of international environmental NGOs provides national groups with access to transnational networks, which offer new sources of knowledge as well as political and financial support to groups whose goals are consistent with their own (Keck and Sikkink 1998). For instance, environmental activists and researchers may play a part in reconfiguring power relations, particularly if and when they place value upon indigenous knowledge and practices and constitute otherwise marginalized groups—and the women within those groups—as important social actors because of their unique knowledge base.[2] In Guatemala, the intersection of globalization and democratization shifts power to NGOs and civil society and creates a political arena characterized by a plurality of social, political, and environmental models, thereby undermining the authority of the state to determine policy.

And yet, even as globalization and democratic reform create new spaces for political mobilization, there are no guarantees that environmental regulations will be implemented through inclusive democratic procedures (Campbell 2000; Neumann 1998; Peluso 1993). Empirical evidence demonstrates that there is no natural congruence between environmentalism and democracy (Desai 1998; Lafferty and Meadowcroft 1996; Midlarsky 1998; Walker 1999). Goodin (1992) explains this tension in theoretical terms by pointing out that environmental regulation is aimed at achieving substantive outcomes, while democracy is about decision making and procedure; conservation can be accomplished without democracy, while democracy may not produce environmentalist policies. In the context of global conservation, the question then emerges: to what extent do newly empowered international environmental NGOs, whose sole purpose is the protection of biodiversity and wildlife habitat, privilege outcomes over procedures?

Moreover, as David Carruthers (2001) and Laura Pulido (1994) argue, the establishment of formal democratic institutions and procedures to accomplish environmental goals does not, in and of itself, bring about a democratization of daily practices. For instance, formal democratic reforms in Guatemala have not transformed the underlying exclusionary social system that has restricted civil rights, political participation and representation, decision making in formal arenas, and therefore citizenship to an elite minority (Sieder 1999). Biological and cultural forms of racism have meant that indigenous people and *ladino*[3] (mixed-race) *campesinos* (peasants) have been considered unfit to take on the rights and responsibilities of citizenship (Cojtí Cuxil 1997). Limits on women's legal rights have been conceptualized in terms of naturalized accounts of biological differences between men and women, particularly in terms of mental capacity and roles within the family (Dore 2000; Thillet de Solórzano 2001). Although feminist and indigenous movements

2. The significance of transnational networks for indigenous movements—not in place of but in addition to preexisting regional or national networks—is increasingly recognized (Van Cott 1994, 2000; Yashar 1998).

3. *Ladino* is the term used in Guatemala to refer to a person of mixed European and indigenous descent; it can also refer to an indigenous person who no longer identifies him- or herself as such.

have made great strides in eliminating structural forms of discrimination in the twentieth century, women and indigenous people are still largely absent from the formal arenas of decision making (Blacklock and Macdonald 2000; Warren 1998; Cojtí Cuxil 1995). To what extent do these histories of exclusion inform the daily practices of conservation?

Democratization and Conservation in the Maya Biosphere Reserve

Implementation of the Maya Biosphere Reserve began in 1990, just four years after civilian elections and democratic reforms in Guatemala. However, the autocratic forms of governance that dominated Guatemala's state bureaucracies initially shaped the new civilian regimes. Indeed, CONAP, the new park administration, held an authoritarian outlook on its mandate to implement the legal statutes creating the reserve. As CONAP's first executive secretary stated, "Our first job in Petén will be to make our presence felt. Once we have laid down precedents for a strict protection of the nuclear zones, then we can negotiate with interested parties for a limited exploitation of the secondary areas" (quoted in Perera 1989). In the early 1990s, a former CONAP leader noted, "the approach was to impose and enforce the law." Those implementing the reserve, he said, "were driven by a TNC [The Nature Conservancy] vision: this is a park and we are going to enforce the law." As an environmental activist and leader of a Central American organization for sustainable development acknowledged, "the Reserve was not implemented in a participatory manner." However, he added, it was urgent to establish the reserve, and the administration would have lost too much time trying to do it in a participatory way. Plus, he said, "people would not have agreed to it anyway." Whether or not they agreed in principle, the USAID and international NGOs were complicit with CONAP's authoritarian methods in the early 1990s, because those practices accomplished desired institutional goals and moral commitments to the protection of nature.

As a consequence of such attitudes and practices, neither municipal and departmental leaders nor local residents were consulted prior to the creation of the reserve. Many individuals found out that they resided within a protected area after being instructed to halt traditional practices in certain areas (e.g., collecting forest products, felling timber for housing, farming). People residing within the reserve demonstrated their disapproval of such exclusionary practices and the potential loss of livelihood by burning down several CONAP checkpoints in 1991 and 1993, and a biological station in 1997 (Cabrera 1991). Also in 1997, a group of officials were kidnapped on their way to meet with and negotiate the removal of a peasant community from the reserve. As a result of one such incident, CONAP abandoned its post in a Q'eqchi' agricultural settlement, although a military encampment eventually replaced it. Indeed, for a time, the military was viewed as a valuable resource in the fight to protect the forest. In the words of one project director, "the military can help give support" because it is "widely respected."

Explaining Coercive Measures

My interviews with governmental and nongovernmental personnel illustrate that they explain such authoritarian and exclusionary measures in a number of ways, ranging from the narrow vision of NGOs, the poor quality of democracy in Guatemala, to the perceived cultural traits of the reserve's inhabitants. In this chapter, I present only a few narratives outlining the primary reason why institutions are unable to accomplish their goals through democratic means.

In answer to a question about why institutions find it difficult to foster participation, the Guatemalan director of a United States–based NGO stated:

> The worst obstacle of all are [sic] the people themselves, the communities don't want to do it. After thirty years of nondeclared war, people's psychology is affected; they make accusations like, are you a communist? Are you a *guerrillero* [guerrilla]? This defines the socioeconomic profile of Guatemala today. We are a country scarred by a nondeclared war. People have come to manipulate the Peace Accords; they think, "now, the government can't do anything to me, the military can't do anything to me."

A CONAP official articulates similar perspectives; he suggested that an important social barrier to participation is education: "Because people don't know how to read and write, so how are we going to involve them? They are thinking principally about making sure they have food for the following day." In this case, illiteracy and poverty are represented as obstacles to reasoning; people are deemed incapable of making decisions about their lives because they cannot read and write and because they are poor. This rationale permits NGOs to explain why projects are announced to and imposed on communities, not negotiated with them. Along similar lines, the Guatemalan director of community development for a Latin American NGO contracted by the US-AID suggested that my study of environmental protection and democratization should begin with an analysis of *campesino* culture. There are several characteristics of *campesino* communities that limit both processes, he said, including "illiteracy, culture, vices, personal interests, and limited capacity to invest," that is, limited capital and limited ability to think ahead. These limitations, he suggested, "often *oblige* the NGOs to act in anti-democratic or even coercive ways" (emphasis added).

In the rationales outlined here, neither the biosphere reserve model nor the means by which it was implemented are mentioned as possible causes of resistance or sabotage; rather, the blame is placed primarily upon the cultural traits of poor Guatemalans.[4] These narratives resonate with elite ideologies rooted in late nineteenth-century Guatemala, which positioned indigenous people as the primary obstacle to modernization (Taracena Arriola 2002). As the rationale went, "the government and plantation owners want to help the indigenous peoples but they resist progress" (González-Ponciano 1999, 18). Similarly, nineteenth- and twentieth-century racial hierarchies have consistently positioned indigenous and mestizo *campesinos* as "racially degenerate" and "unprepared to make use of their civil liberties . . . and for this reason they demand a tough hand to control them" (González-Ponciano 1999, 18). In positioning Guatemalan *campesinos* as obstacles to democratic process in the Maya Biosphere Reserve, conservationists reproduce long-standing racial and cultural hierarchies that have been at the root of Guatemala's violent socio-economic and political system.

In sum, even as the Maya Biosphere Reserve was born at the intersection of global conservation and democratization, the reserve itself was implemented in an exclusionary manner. As I explore in the next section, these processes created a dynamic environment with contradictory implications for the daily practices of conservation.

Negotiating Exclusionary Conservation in the Maya Biosphere Reserve

Within the Maya Biosphere Reserve lie numerous communities that, as mentioned, were not initially consulted during its creation, nor were they included in the reserve's formal power structures. Individuals and collectives—differently situated socially, politically, and geographically—have been differently affected by a shifting political system, new environmental governance strategies, and the presence of international NGOs. In this section, I present two ethnographic vignettes from my research in the reserve outlining how two social groups negotiate conservation and democratization in the context of globalization. In the first example, I focus on San Geronimo,[5] a community of migrant agriculturalists that unexpectedly found itself included within the reserve. In the second case, I concentrate on the "Women Rescuing Indigenous Medicinal Plants" group that emerged in San Timoteo, an indigenous community. In each case, the outcomes are at once uneven, contradictory, and promising; key to these contradictions is the ability of community members to embody the category of political actor, decision maker, and therefore citizen.

4. Doty (1996) argues that such rationales are an important element of colonial and imperial discourses. She identifies various instances whereby British colonial rulers claimed that the cultural traits of "native" peoples compelled them to use authoritarian measures (42, 111).

5. The names of the villages and the organizations have been changed to protect the identity of the inhabitants.

FIGURE 11.2. Migrant farmers make the Maya Biosphere Reserve home (photograph by Kevin Bray)

Migrant Farmers and Conservation in San Geronimo

Since the 1960s, landless or land-poor indigenous and *ladino campesinos* have migrated to the Petén in order to provide their families with what they consider to be a decent life (fig. 11.2). These families see few desirable options in southern Guatemala, where land concentration has been an overriding source of conflict. While statistics vary, Guatemala is said to have the most unequal system of land tenure in Latin America—a significant issue in a country with 60 percent of the population living in rural areas (World Bank 2001). Recent studies suggest that "acute inequality and land poverty are worse" today than in 1979, when 2.5 percent of farms controlled 65 percent of agricultural land, while 88 percent of farms occupied 16 percent of land (Sieder et al. 2002, 49). In addition to economic inequality, landless or land-poor indigenous and *ladino campesinos* historically have faced political inequality, in that property holding and access to education were legal requirements of citizenship and therefore participation in the political system. Although the formal legal system has changed, belief systems that regard illiterate and poor people as mentally inferior continue to determine who counts as a political actor in daily life.

In this ethnographic vignette, I focus on San Geronimo, a community of migrant agriculturalists living within what is now the multiple use zone of the reserve. The eighteen to twenty families had been living in the area for fifteen to twenty years when, in 1991, the Center for Education and Investigation of Tropical Agronomy (CATIE), based in Costa Rica, approached them about initiating a sustainable development project. In 1994, the settlement was granted the first community forestry concession in the reserve, which gives the residents rights to manage a 7,039 hectare area zoned for agriculture, sustainable forestry, and forest conservation (Gretzinger 1998). While the conferring of land rights may suggest a convergence between conservation and democratization, my ethnographic research in this community suggests a more contradictory analysis.

CATIE's management plan conceptualized the project as a mechanism for discovering and teaching the principles of science-based natural resource management. A statement in CATIE's plan is indicative of the envisioned relationship between project staff and locals: the "community will gradually gain the experience needed to ensure the sustainable management of the resources under its responsibility" (CATIE 1994, 1). The plan consistently frames community members as recipients of the knowledge gained by the project's technical staff rather than political actors and environmental decision makers with a stake in the creation of a new resource governance model. Community members do not figure prominently in the management plan or other texts. Information about the residents makes up one-quarter of page 2 and half of page 4; the rest of the twenty-seven-page management plan includes technical information about geography, vegetation, soils, timber inventories, markets, and current land use practices.

My interviews with a majority of the adult residents in 1996 suggest that the social relations outlined in the management plan are reproduced in the project's daily practices. Don Andrés recalled the project's beginning:

> At first, they came and they held meetings and gave us talks and they collaborated with us in everything until they succeeded in convincing us of the forestry concession—because the land wasn't going to be parceled out. In that they told us the truth, although they have tricked us many times. Now we are working on the concession only because they have already involved us, since the benefits from the harvest are minimal.

The president of the Concession Committee indicated that the planning process did not involve locals directly. As Don Juan described it, "we were invited to a meeting and they told us what they were doing and asked if it was good, and we approved." In reference to the construction of a nature trail, Don Andrés said it was built "because CATIE is behind us telling us to do it." Male members of the concession are involved in the project primarily as day laborers and field assistants, carrying out the instructions of the CATIE staff. Interestingly, the project director lamented what he perceived as the community's lack of participation; as he put it, "They just don't have a vision of the project as their own. They just don't have an understanding of this as a business." From his perspective, the inhabitants of San Geronimo are to blame if they do not take ownership of the project, rather than the mechanisms through which the project was carried out and the social relations that shape its daily practices.

Despite the top-down nature of the project, people in San Geronimo positioned themselves as going along with the project because they obtain specific benefits from their participation. The principal benefit is perceived to be land tenure security and the right to plant *milpa,* or corn fields, as well as hunt and collect other kinds of livelihood necessities from the forests. Thus, Don Chema said, "we know that we are renting this land and that they can't remove us, nor can others come in." Similarly, Don Francisco indicated, "the land is ours. We are paying taxes for it and the concession is for San Geronimo [and three other settlements]. So, we are the only ones that have rights to it." Although people in San Geronimo may achieve goals consistent with their own interests, the project positions them, and they position themselves, as manual laborers subject to the authority of others; in contrast, the CATIE staff is positioned as responsible for the intellectual architecture of the project. Indeed, six years into the project, people in San Geronimo did not see themselves as capable of managing the concession alone. When asked about CATIE's impending withdrawal in 1997, most people said that they believed the project would not continue without further assistance. As Juan remarked, "We need help from people that are educated (*preparado*)." This comment was seconded by Chema: "No [we can't continue alone], because there are no educated people here."

This case presents a number of contradictory outcomes. On the one hand, the project's daily practices both reflect and reproduce social hierarchies that position uneducated and economically marginalized people as incapable of making decisions about their future. Community members were unable to enlarge or embody the category of political actor, environmental decision maker, and therefore citizen. Thus, the daily practices of conservation in San Geronimo did little to strengthen or encourage democratic institutional procedures and social relations. On the other hand, the community forestry concession allowed this group of migrant farmers to achieve land tenure and livelihood security in a country wherein land inequality is a primary cause of social and political marginalization. Additional research would be required to shed light on the extent to which this achievement led to reconfigurations in community members' notions of citizenship and democracy.

Gendering Conservation and Citizenship

In my analysis of San Geronimo, I refer to the community as a whole without mentioning the gendered nature of conservation. And yet, in CATIE's project, as in the majority of other projects in the Maya Biosphere Reserve, men are positioned as the primary agents of social and environmental change. Thus, NGOs tend to work with male leaders and male heads of household only. The reasons are many, but all tend to draw from "prevailing cultural notions that assign women to tasks around the house, even though the reality of subsistence life may require them to be in the field or the forest" (USAID 1990, 10). As the Guatemala director of a North American NGO suggested, "even as NGOs try to avoid it, sexism is reproduced": "We do marginalize the position of women. People say that this is a cultural problem, 'we can't change cultural patterns.' We say, 'women don't want to leave their houses.' Even if they don't want to, we also don't search for the mechanisms to find spaces for women." In privileging men and effectively giving them more say over the future of land use in the reserve, conservation projects risk reproducing gender-based inequalities that until the mid-twentieth century were upheld within the legal system in Guatemala, and indeed throughout most of the world (Thillet de Solórzano 2001). And yet, as I outline in my second ethnographic vignette, the intersection of conservation, globalization, and democratization in the reserve also created spaces for new forms of environmental activism and political alliances, which allowed women to disrupt the gender biases at the heart of conservation in the reserve. This case focuses on a group of indigenous women who formed an alliance with a researcher with ties to international environmental NGOs.

The women live in San Timoteo, a historically marginalized indigenous community positioned on the lower rungs of regional and national socioeconomic hierarchies that privilege *ladinos*. As was the case in other communities, the community was excluded from the Maya Biosphere Reserve's formal decision-making circles in the early 1990s. If and when members of the community were invited to the table, men acted as representatives—thereby perpetuating indigenous women's lack of representation and participation in formal political processes.

In 1996, a European botanist funded by a U.S. environmental NGO began working closely with women in San Timoteo to compile basic data on indigenous medicinal plant use and collect recipes for specific remedies. In collaboration with the botanist, several women started the Women Rescuing Indigenous Medicinal Plants Group, a project combining environmental and cultural conservation goals with economic strategies that provide income to women.[6] This idea evolved into an "ethnopharmacy" that would sell traditional remedies gathered and packaged by the women. Rosalia, the group's first president, suggested that the group had a second goal: "that women succeed in forming a group. . . . Women rarely participate, and when you see an organization, it is always directed by a man."

---

6. For an in depth analysis of this group, see Sundberg (2004).

Participation in the group presented most women with a constant struggle to balance their household demands with the desire to expand their reach outside the home, to take responsibility in the group, and to learn new things. For a number of the women, articulating a priority beyond their immediate families had transformative effects on social relations at the level of family and community. As Doña Flor said, "For a long time, I have wanted to do something for humanity." As a young girl, she had wanted to pursue a career, but did not have the opportunity, as she explains:

> We women always occupy ourselves with the house, and only the men leave to work. And a woman can work like men, I mean we don't have the [physical] strength they have, but we have the capacity. . . . At home, when you have finished the housework, you may want to go somewhere, but there is nowhere to go. If I go out in the street, people will say, "that woman just wanders around." . . . But with this responsibility [in the group] I have somewhere to go.

More important, the group gave women the opportunity to speak their minds and to become decision makers; several participants identified a women-only environment as critical to fostering the confidence and assertiveness needed to take such steps. In interviews, the women explained why. In Doña Flor's words, "Between women we talk. With a man there, we will give him the prerogative to speak and he ends up directing." Rosalia, the group's president, concluded that "women are afraid to speak because they have never done so."

The group chose not to work with any of the international NGOs working in the Maya Biosphere Project. However, through the botanist, they found support in EcoLogic Development Fund, a U.S. NGO that offers hands-off financial assistance to grassroots organizations.[7] Since 1996, participation in the group has enabled its participants to take on new kinds of responsibilities, to meet women involved in other grassroots organizations, and to travel to conferences. In addition, the group fulfilled its goal of establishing a plant nursery; they have also been able to building a meeting center. With this infrastructure in place, the group is able to store their materials, produce tinctures and soaps, host events, lead tours, and coordinate volunteers (primarily Spanish-language students). Furthermore, the support from outsiders has given the medicinal plants group a certain legitimacy and status in the community that extends to the individual participants (fig. 11.3).

This case highlights the more promising side of global conservation, in that the presence of international NGOs enabled a transnational alliance between a researcher and a group of previously marginalized indigenous women. In placing value upon women's knowledge and use of medicinal

---

7. Visit EcoLogic Development Fund's website at http://www.ecologic.org.

FIGURE 11.3. Drying medicinal plants (photograph by the author)

plants, this alliance disrupted existing power structures and created a space for a group of indigenous women to emerge as political actors and participants in civil society. The formation of the Women Rescuing Indigenous Medicinal Plants Group also had implications on a more intimate level, as each individual woman as well as the group as a whole took significant steps to speak, assume responsibility, and make decisions, thereby challenging gender relations as well as gendered notions of citizenship. While democratization is usually measured in terms of formal institutions, this case points to the importance of analyzing changes in everyday practices that enlarge and transform the category of citizen within the context of conservation and globalization.

Concluding Discussion and Recommendations

My research on the daily practices of conservation in Guatemala reveals the contradictory articulations between conservation, globalization, and democracy. On the one hand, globalization and democratization contributed to the emergence of an environmental movement in Guatemala as well as the creation of new conservation policies and conservation territories such as the Maya Biosphere Reserve. On the other hand, the Guatemalan state and its international partners took an authoritarian approach to environmental protection, thereby reproducing entrenched social inequalities in Guatemalan society that restrict who counts as a political actor. Deemed incapable of participating in environmental decision-making, the reserve's resident population were said to "oblige the NGOs to act in anti-democratic or even co-ercive ways."

The implications of these contradictory tendencies created a challenging environment for people residing within the reserve. My ethnographic research in two communities demonstrates how different groups dealt with these circumstances. In San Geronimo, migrant *campesinos* were excluded from participating in the intellectual architecture of the project. However, community leaders supported the plans proposed by an NGO because it guaranteed them land tenure and livelihood security in a country wherein a majority of the population faces economic hardship and, by extension, political marginalization. The long-term political implications of this project have yet to be revealed. My analysis of the Women Rescuing Indigenous Medicinal Plants Group suggests that the presence of conservation projects in the area disrupted local power structures, while a researcher provided access to new networks of political and financial support. In this case, women became decision makers and participants in civil society.

In Guatemala, as in other Latin American countries, conservation needs and priorities are often considered in isolation from other social processes. For instance, when asked if processes of democratization are considered in the design of conservation projects in the Maya Biosphere Reserve, the Guatemalan director of a United States–based international NGO replied, "No, because all the NGOs are sectorialized; we only do our thing. Human rights, which include respect for others, do include processes of democratization. But development only includes things like clean water, and environmental NGOs only focus on environment; democratization is not a variable." He added, "We [NGOs] don't approach things in an integrated fashion. Our challenges oblige us to pursue specific goals."

As outlined in this chapter, however, the creation of new conservation territories in Guatemala is intimately connected to global conservation organizations, neoliberalism, and the ongoing process of democratic reform. Thus, it is incumbent upon environmental activists, agents of the state, and NGO personnel to consider how the daily practices of conservation intersect with and are informed by other social and political processes. In Guatemala and other Latin American countries with histories of systematic social and political exclusion, it is particularly important to consider if and how conservation projects facilitate processes of democratization. On the one hand, this implies reconceptualizing who counts as a citizen, political actor, and environmental decision maker in specific social and geographical contexts. This also means that human rights are a critical variable in conservation; although previously considered separately, these two issues must be thought through and practiced together. On the other hand, conservation policies must include forms of governance that employ, but also foster, democratic social relations and procedures. If state and NGO conservationists are seen to be supporting authoritarian measures, their policies may be met with resistance and their tactics may be counterproductive. Moreover, given that both international and national environmental activists tend to come from the professional and elite classes, coercive conservation may be seen as contributing to social authoritarianism.

"The foreigners are giving money, but they themselves are eating it; those who live here are not earning anything," said a man in San Geronimo. This comment—repeated to me in various ways by other people—reveals the extent to which conservation projects are associated with globalization as well as the reproduction of inequality in Guatemala. Changing this image of conservation in Guatemala and other Latin American countries depends upon management policies that are attentive to human rights and governance strategies that foster democratization.

Acknowledgments

This chapter would not have been possible without the individuals and communities in Guatemala, who gave of their time and energy to answer my questions and allowed me to observe and participate in their lives. In addition, I am very grateful to Karl Zimmerer for inviting me to participate in the University of Wisconsin's Globalization and Geographies of Conservation Conference, where I presented a previous version of this chapter. The workshop participants provided useful feedback, for which I am thankful.

References

Berger, S. 1997. Environmentalism in Guatemala: When fish have ears. *Latin American Research Review* 32 (2): 99–115.

Blacklock, C., and L. Macdonald. 2000. Women and citizenship in Mexico and Guatemala. In *International perspectives on gender and democratisation*, ed. S. Rai, 19–40. London: Macmillan.

Bulmer-Thomas, V. 1996. *The new economic model in Latin America*. London: Institute for Latin American Studies/Macmillan.

Cabrera, M. 1991. Informe del viaje realizado a la comunidad de El Naranjo-Frontera, Municipio la Libertad, El Petén, Durante las fechas 28 febrero al 2 de marzo 1991. Internal Report. Guatemala City, Guatemala: CECON.

Campbell, L. 2000. Human need in rural developing areas: Perceptions of wildlife conservation experts. *The Canadian Geographer* 44 (2): 167–81.

Carruthers, D. 2001. Environmental politics in Chile: Legacies of dictatorship and democracy. *Third World Quarterly* 22 (3): 343–58.

CATIE (Center for Education and Investigation of Tropical Agronomy). 1994. *Plan de manejo forestal para la unidad de manejo San Miguel, El Petén, Guatemala*. Turrialba, Costa Rica: CATIE.

Christen, C., S. Herculano, K. Hochstetler, R. Prell, M. Price, and J. T. Roberts. 1998. Latin American environmentalism: Comparative views. *Studies in Comparative International Development* 33 (2): 58–87.

Cojtí Cuxil, D. 1995. *Configuración del pensamiento político del Pueblo Maya, segunda parte*. Guatemala City, Guatemala: Editorial Cholsamaj-SPEM.

———. 1997. *El movimiento Maya (en Guatemala)*. Guatemala City, Guatemala: Editorial Cholsamaj.

CONAP (Consejo Nacional de Áreas Protegidas). 1996. *Plan maestro reserva de la biósfera Maya*. Turrialba, Costa Rica: Center for Education and Investigation of Tropical Agronomy.

Desai, U., ed. 1998. *Ecological policy and politics in developing countries: Economic growth, democracy, and environment*. Albany: State University of New York Press.

Dore, E. 2000. One step forward, two steps back: Gender and the state in the long nineteenth century. In *Hidden histories of gender and the state in Latin America*, ed. E. Dore and M. Molyneux, 3–32. Durham, NC: Duke University Press.

Doty, R. 1996. *Imperial encounters: The politics of representation in North-South relations*. Borderlines 5. Minneapolis: University of Minnesota Press.

González-Ponciano, J. 1999. Esas Sangres No Estan Limpias, Modernidad y pensamiento civilizatorio en Guatemala (1954–1977). In *¿Racismo en Guatemala? Abriendo el Debate sobre un tema tabú*, ed. C. Arenas Bianchi, C. Hale, and G. Palma Murga, 1–46. Guatemala City, Guatemala: AVANSCO.

Goodin, R. 1992. *Green political theory*. Cambridge, MA: Polity.

Gretzinger, S. 1998. Community forestry concessions: An economic alternative for the Maya Biosphere Reserve in the Petén, Guatemala. In *Timber, tourists, and temples: Conservation and development in the Maya Forest of Belize, Guatemala, and Mexico*, ed. R. B. Primack, D. Bray, H. Galletti, and I. Ponciano, 111–24. Washington, DC: Island.

Haraway, D. 1991. Situated knowledges: The science question in feminism and the privilege of partial perspective. In *Simians, cyborgs, and women: The reinvention of nature*, D. Haraway, 183–201. New York: Routledge.

Keck, M., and K. Sikkink. 1998. *Activists beyond borders: Advocacy networks in international politics*. Ithaca, NY: Cornell University Press.

Lafferty, W., and J. Meadowcroft, eds. 1996. *Democracy and the environment: Problems and prospects*. Brookfield, VT: Edward Elgar.

Midlarsky, M. 1998. Democracy and the environment: An empirical assessment. *Journal of Peace Research* 35 (3): 341–61.

Moore, D. S. 1998. Clear waters and muddied histories: Environmental history and the politics of community in Zimbabwe's eastern highlands. *Jouirnal of Southern African Studies* 24 (2): 377–403.

Mumme, S., and E. Korzetz. 1997. Democratization, politics and environmental reform in Latin America. In *Latin American environmental policy in international perspective*, ed. G. MacDonald, D. Nielson, and M. Stern, 40–57. Boulder, CO: Westview.

Neumann, R. 1998. *Imposing wilderness: Struggles over livelihood and nature preservation in Africa*. Berkeley: University of California Press.

Peluso, N. L. 1993. Coercing conservation: The politics of state resource control. *Global Environmental Change* 3 (2): 199–217.

Perera, V. 1989. A forest dies in Guatemala. *The Nation*, 6 November, 521–24.

Pulido, L. 1994. Restructuring and the contraction and expansion of environmental rights in the United States. *Environment and Planning A* 26 (6): 915–36.

Sieder, R. 1999. Rethinking democratisation and citizenship: Legal pluralism and institutional reform in Guatemala. *Citizenship Studies* 3:103–18.

Sieder, R., M. Thomas, G. Vickers, and J. Spence. 2002. *Who governs? Guatemala five years after the peace accords*. Cambridge, MA: Hemisphere Initiatives.

Sundberg, J. 1999. *Conservation encounters: NGOs, local people, and changing cultural landscapes*. PhD diss., University of Texas, Austin.

———. 2004. Identities-in-the-making: Conservation, gender, and race in the Maya Biosphere Reserve, Guatemala. *Gender, Place, and Culture* 11 (1): 43–66.

Taracena Arriola, A., with G. Gellert, E. Castillo, T. Paiz Sagastume, and K. Walter. 2002. *Etnicidad, estado y nación en Guatemala, 1808–1944, Volumen I*. Guatemala City, Guatemala: Nawal Wuj.

Thillet de Solórzano, B. 2001. *Mujeres y percepciones políticas.* Guatemala City, Guatemala: FLACSO.

USAID (United States Agency for International Development). 1989. *Project paper: Maya Biosphere Project.* Washington, DC: USAID.

———. 1990. *Social soundness analysis, Maya Biosphere Project.* Washington, DC: USAID.

Van Cott, D. L. 1994. Indigenous peoples and democracy: Issues for policy makers. In *Indigenous peoples and democracy in Latin America,* ed. D. L. Van Cott., 1–27. New York: St. Martin's.

———. 2000. *The friendly liquidation of the past: The politics of diversity in Latin America.* Pittsburgh, PA: University of Pittsburgh Press.

Walker, P. 1999. Democracy and environment: Congruencies and contradictions in southern Africa. *Political Geography* 18 (3): 257–84.

Warren, K. 1998. *Indigenous movements and their critics: Pan-Maya activism in Guatemala.* Princeton, NJ: Princeton University Press.

World Bank. 2001. *Guatemala data profile.* Available at http://www.worldbank.org. Accessed 27 February 2003.

Yashar, D. 1998. Contesting citizenship: Indigenous movements and democracy in Latin America. *Comparative Politics* 31 (1): 23–42.

Zimmerer, K. S., and E. D. Carter. 2002. Conservation and sustainability in Latin America. *Yearbook of the Conference of Latin Americanist Geographers* 27:22–43.

12 Decentralization, Land Policy, and the Politics of Scale in Burkina Faso

LESLIE C. GRAY

The balance of international environment and development efforts is tipping toward the local scale. With agendas of community-based natural resource management, participatory development, decentralization, and other nature-society hybrids, international donors have recognized the local scale as the place where development and conservation can happen. The World Bank (1999, 112) argues that the scale of a policy issue "dictates the level of government best equipped to tackle it." The recognition that local participation in environmental governance is necessary for successful conservation and environmental management has led to an international push for decentralizing environmental management to the local scale (state or municipal governments and local communities).

There are several reasons why international institutions and governments believe that local communities are the best scale for designing environment and development policy. A major impetus is the generalized disappointment with top-down models of development (Schroeder 1999) and a global shift away from state-centric development (Toulmin 2000). Decentralized development is intended to promote good governance and democracy, with the hope that the majority of resource users will be represented in the decision-making process (World Bank 1997; IIED 1999). Finally, knowledge about local needs, and resource use patterns is better understood at the local scale, potentially making local-scale development efforts more responsive and less expensive (World Bank 2000).

As with many countries in West Africa, the government of Burkina Faso, under pressure from international institutions such as the World Bank, is starting to decentralize land management policy and proposing to decentralize many other rural services in the coming years. The predominant policy has been land zoning, using the *gestion des terroirs* (village land management) approach, whereby local communities clarify land rights and zone land into specific land use categories. The approach has multiple goals common to other community natural resource management programs (e.g., Kellert et al. 2000), among them sustainable resource use, equity, participation, and more secure land rights. The approach also reflects one of the major themes of this edited volume: the globalized trend of applying conservation techniques to agricultural zones. In West Africa, international donors and governments perceive that population growth and migration put these types of land use zones at risk.

This chapter examines decentralization in the context of land and environmental policy in Burkina Faso, arguing that the *gestion des terroirs* approach has met neither the equity nor environmental conservation goals put forth by programs. Programs have generally not met equity goals because they have devolved natural resource control to community actors that are unrepresentative and unaccountable (e.g., Ribot 1999). As a consequence, many programs have been fraught with conflict and competition and ultimately abandoned, thus failing to meet their environmental conservation goals. This reflects a common experience of decentralization and community natural resource management programs in general in West Africa, where programs have not been as successful as hoped (Benjaminsen and Lund 2001; Leach, Mearns, and Scoones 1999; Ribot 1999; Turner 1999). The discussion here will explore what Schroeder (1999) calls "the politics of scale" in environmental planning efforts. In particular, it will highlight concerns about whether externally conceived development projects can become "owned" by local communities, who have the power to make decisions at the local scale, and whether the local scale is the right place to undertake land policy (Lavigne Delville 2000; Peters 1996; Ribot 1999).

A primary weakness of the *gestion des terroirs* approach is that what is portrayed as decentralized development is essentially top-down development, imposed by international donors and imperfectly implemented by governmental and nongovernmental organizations. Decentralization, in theory, should cede powers to local-level actors and institutions in a way that supports greater efficiency and equity in the delivery of services and management of natural resources. Agrawal and Ribot (1999) argue that downward accountability is key in this rubric. The *gestion des terroirs* approach in Burkina Faso and elsewhere in West Africa has done just the opposite by empowering actors who are not representative of their populations on the whole and who are not accountable to them in any significant way. Generally officials are instead upwardly accountable to national government officials and representatives of nongovernmental organizations. Indeed, locals who were interviewed did not see that they had any ownership over projects, but rather saw the exercise of zoning as something that was externally directed and controlled.

Because projects devolve control to authorities (local customary authorities and representatives of dominant ethnic groups, all generally male) that are neither democratic nor representative, conflicts abound. The case studies presented in this paper will illustrate how local farmers have used the rhetoric of the *gestion des terroirs* program to try to expel migrant farmers from their land. Migrant farmers, who make up more than half of the population and, in many cases, have lived in the region for decades, have little power in designing zoning programs, despite the rhetoric of participatory development. Because they have little power, land zoning is generally undertaken in ways that disadvantage them. So what begins as an internationally conceived project to increase participation in environmental decision making is effectively captured by actors at the local scale and used in ways that reinforce existing unequal power relations. Devolving resource control should be about communities participating in determining the rules of access, yet in practice decentralization has resulted in exclusion of key parts of the population. The experience of this program mirrors the finding by Sundberg (chap. 11 in this volume) that the goals of conservation and democratization do not always coincide.

Problems result, in part, from a spatial mismatch between the respective interests of international institutions and national governments. International institutions, as exemplified by the World Bank and the Caisse Centrale de Coopération Economique in this case study, give money for projects to devolve control over resources. States comply, not because they necessarily believe in the project, but because conditionality forces them to do so. This forces a serious contradiction because most African state actors, operating under neopatrimonial and undemocratic regimes, are interested in keeping power as centralized as possible—any real devolution of power has the potential to threaten vested interest groups and regime survival (Boone 2003). Donors therefore implement programs in situations where there are no downwardly accountable authorities to devolve power to. Therefore, what emerges is a false decentralization, one intended to please international donors, but one that also keeps real power within the state and away from local communities.

Another mismatch lies in how international and national governments view the local scale and local realities on the ground. For example, the globalized discourses that justify interventions revolve around ideas that traditional tenure systems are breaking down under population pressure, resulting in land degradation and conflicts over natural resources (PNGT 1989; Toulmin 1994). However, a number of village-level studies counter this neo-Malthusian vision of land tenure breaking down under population pressure, showing land tenure relations to be far more adaptive to changing socioeconomic, demographic, and environmental conditions (Benjaminsen and

Lund 2003; Braselle, Gaspart, and Platteau 1998; Gray and Kevane 2001; Paré 2001). The basic premise that current tenure relations are resulting in widespread land degradation is not at all confirmed by empirical research, but nonetheless this idea is widely accepted in international policy circles.

This does not mean that land tenure systems are unproblematic. On the contrary, concerns about conflict and equitable distribution of land resources abound (Benjaminsen and Lund 2003; Turner 1999). The problem lies in the solution. Instead of building upon local innovations and adaptations, development programs impose standardized solutions to land problems, in a top-down fashion. These standardized solutions to local-scale issues reflect globalized policy discourses about community participation and management of common property resources, instead of localized realities. In other regions of the world, the standardized solution involves privatization, but in much of Sub-Saharan Africa, which has a history of failed titling schemes, solutions involve a reimagination of customary land tenure systems. Part of the problem is that the standardized solution and the idealization of communities as consensual entities (Agrawal and Gibson 1999) downplays the role of power in determining resource access and the role of conflict in struggles over resources. This clouded, apolitical view of reality ultimately results in more conflict.

The following sections will examine the problematic nature of decentralization of land management in West Africa, explaining why policies have generally failed. Two case studies will illustrate the politics of changing land rights in several villages in Burkina Faso, one examining the outcomes of land zoning under the *gestion des terroirs* approach and another looking at a local-level land conflict. These case studies are based on fieldwork in several villages in southwestern Burkina Faso (fig. 12.1). While much of this paper is a critique of the *gestion des terroirs* approach in Burkina Faso, the last section will discuss ways to rethink the approach, particularly in ways that ensure both the environmental and equity goals of the program. Suggested solutions include mechanisms for conflict resolution, devolution of powers to accountable authorities, and participation of local people in determining the rules of access to natural resources.

Decentralization, Land, and the Politics of Scale in West Africa

When international donors talk about decentralization, they are generally referring to a shift of power away from the center state (Carney and Farrington 1998). Decentralization within government structures—from center to periphery—is generally referred to as *deconcentration*. Political decentralization is when power is devolved to local-level actors and institutions, generally in the hopes that these are representative and accountable bodies (Agrawal and Ribot 1999). Natural resource decentralization generally has both equity

FIGURE 12.1. Map of study area

and environmental goals. Programs intend to give communities control over their natural resources in a way that represents different community interests and at the same time creates management strategies that will conserve natural resources.

Donors are particularly interested in land tenure because they view customary tenure systems as a constraint to agricultural development. This belief first emerged from colonial administrators concerned that customary tenure systems would prohibit farmers from investing in land improvements, because of the potential for multiple claims upon their land, and from obtaining loans because land could not used as collateral (Bassett 1993). Many of these critiques of customary tenure have since been discredited: customary systems have been shown in many circumstances to be quite secure and amenable to commercial agriculture (Atwood 1990; Lund 2001). Yet the idea that customary tenure systems constrain development persists in policy circles. Policy interventions such as land titling are widely regarded as failures (Platteau 2000), so now donors attempt to design policy to strengthen customary tenure systems. Decentralization programs attempt to do this by giving communities legal control over common property resources such as land, forests, and fallow lands.

One part of the decentralization of land management has focused on programs to clarify land rights. Land has become a significant site of conflict among many groups, including migrants and locals, farmers and herders, intergenerational groups, and neighboring villages (IIED 1999). To counter perceptions of degradation and solve conflicts, donor organizations have implemented projects to zone land, hoping in the process to improve the management of common property resources. Zoning is undertaken as a participatory process where program staff and village committees cooperatively map, delineate, and divide village land into land use zones.

Critiques of decentralization programs in West Africa abound; they are generally focused on whether programs are meeting goals of greater equity and participation. Indeed, several studies show that decentralized resource control is proving to be problematic (Benjaminsen and Lund 2001; Ribot 1999). Land conflicts have emerged when projects attempt to clarify rights. Part of this has to do with devolving resources in an arena of complex and bundled rights, where redefinitions of rights often result in exclusion. Resource boundaries that have been historically flexible and negotiated are replaced with more defined boundaries (Brosius, Tsing, and Zerner 1998), which may lead to certain people losing rights in the process. People with secondary rights—for example, a woman who has rights to collect tree products—are often most negatively affected. This is one example of the many problems that emerge in the way that external projects view the "local" scale, particularly how social relations are constructed.

Another question concerns the political inequalities of village societies and the extent to which external donors reinforce unequal access to resources. Local communities, rather than being homogeneous social units that act cooperatively, are made up of multiple actors with competing political agendas. This means that village social relations, rather than being based on cooperation and consensus, are instead based on conflict and competition (Agrawal and Gibson 1999). Despite the cooperative rhetoric of participatory development, externally funded projects produce intracommunity competition for resources. This competition is not necessarily a bad thing, depending on how the rules of engagement and negotiation are structured. A main critique of models of decentralized development is that authority is devolved to unaccountable community structures (Neumann 1997). Many institutions in place for decision-making at the community level are inequitable and nonparticipatory (Engberg-Pedersen 1995). Projects, in the name of participatory development, reinforce local and generally unaccountable decision-making bodies. Ribot (1996) argues that this results in participation without representation. Overlooking the inequitable nature of village social relations then frequently results in increased conflict and competition (Leach, Mearns, and Scoones 1999).

A final question concerns the basic nature of state-society relations in African societies and whether a centralized state will decentralize. Decentralization in Sub-Saharan Africa takes place in the context of state structures that are highly centralized and neopatrimonial. This differs from Latin America, where decentralization largely takes place within the context of elected municipal governments (Larson 2002). Undemocratic African governments have been wary of empowering any actors that will challenge their power; thus, Sub-Saharan Africa countries have few directly elected local governments. Policy at the local scale is directly controlled by the central state, meaning that local government is in effect local administration. Local communities have little say in actually making policy (Thomson 2000). Because of the absence of local representative governments, donors have attempted to create village institutions that are ostensibly representative (village committees in the case of *gestion des terroirs* projects), or they have worked through customary authorities. Therefore, Agrawal and Ribot (1999) ask whether many of the reforms that take place in the name of decentralization really constitute decentralization. Real decentralization, they argue, should be representative and downwardly accountable: actors should be able to participate in and modify the institutions that they are involved with. Instead, "governments often perform acts of decentralization as theater pieces to impress or appease international donors and nongovernmental organizations or domestic constituencies" (Agrawal and Ribot 1999, 474). The World Bank has recognized the problematic nature of decentralization and devolution of authority under these conditions and is proposing and funding the creation of elected rural municipalities throughout the developing world. Whether national governments will give up real power to the local scale remains to be seen.

Land Rights and the Decentralization Experience in Burkina Faso

Land law in Burkina Faso is ambiguous. Formal law declares that land belongs to the state, although new laws suggest that under the *gestion des terroirs* approach, villages can control their land resources. In general, though, there is no private property for agricultural land, a situation similar to that of many African countries (Benjaminsen and Lund 2003). Therefore, customary systems of land tenure dominate how land is allocated and controlled. In Burkina Faso, land holding is complex and multilayered, reflecting the generalized situation of land tenure in Sub-Saharan Africa, where people generally have many levels of access to land with multiple and overlapping rights, claims, and authorities (Haugerud 1989; Shipton and Goheen 1992). Historically, land use has depended on membership in a corporate group. Individuals who were members of groups had rights to use uncultivated land belonging to the broader kinship or residence group. Their rights extended to land under cultivation, but as soon as the field was left fallow, the field would return to the pool of land controlled by the collective (Boutillier 1964). This was not the only way of obtaining land: borrowing land that belonged to other groups was and is extremely common, particularly if the borrower is a migrant or a member of a land-poor lineage. Rights to borrowed land often became permanent over time, especially as land was passed down through generations.

In southwestern Burkina Faso, land rights have recently become much more complicated due to demographic and socioeconomic changes. The largest demographic change has been the increased influx of Mossi migrants into the region. This is a pattern that has been historically important (Cordell, Gregory, and Piche 1996) but accelerated dramatically during the 1970s and 1980s when people moved from the heavily populated northern and central regions to the southern part of the country, where good quality land has historically been abundant and population densities low. This migration led to a doubling of the population of the province of Houet, the heart of the study area in southwestern Burkina Faso, between 1975 and 1985. When Mossi initially arrived in the southwest, borrowing land from local Bwa farmers was relatively easy. Over time, as the perception that migration led to land shortages became more commonplace, borrowing land has become more difficult. Mossi migrants' borrowing rights are limited to use rights, and as a result many migrants fear that their land could be taken away from them, which in practice has been rare. Bwa farmers openly state that Mossi migrants can never own land, though in villages where cultivation is essentially permanent, migrants have developed very strong rights to cultivated land.

During the same time frame that migration has increased, production practices have changed radically in southwestern Burkina Faso. Farmers enthusiastically adopted both cotton and animal traction into their farming systems. Cotton grain production increased more than sevenfold from 1970 to 1990, due to a host of institutional changes in commodity pricing, agricultural marketing, and financial support (Schwartz 1991). Both of these trends, population increase and new production patterns, have led to greater land scarcity, and in response farmers have developed new adaptations to land management. One of the most prominent strategies has been agricultural intensification. Because land of good quality is perceived as scarce, farmers have been hesitant to leave their land fallow for fear that it will be taken away from them. Instead, farmers use different techniques to invest in the soil quality of their agricultural fields, enabling them to stay on fields for longer periods. As a result, land management in southwestern Burkina Faso appears to be evolving toward more individual control with higher levels of investment in soil quality (Braselle, Gaspart, and Platteau 1998; Gray and Kevane 2001; Paré 2001).

It is in this context that in 1998 Burkina Faso passed a decentralization law that formally recognized local level institutions. The law allows groups of villages to federate themselves as a political entity. These groups can then request formal status as a *commune rurale* from the government (Donnelly-Roark, Ouedraogo, and Ye 2001) and enter into agreements with governmental and non-governmental organizations to provide different types of village projects. As of yet, these local decentralized governmental structures do not exist in most rural areas, but the World Bank plans to begin the process of building up representative local municipalities in the next five years. In the meantime, the focus is on nongovernmental or national-level governmental organizations working with different local community actors (World Bank 2001).

The main experience that Burkina Faso has had with decentralization is through the use of the *gestion des terroirs villageois* approach. The *gestion des terroirs* program aims to improve land management at the local level by creating institutions for managing land in a sustainable manner (Toulmin 1994). Both governmental and nongovernmental organizations are attempting to give local communities more control over land resources using this approach. Different projects tend to emphasize different elements depending on the preferences of individual donors (Engberg-Pedersen 1995). Many projects implement programs that promote soil and water conservation, intensification techniques, community development projects, and land zoning.

Of these programs, by far the most controversial has been land zoning. The goal of this strategy is to empower villages to manage common property resources by zoning land into different land use categories. Projects to zone land are generally undertaken in a participatory manner, as project staff work with villagers to create land management committees that are officially known as the Commission Villageoise de Gestion des Terroirs (CVGT). These committees, along with project staff, work to inventory natural resources, draw maps, define resource use boundaries, and decide land use categories. These CVGT committees have official authority: under the agrarian reform law of 1991, the committees were given legal rights to manage land at the village level, changing the 1984 law that designated all land as state property. Giving official power over allocation of land to CVGT means that control over resources is effectively being given to local communities.

Projects are committed to local control and bottom-up participatory development. The CVGT committees are intended to represent the views of different community members. The goal is to make decisions through consensus. Discussions are held with different parties who have interests in the village territory, who then make up the CVGT. These groups include leaders of different village community groups, such as local ethnic groups, migrants, pastoralists and village organizations, but also try to include representatives of groups that are often ignored in official decision-making circles, such as women and young men (Faure 1992). Village committees meet with project staff to draw up a zoning plan. Zoning projects are different in different areas; some are broad reallocations of village land to different land use categories, generally fallow, forest, and agriculture. Other plans are more modest—for example, negotiating a livestock path through farmer fields.

In practice, these plans are supposed to result in better resource management. The next section will illustrate how local politics often subvert attempts to zone land, as projects become caught up in local material and symbolic political struggles over land.

Case Studies of *Gestion des Terroirs* and Conflict in Southwestern Burkina Faso

This section will examine both the implementation of the *gestion des terroirs* approach and land conflict in several villages (Kassaho, Dimikuy, and Sara) in southwestern Burkina Faso, an area of rapid population growth and agricultural change in an environment of high rainfall (800–1,000 mm/year) and fairly fertile soils. In the village study area, two main organizations implement the *gestion des terroirs* program. The World Bank funds the Programme National Gestion des Terroirs (PNGT) and is involved in several villages in the region. Its efforts initially concentrated on improving agricultural practices, but it is now involved in zoning land as well. The second organization using the *gestion des terroirs* approach in the region is Programme de Développement Rurale Integreé (PDRI), which is funded by the Caisse Centrale de Coopération Economique, a French governmental economic cooperation organization. The PDRI has been intensively involved in zoning programs in the southwest. It is also involved in community development projects, providing funds for different activities, such as schools and women's groups. As the case studies will demonstrate, local-level political processes are vital in determining the course of such projects.

Both Kassaho and Dimikuy are villages where the population is split between locals and migrants (Gray 2002). In Kassaho, attempts to zone land involved lengthy meetings with committees of Mossi, Fulani, and Bwa families, involving them in every stage of the zoning plan, which in this case would divide village land into pasture, fallow, and cultivation zones. Project staff acknowledged that within the zoning project there would be winners and losers, as some families would have their allocation of land reduced when they moved to new fields. Any loss, though, would theoretically be offset by in-tensification. Tensions over zoning land were very high, both between elders and younger people and between Mossi and Bwa farmers. This tension was evident despite the use of participatory development methods, which gave many farmers input in the process. Part of the problem was that in the zon-ing process, local Bwa villagers decided that most of the Mossi migrant farm-ers' fields would be put into either fallow or forest and then they would move into a new area where most of the Bwa farmers were already farming. Even though migrant Mossi were on the village committee to zone land, their sta-tus as migrants precluded their having any meaningful input on the zoning project. In the end, it was only migrant farmers who had to move fields, and consequently only migrants lost land in the zoning process.

Most Mossi agreed to follow the zoning plan, but when it came time to abandon fields, five Mossi families refused to move. This caused extreme ten-sion in the village, with Bwa farmers threatening to evict Mossi farmers from their fields. In the end, the local prefect decided that the Mossi families could remain on their fields but would have to move the next season. Most Bwa farmers were very unhappy with this decision; accusations of bribery and misconduct were widespread. Due to these conflicts, PDRI abandoned zoning in Kassaho. Since that time, villagers from two neighboring villagers, most of them Bwa, have illegally squatted on the land that was left by the bulk of Mossi farmers during the zoning process. This has led to new intervillage con-flicts between Bwa farmers that have brought complaints all the way up to the level of the *haut commissaire,* equivalent to a provincial governor, illus-trating the intensity of conflicts over land rights in the region. Most villagers feel that higher authorities are unwilling to intervene in any meaningful way, mainly because they are fearful of stoking the flames of conflict.

In the village of Dimikuy, implementation of the same program ended in similar conflicts. In Dimikuy, PDRI officials began meeting with members of different village groups (Bwa, Mossi, elders, juniors, women, and men) in 1995 and 1996. As in Kassaho, the goal of the PDRI representative was to re-duce the area that all farmers were cultivating, by substituting inputs for land. Local Bwa farmers, however, interpreted the program based on their own political agendas and knowledge of the divisive experience with zoning in Kassaho, and came to believe that the primary goal of the program was to reduce the number of fields that Mossi families farmed. Young Bwa men, in particular, wanted the PDRI to take back land from Mossi migrants because they were cutting down trees. Several young men decided that the PDRI would support them in their goal of taking land back from Mossi farmers be-cause "the Mossi have come and destroyed our land." Farmers manipulated the rhetoric of project planners, arguing that the poor environmental prac-tices of the Mossi had given the Bwa special rights over the land.

As a result of their meetings with the PRDI, the young men's group decided that it would inform the Mossi that they needed to reduce their field numbers. Most of this was predicated on the notion that Mossi farmers were poor land managers and had a penchant for cutting down trees. Indeed, these young men were quite adamant in their belief that the PDRI was behind them in this effort to reduce the fields of Mossi farmers. There were conflicts between the younger Bwa and their elders, who also had power to allocate land but were not as enthusiastic about the process of evicting Mossi migrant farmers. This was part of a greater resentment between elders and juniors. Junior men blamed the elders for giving so much of the land away to migrant farmers in the first place, often for some ritual, though by no means small, remuneration.

Unfortunately for the Bwa, while they were longer established in Dimikuy, they had become far outnumbered by the Mossi. In reaction to the aggressive rhetoric of young Bwa farmers, the Mossi naturally became suspicious that the ultimate motive of the project was to evict them from their land. Mossi farmers came to view the PDRI with suspicion and rejected participating in any type of project. In Dimikuy, as in Kassaho, the PDRI ultimately abandoned the zoning project. In both villages, *gestion des terroirs* initiatives failed because they exacerbated already volatile conflicts between ethnic groups.

Tionkono, an agricultural area near the village of Sara, has become the site of bitter conflict over land between Mossi migrants and Bwa farmers. This area of large forests and fertile soil is one of the few remaining forested areas near the village. The soil is viewed as extremely high quality. This area is controlled by one large family, the Tamini, who left the land in 1975 hoping to return to it sometime in the future. Bala, an area contiguous to Tionkono, has been farmed continuously since Mossi migrants first started arriving in the 1960s. By planting trees in Bala, Mossi migrants effectively placed a claim on the land. When Mossi farmers began moving from Bala into Tionkono, conflicts over land arose. To the Bwa, there seemed to be no limits to Mossi encroachment. As one Bwa man in Tionkono noted, "the Tamini might have only given land to five Mossi, but each Mossi has ten behind him."

To most Bwa, it was the perceived Mossi invasion that led to the conflict at Tionkono. Mossi farmers, however, saw the situation quite differently. They contended that the conflict hinged on a tale of a snatched and barbecued sheep. Apparently, a group of hungry young Bwa men caught a sheep running around in the bush, thought it was lost, and slaughtered and ate it. It turned out the sheep was wayward, but not abandoned. It belonged to the younger brother of the wealthiest Mossi farmer in the village. The Mossi believed that the argument between the younger brother and the Bwa who had slaughtered the sheep was at the bottom of the land conflict. It was after this incident that Bwa began to take land away from Mossi. The Bwa *responsable administratif* of Sara, when asked if there was any truth to this allegation, agreed that the conflict over compensation for the sheep might have been the breaking point in a general situation of escalating tension over land.

After the sheep incident in 1993, the Tamini held a family meeting, delegating people to go to all the Mossi farming in Tionkono to tell they had to leave their fields after that year's harvest. This included Mossi from the village of Sara, who had farmed there for more than two decades, as well as farmers from the nascent village of Bala, and those who had built houses and were beginning to settle in Tionkono. The first time a small group of between two to four people Bwa men alerted the Mossi directly. Then, in November, as the harvest neared completion, about twenty young Bwa with shotguns and axes went from house to house, informing Mossi farmers that they would have to leave after the harvest.

After this experience, the Mossi met and went together to the prefect to explain what had happened. The prefect convoked representatives of the Tamini family to the prefecture and asked them to let the Mossi cultivate. They unequivocally refused. The next year, young men again went to Mossi families bearing rifles and axes to tell them to leave their fields. Again, local government officials intervened unsuccessfully. In 1995, the local prefect responded by calling a meeting with the concerned Mossi and the Tamini family in Sara, but informed them that he would be coming with representatives of the local military police force. This local show of force was meant to convey to the Tamini family that they had better allow the Mossi to cultivate this next season. This, in fact, was the policy of the local government, which was very sensitive to ethnic conflicts over land that loomed large over most of the province. When he arrived with six military policemen, the prefect informed the Tamini family that the problem had reached "important people" in the capital (Ouagadougou), who had asked that the Mossi be allowed to cultivate one more season, after which they would come and resolve the problem. The Tamini family, in a general family meeting, delegated someone to go to the prefect. They had decided that if the prefect didn't say anything or refused to listen to the Bwa, then people would find out if the "ancestors [had] force or not." If they were forceful, then the Mossi would leave. If they were not, then they would stay. As they predicted, the prefect had not changed his position.

To find out whether the ancestors had force or not, the family dispatched several young men to go to a specific tree in the forest, to cut branches, dig holes, and put the branches into the ground like stakes in all of the Mossi fields in Tionkono. Several Bwa explained that if anyone attempted to cultivate the field without conducting the appropriate sacrifices, they would die. When the Mossi farmers saw this, which they considered sorcery, they went directly to the police, who called yet another meeting between the Mossi and Bwa families. The police explained that the Bwa would go to prison unless they removed the stakes and let the Mossi cultivate their fields. The Tamini family then removed the stakes and conducted sacrifices in the village. By that time, however, most Mossi in Tionkono had left their fields and Bwa farmers had moved in to reclaim them. The use of sorcery had worked to expel Mossi farmers who were too afraid to return to their abandoned fields even though the authorities urged them to.

Since that time, the conflict has died down, but tensions simmer below the surface. They reemerged in 2000 when the PNGT project was in the midst of creating a livestock route from the village of Sara to pasture areas. Mossi farmers were very much against the project because it seemed to go through only their fields. Bwa farmers dominated the zoning committees, and as a result only Mossi farmers were required to abandon fields to facilitate the livestock path. At the time I left, the zoning plan had still not been completed, but Mossi farmers were extremely unhappy and felt they had been targeted by the committee.

Discussion and Recommendations: Rethinking the *Gestion des Terroirs* Approach in Burkina Faso

Each of the case studies illustrates the problematic nature of land relations in Burkina Faso. The first case study illustrated that the *gestion des terroirs* approach as applied in southwestern Burkina Faso has in general failed to achieve its goals of conserving natural resources or making natural resource governance more equitable. Attempts to devolve control have been fraught with conflict and have generally been abandoned. The second case study demonstrates the conflictual nature of land relations in southwestern Burkina, where conflicts between locals and migrants have made any sort of *gestion des terroirs* intervention difficult.

The case studies illustrate that many of the customary institutions for decision making at the local level, such as councils of elders, are based on inequitable and nonparticipatory principles. The authority of decisions made in such a manner usually derives from the status of the decision maker rather than from compliance with rule-based, democratic processes. As a result, such decisions seldom reflect a consensus of opinion within the community. Just because *gestion des terroirs* projects designate committees as "participatory" does not mean that they are legitimate in the eyes of villagers (Engberg-Pedersen 1995, 15). Thus, policy made by these committees, rather than resolving conflicts through debate, compromise, and consensus, actually allows animosity to persist and fester.

In the case study area, local norms about land give clear rights to Bwa farmers and lineage elders. Mossi farmers, women, and other community members generally are not viewed as having rights to land. This is reflected in each of the examples presented earlier: despite the fact that Mossi farmers were represented in committees, and in most cases outnumber the Bwa, they ultimately had little influence over the final decisions on zoning land. Thus projects, despite attempts to use participatory methods of planning, should be aware that they might be empowering institutions that are not based on democratic principles. Some villagers may have a greater voice in the decision-making process and see greater benefits of projects.

The case studies also remind us that participatory, community-based management schemes such as the *gestion des terroirs* approach do not exist in a political vacuum. Local institutions for resource management do not replace or supersede the institutions of the central state. In the cases discussed above, Bwa and Mossi farmers, dissatisfied with the outcomes of village-level management and preoccupied with increasing social tension, appealed to state agents such as the prefect and provincial governor; in the latter case, even the military police became involved. In all cases, state agents made arbitrary judgments or followed the directives of their superiors, consistently choosing the path of least resistance, thereby leaving ethnic tensions to fester in the long run, presumably to be dealt with by some other unlucky bureaucrat. In comparison to state institutions in Burkina Faso, village-level management seems enlightened, democratic, and participatory; unfortunately, when decision making breaks down at the local level, villagers can appeal only to the apprehensive, arbitrary administrators of the central state. While recent critiques of village-level management have focused on problems of downward accountability (Agrawal and Ribot 1999), attention must also be paid to the quality of governance and leadership at higher levels.

The agenda of managing natural resources by zoning is not an idea that has emerged from local communities. A contradiction emerges because the push for both village participation and decentralization arises from the perception of international donors that peasants know how to manage their natural resource base better than government officials. Unfortunately, this is frequently contradicted by project staff and government officials, who, though schooled in the language of participatory development, blame peasant farmers for degradation. This commitment to indigenous knowledge on the one hand, and a criticism of local practices as degrading, sets a contradiction in motion whereby projects that purport to be participatory actually rely on top-down models of knowledge extension. Batterbury (1998, 893) illustrates this in his study of the *gestion des terroirs* approach in northern Burkina Faso, concluding that the approach is a "'second-best' form of community development, because it is initially managed 'from above' despite a populist framework and strong local input."

What are the policy implications of this study? A key question is whether the local scale is the correct place to undertake land policy in Burkina Faso, particularly when local-level political struggles over resources may drive the resource allocation process. The complex nature of local land systems is further complicated in West Africa by the fact that dual systems of formal and customary control over land coexist, neither of which is totally dominant (Lavigne Delville 2000). National laws generally assert that the state owns all land, but effective control is held at the local level. Local programs that attempt to clarify land rules by delimiting rights provide yet another level of authority to manipulate by local political actors. Nongovernmental institutions are not viewed with legitimacy in resolving local conflicts, and local representatives of the central government are also unclear about where authority lies. Due to the multiplicity of authorities arbitrating land and resource conflicts, the system is easily manipulated (Lavigne Delville 2000).

It remains to be seen whether national governments and local peoples will assume "ownership" over externally conceived decentralization, particularly in the context of local governance systems that are neither representative nor accountable. Much of the effort to decentralize land tenure has emerged from external agendas and is implemented under the conditionality of structural adjustment packages. Research indicates that many locals have not taken ownership over decentralization efforts, but see them as managed by outside interests (Engberg-Pedersen 1995; Gray 2002). Villagers have little say about agenda setting or problems that they perceive as most problematic. Thus, what is purported to be a bottom-up exercise ends up being perceived locally and at other scales as top-down (Batterbury 1998). The case studies show that local actors do not just passively accept project goals but manipulate projects to achieve their own political goals. When conflicts erupt, projects have generally reacted by abandoning projects, contributing to the sense that decentralization is something imposed by external actors.

The large question that must be resolved is one that cannot be solved at the local scale: what are the rights of migrants to land? This issue is one that looms large in much of Sub-Saharan Africa, where customary rights do not guarantee permanent rights, particularly to migrants. Governments have generally not wanted to address this issue because of the potential for more overt conflict and violence and because granting rights to one group may mean disenfranchising another. Instead, what generally happens is that local-level conflicts create reality on the ground, a process that does not generally guarantee fair outcomes.

How could a better *gestion des terroirs* approach be created? One of the first recommendations would be to give communities real power by creating locally elected institutions that are seen as legitimate and representative. As it stands, most land zoning efforts are guided by nongovernmental organizations that locally have little legitimacy. Local government in Burkina Faso, despite moves to decentralize services, is little more than local administration, representing an appointed arm of the central government.

Second, conflict resolution must be part of any program. Given the likelihood of conflict in any attempt to restructure resource access, programs must see villages as essentially conflictual territories and think about ways to create conflict resolution. One possible avenue for reconciliation would be to create authorities that have the right to settle disputes and enforce outcomes. These might be in the realm of customary authorities or formal authorities. But a key element of this is to adjudicate in a way that is both transparent and accountable (IIED 1999) and is ultimately perceived to be fair. One of the problems with governance structures is that formal authorities are largely unaccountable to local populations, while local authorities are often seen to be biased against the interests of migrants. For any process to be seen as fair, it will have to overcome these challenges. Clearly, successful conflict resolutions mechanisms are linked to increased democratization and representation in the powers that determine resource allocation.

Third, people need to know that they can participate in changing rules for natural resource management as long as they can do so in a way that does not intentionally disenfranchise them. Real ownership over resources can lead to better management. Finally, projects need to build on indigenous innovations for resource management. It is clear that projects attempting to idealize property rights as communal and unchanging are failing. The state must recognize growing individual control over land. However, attempts to formalize tenure through titling or registration have generally failed, resulting in greater insecurity. In West Africa, there is a growing movement to formalize informal land transactions in an incremental way. In several regions in Sub-Saharan Africa, local peoples are creating and demanding written contracts. Mathieu (2001) argues that under conditions of uncertainty and fast-paced agrarian change, land users use these contracts to guarantee their rights to land and forestall counterclaims. This informal recognition of land transactions—which in essence clarifies the terms of the agreement—might reduce conflicts since the terms of the agreement are laid out and often witnessed and signed. Because these are done at the local level, the administrative burden is minimal. This could potentially lay the groundwork for more institutional capacity and markets for land transactions (Lavigne Delville 2000).

References

Agrawal, A., and C. C. Gibson. 1999. Enchantment and disenchantment: The role of community in natural resource conservation. *World Development* 27 (4): 629–49.

Agrawal, A., and J. Ribot. 1999. Accountability in decentralization: A framework with South Asian and West African Cases. *Journal of Developing Areas* 33:473–502.

Atwood, D. A. 1990. Land registration in Africa. *World Development* 18 (5): 659–71.

Bassett, T. 1993. Introduction: The land question and agricultural transformation in Sub-Saharan Africa. In *Land in African agrarian systems,* ed. T. Bassett and D. Crummey, 3–31. Madison: University of Wisconsin Press.

Batterbury, S. 1998. Local environmental management, land degradation and the "gestion des ter-roirs" approach in West Africa: Policies and pitfalls. *Journal of International Development* 10:871–98.

Benjaminsen, T., and C. Lund, eds. 2001. *Politics, property and production in the West African Sahel: Understanding natural resources management.* Uppsala, Sweden: Nordiska Afrikainstitutet.

——, eds. 2003. *Securing land rights in Africa.* London: Frank Cass.

Boone, C. 2003. Decentralization as political strategy in West Africa. *Comparative Political Studies* 36 (4): 355–80.

Boutillier, J. 1964. *Les structures foncières en Haute-Volta.* Ouagadougou, Burkina Faso: Centre IFAN-ORSTOM.

Braselle, A. S., F. Gaspart, and J. P. Platteau. 1998. Land tenure security and investment incentives: Further puzzling evidence from Burkina. Namur, Belgium: Facultes universitaires Notre Dame de la Paix, Faculte des sciences economiques, sociales et de gestion, Centre de recherche en economie du developpement (CRED).

Brosius, J. P., A. L. Tsing, and C. Zerner. 1998. Representing communities: Histories and politics of community-based natural resource management. *Society and Natural Resources* 11 (2): 157–69.

Carney, D., and J. Farrington. 1998. *Natural resource management and institutional change.* London: Routledge.

Cordell, D., J. Gregory, and V. Piche. 1996. *Hoe and wage: A social history of a circular migration system in West Africa.* Boulder, CO: Westview.

Donnelly-Roark, P., K. Ouedraogo, and Xiao Ye. 2001. *Can local institutions reduce poverty? Rural decentralization in Burkina Faso.* Washington, DC: Environment and Social Development Unit, Africa Region, World Bank.

Engberg-Pedersen, L. 1995. *Creating local democratic politics from above: The* gestion des terroirs *approach in Burkina Faso.* London: International Institute for Environment and Development.

Faure, A. 1992. *Perceptions de l'approche gestion des terroirs par les populations rurales au Burkina Faso.* Paris: Caisse Centrale de Coopération Economique.

Gray, L. C. 2002. Environmental policy, land rights and conflict: Rethinking community natural resource management programs in Burkina Faso. *Environment and Planning D: Society and Space* 20 (2): 167–82.

Gray, L. C., and M. Kevane. 2001. Evolving tenure rights and agricultural intensification in southwestern Burkina Faso. *World Development* 29 (4): 573–87.

Haugerud, A. 1989. The consequences of land tenure reform among smallholders in the Kenya highlands. *Africa* 59:62–90.

IIED (International Institute for Environment and Development). 1999. *Land tenure and resource access in West Africa: Issues and opportunities for the next twenty-five years.* London: IIED.

Kellert, S., J. Mehta, S. Ebbin, and L. Lichtenfeld. 2000. Community natural resource management: Promise, rhetoric and reality. *Society and Natural Resources* 13:705–15.

Larson, A. 2002. Natural resources and decentralization in Nicaragua: Are local governments up to the job? *World Development* 30 (1): 17–31.

Lavigne Delville, P. 2000. Harmonizing formal law and customary land rights in French-speaking West Africa. In *Evolving land rights, policy and tenure in Africa,* ed. C. Toulmin and J. Quan, 97–122. London: International Institute for Environment and Development.

Leach, M., R. Mearns, and I. Scoones. 1999. Environmental entitlements: Dynamics and institutions in community-based natural resource management. *World Development* 27 (2): 225–47.

Lund, C. 2001. Questioning some assumptions about land tenure. In *Politics, property and production in the West African Sahel: Understanding natural resources management,* ed. T. Benjaminsen and C. Lund, 144–62. Uppsala, Sweden: Nordiska Afrikainstitutet.

Mathieu, P. 2001. Transformations informelles et marches foncieres emergents en Afrique. In *Politics, property and production in the West African Sahel: Understanding natural resources management,* ed. T. Benjaminsen and C. Lund, 22–39. Uppsala, Sweden: Nordiska Afrikainstitutet.

Neumann, R. 1997. Primitive ideas: Protected area buffer zones and the politics of land in Africa. *Development and Change* 28:559–82.

Paré, L. 2001. *Negotiating rights: Access to land in the cotton zone, Burkina Faso.* London: International Institute for Environment and Development Drylands Programme.

Peters, P. 1996. Who's local here? The politics of participation in development. *Cultural Survival Quarterly* 20 (3): 22–25.

Platteau, J. P. 2000. Does Africa need land reform? In *Evolving land rights, policy and tenure in Africa,* ed. C. Toulmin and J. Quan, 51–74. London: International Institute for Environment and Development.

PNGT (Programme National de Gestion des Terroirs). 1989. Rapport de synthese et d'analyse des experiences pilotes de gestion des terroirs villageois. Ouagadougou, Burkina Faso: Cellule de Coordination du Programme National de Gestion des Terroirs Villageois.

Ribot, J. 1996. Participation without representation: Chiefs, councils and forestry law in the West African Sahel. *Cultural Survival Quarterly* (Fall): 40–44.

———. 1999. Decentralisation, participation and accountability in sahelian forestry: Legal instruments of political-administrative control. *Africa* 69 (1): 23–65.

Schroeder, R. A. 1999. Geographies of environmental intervention in Africa. *Progress in Human Geography* 23 (3): 359–78.

Schwartz, A. 1991. L'exploitation agricole de L'aire cotoniere Burkinabe: Caracteristiques sociologiques, demographiques, economiques. Document de Travail. Ouagadougou, Burkina Faso: Office de la Recherche Scientifique et Technique Outre-Mer (ORSTOM).

Shipton, P., and M. Goheen. 1992. Understanding African land-holding: Power, wealth and meaning. *Africa* 62:307–25.

Thomson, A. 2000. *An introduction to African politics.* London: Routledge.

Toulmin, C. 1994. *Gestion de terroir: Concepts and development.* London: International Institute for Environment and Development.

———. 2000. Decentralization and land tenure. In *Evolving land rights, policy and tenure in Africa,* ed. C. Toulmin and J. Quan, 229–45. London: DFID/International Institute for Environment and Development/NRI.

Turner, M. 1999. Conflict, environmental change, and social institutions in dryland Africa: Limitations of the community resource management approach. *Society and Natural Resources* 12 (7): 643–58.

World Bank. 1997. *World development report: The state in a changing world.* New York: Oxford University Press.

———. 1999. *World development report: Knowledge for development.* New York: Oxford University Press.

———. 2000. *World development report: Attacking poverty.* New York: Oxford University Press.

———. 2001. *Burkina Faso: Poverty reduction support credit.* Washington, DC: World Bank.

13 Fences, Ecologies, and Changes in Pastoral Life: Sandy Land Reclamation in Uxin Ju, Inner Mongolia, China

HONG JIANG

Decentralization has been an important global trend in recent decades, driven by the neoliberal economic agenda that promotes market economic incentives of decentralized management. In tandem with this economic logic is the conviction that devolution of decision making in resource use will promote environmental sustainability (World Bank 1997). For these reasons, in developed countries a trend toward decentralization in the provision of public services has intensified (Bennett 1990), and the community-based conservation movement has been heralded as a grand vision (Weber 2000). In developing countries, especially in Africa and Latin America, the World Bank and other international organizations have promoted decentralization as a prerequisite for sustainable resource use and environmental conservation (World Bank 1997). While decentralization generally devolves power from the central government to local administrations and communities, in more drastic changes, it has privatized land and resources previously held by the state and collectives (Turner 1999).

Perhaps nowhere else is the global trend of decentralization as prominent as in postsocialist regimes, as an element of neoliberal "structural adjustment" programs to fix decrepit economies and degrading environments (Archibald, Banu, and Bochniarz 2004; Gutner 2002; Pavlínek and Pickles 2000). In China, for example, active engagements with the market and decentralization of governance have accompanied the weakening of Marxist ideologies since the 1980s (Lavigne 1999; McMillan and Naughton 1996; Turner 1999). China's decentralization has several unique characteristics. First, in terms of rural land use, instead of community-level management, which has been a common form of decentralization globally, China has drastically devolved decision-making power to the level of households. Most cropland and much pastureland have been contracted to households through a Household Responsibility System (Unger 2002). Although the state retains ownership of land resources, households have been given long-term rights to use the land; thus, the contract system has been described as a process of privatization, which is largely shared by other postsocialist countries (Turner 1999).

Second, various levels of the government have played crucial roles in the decentralization process. Decentralization was an important part of the national policy shift from revolutionary politics to economic growth, and various levels of the state continue to control land use in important ways. This stands in sharp contrast to much of Africa and Latin America, where state involvement is weak and it is mainly international agencies, donors, and nongovernmental organizations (NGOs) that have provided the engine for decentralized resource use (Schroeder 1999; Sundberg 1999). While decentralization has been connected with democratic forms of government, China sets itself apart from the widespread postsocialist democratic transition, since it has remained a one-party authoritarian state in which decentralization is promoted as a governmental policy of economic growth. In comparison, in Russia and Eastern Europe, democratic politics and the weakening of authoritarian control have been an important part of power devolution and resource privatization (see, e.g., Alm and Buckley 1994; Pickvance 1998). Third, as is common in postsocialist transition, China's decentralization has come with an economic reform policy that aimed to reverse the economic failure of the former collective management (Jia and Lin 1994; Lin 1999). This increasing drive for development and growing reliance on the market have resulted in environmental issues' being viewed through economic lenses (Jiang 1999).

The unique Chinese experience provides an important sociopolitical context in which to explore decentralized pastureland use in this paper. The complicated results of decentralization have been widely acknowledged (Lutz and Caldecott 1996; Ribot 1999). While some champion decentralization's positive effects on environmental sustainability (e.g., Bakir 2001; Oliveria 2002), many studies reveal the unfulfilled promises of decentralization. Sundberg (chap. 11 in this volume) and Gray (chap. 12 in this volume) suggest that community-level resource use can often exacerbate existing inequalities within the community, and consequently there is no guarantee for equitable use; Lane (2003) warns of the manipulation of decentralized resource management by local power holders; Nemarundwe (2004) argues that overlapping authority at the community level can serve to frustrate, rather than clarify, resource use; and Williams and Wells (1996) and Zhang (2000) report cases of reduced funding and concern for conservation as the power is decentralized and responsibility for environmental initiatives is placed on regional and local governments. While political, administrative, and financial difficulties can affect decentralized resource use, ecological dynamics add another level of complexity (Leach, Mearns, and Scoones 1999). As Assetto, Hajba, and Mumme (2003) indicate, decentralization is a necessary but not a sufficient condition for environmental protection and equitable resource use.

This chapter joins in the study of decentralized resource use by analyzing the impact of household-based pastureland management in Uxin Ju, a community in western Inner Mongolia, China. After pastureland is distributed and fenced, sandy land reclamation becomes a growing concern as degradation rises. While household-based pastureland use and reclamation have brought economic growth to the area, I argue that their effects are not uniformly positive when cultural change and ecological processes are taken into consideration. Instead of environmental conservation within designated protected areas, I discuss pastureland use that is an integral part of rural livelihood. Issues to be explored here are relevant to the third wave of conservation that emphasizes conservation with sustainable use (Brandon, Redford, and Sanderson 1998). While much of this chapter will discuss more aggressive approaches to resource use that prioritize economy over ecology, toward the end I will discuss a renewed need for stricter conservation precisely because of this aggressive use.

This chapter proceeds as follows. In the first section, I will introduce how decentralization is implemented in Inner Mongolia, and its forms on the landscape in Uxin Ju, particularly the building of fences. The second section examines ways in which sandy land is reclaimed inside fences. The third section examines the impact of fencing on the Mongolian lifestyle, asserting the far-reaching cultural implications of resource privatization. While in general fences have led the Mongols away from traditional nomadism, ecological processes of pastureland management have resulted in uneven landscape changes and concomitant inequality among the Mongols. The next two sections explore such landscape and socioeconomic differentiation, and discuss the need for a less aggressive approach to pastureland conservation. This paper concludes with a critique of household-level resource management from the perspectives of ecological processes, cultural change, and management scales.

Household-Based Pastureland Management

Uxin Ju is a Mongolian township situated in Uxin banner, Ih Ju league in Inner Mongolia Autonomous Region, China. It has four administrative villages (or *gacha* in Mongolian): Uxin Ju, Bayintaolegai, Chahanmiao, and Buridu (see fig. 13.1). Its population is dominated by Mongols, and its economy is based on sheep and cattle husbandry. Much of its 1,744 km^2 (or 2.6 million *mu*) is dominated by sandy land (83 percent), and land forms include upland (usually sandy), lowland, and sand dunes in between. The sandy upland and dune areas, if vegetated, are covered by *Artemisia ordosica* and *Caragana* spp. In between the moving sand dunes are small depressions where water conditions

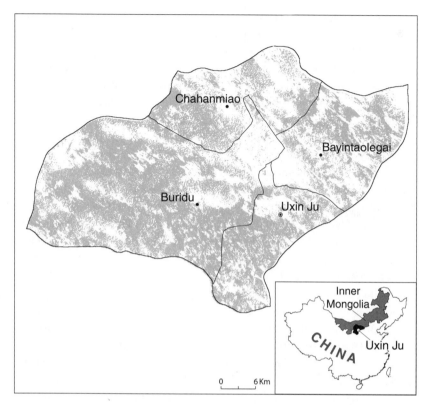

FIGURE 13.1. Map of study area: Uxin Ju

are favorable and willow shrubs (*Salix psammophila*) can take root. The low-land, which has the most favorable water conditions, is dominated by grassy species *Carex duriuscula* and *Achnatherum splendens* and provides the best quality pastures in Uxin Ju.

China's economic reform has brought about profound change to the landscape of Uxin Ju. Faced with the failure of the collective economy, the Chinese central government launched an economic reform policy in the late 1970s after the death of Mao. The reform started in the farming areas, where, under a Household Responsibility System (HRS), cropland and other production materials, previously managed by the communes, were distributed to the households. The HRS soon spread to the pastoral areas, and the Inner Mongolia regional government took the lead in devising and implementing a two-tier HRS: both livestock and pastureland would be distributed to households. While the livestock was sold to households as private property, the state retained ownership of land, the use rights to which were contracted to the households. The distribution of pastureland has been subject to periodic adjustment based on household demographic change. In Uxin Ju and many areas of Inner Mongolia, pastureland distribution was twice adjusted, in 1991 and 1997; the last adjustment also extended the contract terms to thirty to fifty years from an unspecified period (Jiang 2006).

The household contract has the de facto effect of privatization, albeit a limited one, since land cannot be bought or sold but only have its use rights transferred. Williams (2001) traces the shift to household-based management to the global distrust of common resource institutions and the common belief that privatization will promote ecological protection. Decentralization was widely acclaimed by the Chinese government as a cure for the economic and environmental problems of the collective management in the Mao era. The local area newspaper in Uxin Ju, *The Ordos Daily* (2 February 1989), celebrated the pastureland distribution with an article entitled "Now the Pasture Has Masters," which asserted that this decentralization is the best way to care for the land and to develop the economy. Consequences of privatization will be discussed in greater detail later, but, in brief, the acclaimed promise of household-based pastureland use has been undermined by ecological, cultural, and economic problems.

Not all of China's pastoral areas have implemented the two-tier HRS. Some have kept the pastureland as a common resource (Banks 2001), or demarcated household pastures only on paper (Thwaites et al. 1998), or fenced the pastureland only partially (Williams 2001). But in Uxin Ju, pastureland was fenced soon after its distribution to households in 1984–85. Such a wide variation in pastureland decentralization is attributed to differing priorities of regional and local governments, which are responsible for the actual implementation of many of the reform policies (Jiang 2006). In Inner Mongolia, while the regional government issued policies regarding HRS, methods of implementation became more detailed at the lower levels of government. The actual pastureland distribution was carried out by village (*gacha*) committees, which stipulated distribution rules. In the communities that I have investigated around Uxin Ju, two types of distribution rules were employed: one, based on strict equality of all people, was to distribute pastureland according to the number of persons in a household; the second was to take into consideration both the number of persons and livestock in a household. Uxin Ju followed the first rule. To minimize travel distance, distribution plans attempted to assign pastureland to the nearest households (instead of giving each household a piece of each type of land, as in farming areas; see Kung and Liu 1997). To account for variations in species, coverage, and productivity, all pastureland was assessed according to usable biomass and converted to "standard pasture," each *mu* corresponding to 100 kg of dry forage. The built-in structural equality in pastureland distribution comes from the Mao era's ethos (Kung and Liu 1997); such equality, although subject to cadre manipulations, helps minimize the struggle for access that is common in the decentralized resource use in Africa (Gray, chap. 12 in this volume) and Latin America (Sundberg, chap. 11 in this volume). In Uxin Ju, however, initial conditions of equality have been undermined by socioeconomic stratifications resulting from uneven landscape change, as will be discussed later.

Barbed wire fences quickly followed pasture distribution, creating enclosures called *kulum* by the Mongols. From the experience of eastern Inner Mongolia (Thwaites et al. 1998), it is clear that unless boundaries are demarcated with fences, pastureland distribution cannot be adhered to, since livestock will roam into unfenced pastures. In Uxin Ju, the people's ready acceptance of the concept of fences came from the early socialist period when they built large enclosures in order to protect planted trees and crops, as well as to conserve pasture for the cutting of hay. In the 1980s, after the pastureland distribution, the practical need for fences quickly became clear since households had to protect their allotted pastures from turning into commons. The government encouraged fencing by providing subsidies. Data from Ih Ju League show that from 1978 to 1996, government funds accounted for 23.3 percent of total investment in pasture enclosure (Liu and Wang 1998). The rest of the funds came from individual households. In a 1998 interview, Bayingsonbuer, a Mongol living in Bayintaolegai *gacha,* said that since 1985, his household had spent over 20,000 yuan (approximately US $2,400) on fences. This is a hefty sum given that the 1998 per capita income was only 1,952 yuan (US $235). By the late 1990s, all contracted pasturelands in Uxin Ju, including those assigned to the poor, had been fenced.

Sandy Land Reclamation
With fences came more aggressive use of the pastureland as well as increased emphasis on sandy land improvement. Since the implementation of economic reform, China has paid more attention to rural environmental issues and has attempted to use ecological science to guide environmental management. The term *ecological construction* came into wide use in the 1990s, referring to the repair of land degradation using ecological principles (Jiang 2006). In reality, however, the implementation of ecological programs by all levels of the government has focused on aggressive approaches that undermine ecological processes. In Uxin Ju, for example, instead of protecting or recovering the original species, improvement efforts only considered biomass. Trees, which were nonexistent on the natural landscape, have been planted, along with exotic shrub and grass species. Even the expansion of irrigated cropland has been seen as sign of ecological improvement. Concern for sustainability has been tilted toward the economy, while ecology has largely been seen as providing a necessary service.

Sandy land reclamation entails the conversion of degraded sandy pastureland and moving sand into usable pastures, and it is achieved through two kinds of efforts: transplanting and seeding, both followed by a prohibition of grazing in enclosures, at least in the summer, and sometimes year-round, for three years. The timing and length of these grazing bans have been determined and monitored by the local government, and violators are fined. Transplanting uses seedlings of trees and shrubs, and is practiced in enclosures closer to houses. In recent years, the total transplanted area per year has reached 30,000 *mu* or more. The most commonly transplanted tree species is the willow (*Salix matsudana*), and shrubs include local species such as *Salix psammophila, Artemisia ordosica,* and *Caragana* spp., and nonlocal species such as *Hedysarum mongolicum* and *H. scoparium.* Methods of sand dune reclamation, which have been developed locally since the 1960s, are described as "blocking the front and hauling the back; boots first, then gowns and lastly hats." That is, trees and shrubs are planted first on the lower part of leeward (front) and windward (back) sides. As the vegetation develops at the bottom of the sand dune like "boots," the movement of sand is constrained, and the sand dunes are lowered. Planting slowly moves up to the sand dunes and dresses them in green "gowns" and finally covers them with "hats." This method has worked well in some locations and has converted sandy land into usable pastures. It meets great difficulty, however, at the landscape scale, as will be discussed later.

Seeding is done with the help of airplanes that are operated by the local government. In 2001, Uxin Ju seeded 122,000 *mu* of sandy land. Airplane sowing is a labor-saving method of land improvement used mostly in large continuous areas of sandy, semivegetated, and low-lying land. In addition to shrubs, grasses such as *Melilotus albus* and *Medicago sativa* are also seeded. Since the area to be sown has to be large enough to justify the operation cost, airplane seeding goes beyond the boundaries of fences and requires coordination among multiple households. Since decentralization, this is one of the ways in which collective management maintains great importance. Funds for seeding are allocated from the government to the village. Once an area is chosen for operation by the technical staff in Uxin Ju, the households involved are expected to participate. If no rain follows seeding in ten to fifteen days, the operation will most likely fail. In most cases, even if seeds germinate, success is only partial because subsequent droughts may cause the young seedlings to die. The government reports that the average success rate for airplane seeding is 80 percent, but my interviews suggest that the actual seedling survival rate averages 50 percent or lower. In the recent drought from 1999 to 2001, most seeds failed to germinate. Regardless of seeding success, the participating households are obligated to follow the grazing ban for a three-year period after seeding. From the fourth year on, they can graze only in the cold season. Most households prefer to have their pasture seeded because of potential near-future gain, so there have not been major conflicts between the participating households and the government. However, there are cases when households are poor and have difficulty paying their share of the seeding cost. After their pasture is seeded and closed, most of these households have to rent pastureland from other households at 0.25 yuan per sheep per day; this only increases the financial burden on poor households.

Reclamation is a joint venture between the government and individual households. Although land use decisions are made by the households, the government still plays a crucial role in guiding land use through managerial and financial means. All government leaders are allocated responsibility to help a certain number of households in the reclamation process. Each year, Uxin Ju has spring and fall planting sprees during which all leaders go to their selected households to participate in planting. Each *gacha* also designates several large areas for concentrated efforts, and households that contracted the pasture of these areas are expected to plant trees there during the planting seasons. Government funding for planting and seeding is available through specific programs such as the North China Revegetation and Return Cultivation to Trees programs from the national government and the Family Pasture program from the regional government. Specific programs change over time, but overall, most of the seedlings and seeds are purchased with government funds, and labor input comes from households. In airplane seeding, the banner government paid 100 percent of the operational cost and seeds before 1995; now, it pays only slightly more than half, while requiring that households pay 2.5 yuan per *mu*.

Most households actively participate in sandy land reclamation in order to augment household pastures. A sample of the household surveys that I conducted in 2001 shows that 48 percent of the sandy land had been transplanted or seeded by the households, although not all efforts were successful. In addition, the number of trees (willow and poplar) has increased drastically. Official statistics show that the number of trees in the late 1960s totaled 80,000, with each household averaging about ninety trees. By 2001, my sampled households each owned, on average, 2,800 trees. These trees, along with irrigated cropland and some planted shrubs, are located in what is called "cultivation enclosures." Although not used for grazing, this type of enclosure is the most productive and also the newest modification to the landscape that makes the pastoral area most farmlike.

Other than the regularly contracted pastureland, there are also "wastelands" that were not initially distributed in 1984. These are moving sand areas or sand dunes, called "sandy waste." Instead of allowing these lands to lie "waste," the local government contracted them out to households on a voluntary basis and encouraged households to reclaim them. According to the local policy, any wasteland reclaimed will be awarded to the households that improved upon it as additional tax-free pasture in addition to their original allocation. Since all usable pastureland had been distributed, the only way to increase pastureland was to reclaim these wastelands. By 1991, most waste-lands had been contracted. In most cases, large areas of sandy wasteland were contracted by multiple households. Some wastelands in remote locations were rented to outside groups (such as the First Chemical Engineering Plant of Ih Ju league). My interviews suggest that some households have already im-proved upon these wastelands with fences and planting; others have started to make plans for reclamation.

Changing Pastoral Life under Fences

Land use concerns are not only economic but also cultural. As Turner (2002) points out, ways of making a living are closely related to ethnic identity. The majority (90 percent) of those managing Uxin Ju's rural landscape are Mon-gols, whose traditional lifestyle was nomadic grazing. As pastures and sand dunes become constrained by fences, mobility for both livestock and pastoral-ists ends. In livestock raising, increases in cropping and planting have shifted the past reliance on natural pasture to increasing dependence on human-produced biomass. On the positive side, planted trees (leaves), shrubs, and grasses as well as crops have provided additional forage and feed, and this has made livestock grazing more resistant to drought and winter shortage. The pastoral economy has thus been improved. But economic benefits come with costs and consequences. First, as a result of pasture improvement and pro-tection, goats, a traditional livestock, have been eliminated since they "de-stroy the planted trees" and "browse bare the pastures" (Wushen Qi 2001). Goats' durability and adaptability to the dry, shrubby vegetation has now be-come their liability. After several decades of reduction, by 1999, the number of goats had been reduced to 73 from 40,000–60,000 in the 1970s. Goats may indeed be maladapted to the newly created landscape, yet the impact of their removal on local diet and health warrants careful research. A second conse-quence is that rotational grazing, practiced over large areas in the past, has now become impossible. While livestock privatization encourages the growth of sheep numbers, sheep are now grazed for longer periods of time (two months at a minimum) on one piece of pasture within fences.

Rotational grazing in enclosures is the Mongols' way of adapting to the new social and environmental situations. Traditional nomadic grazing in-volves frequent movement of herds on large open pastureland. Uxin Ju people call it *zouchang* (walking the field). This extensive mobility not only protects the pastureland from overuse, but also encourages more efficient use of pastureland resources (Humphrey and Sneath 1999). Nomadic grazing started to be limited after the 1600s when the Qing court restricted Mongol princes to fixed banner lands (Barfield 1989). With the establishment of com-munes in 1958 under the socialist regime, land was further divided according to the basic production unit of a brigade (about twenty to thirty households). Limited long-distance movement was still possible as Uxin Ju swapped land use with Otog banner to the west and grazed on their highland stony shrub in the summer. Now, with fences and household management, pastureland is too divided to allow for any extensive movement of livestock. Pastoralists have adjusted to this new pastoral condition by rotational grazing on their small patches of pasture inside fences. Grazing on other households' pastureland is still practiced, but the relationship is that of land rent rather than land use swapping. Some households rent pastureland from others at a fixed fee and use it more intensively; some pay according to sheep number and length of grazing to graze their sheep on others' allocated pastureland. However, neither of these methods helps protect the pastureland as mobility did in the past. It is not surprising that Williams (2001) suspects that pastureland distribution might have marked a greater transformation of pastoral life than collectivization in 1958.

With fences, the migratory routine of human life is disappearing, and in its place is the settled routine of a busy life. Although livestock can be left to graze in fenced enclosures without much supervision, thus reducing labor requirements for grazing, overall labor demands have drastically increased. With trees to plant, tree leaves to cut, crops to irrigate and cultivate, fences to mend, and more work required to feed the improved sheep variety and to prevent disease, pastoralists' daily routines have become much more labor intensive. To illustrate, Bayinsongbuer complained that he had to spend two to three months each year mending fences. The cropping season is especially labor intensive for pastoralists who are not used to the rhythm of farming. It is no wonder that my interviewees commented about the loss of leisure life. When asked whether they would trade their current busier but materially richer lifestyle for their past leisure, however, most of my interviewees indicated that they would not. This juxtaposition of complaint and acceptance points to the pastoralists' internal conflicts between tradition and change.

Williams (2001) maintains that pastureland privatization represents Chinese cultural and political imposition. Household use is a familiar relationship to the land for the Chinese farmers, and the tendency to fix people to particular locations comes from the farming tradition. But for the Mongols, pastureland has traditionally been used as a common resource. As a traditionally Mongolian area, Uxin Ju has long been under the influence of Chinese culture and politics, and change in land use is closely related to the process of Sinicization. I would submit that in Uxin Ju, such influence is not as hegemonic as Williams suggests. Many Mongols have learned to adapt to the lifestyle change under fences and have started to embrace planting and cropping as an integral part of Mongolian identity (see Jiang 2004 for details). Ironically, the Mongols' acceptance of sedentarized agropastoralism, while culturally empowering, seems only to accelerate the loss of traditional no-madism.

Ecological Processes and Uneven Landscape Consequences of Reclamation
Unlike flat land, sandy land is undulating and rolling; it is a landscape of elas-ticity created by the movement of sand and pasture. As sand dunes move with the direction of the wind, new pastures are created on locations the dunes have just vacated. This dune land remains resilient to moving sand, in con-trast to the common view that this land is fragile and easily sandified (e.g., Sun 1990). Bare sand dunes serve important roles in supporting functions of the landscape. Precipitation quickly filtrates into the ground to supplement the groundwater, which in turn supplies water for the lowland. Moving sand dunes are considered areas of "water provision." Instead of being "waste-lands," moving dunes are a crucial component on the landscape that nur-tures the key lowland pastures. Research done by the Chinese Academy of Science led by Zhang Xinshi shows that if planting on sand dunes is too dense, not only will the established vegetation die off after a few years, but the adjacent lowland will also be adversely affected (X. Zhang, personal com-munication, 2001).

Two reasons explain why moving sand is now considered a problem to so-ciety. First, pastureland overuse has caused more sand to be exposed. Not only has the number of livestock increased but grazing intensity has also been heightened with fences. In 1998, official statistics showed that about 50 per-cent of Uxin Ju was occupied by moving sand, much of which was considered to be caused by human misuse. Many scholars emphasize the human causes and human control of sand expansion (Huang and Song 1986; Sun 1990), and most people I interviewed believe that much of the sandy land can be im-proved. Expansion of sand is seen as a major problem that reduces usable pas-tureland. Instead of seeing moving sand as a natural function of the land-scape, scholars and local people use it as a measure of degradation. Once put in the category of degraded land, sandy land has to be reformed. Second, the mobility of sand challenges sedentary lifestyles. In nomadic societies, as sand moved, so did people, and thus the human relationship with the sandy land environment was more harmonious (Chen 2001). Such mobility, however, was only possible when intensity of use was low. Now that people are more densely settled, mobility of sand has become hazardous to houses, animal shelters, and roads (Huang and Song 1986). The elasticity of the dune land-scape contradicts with the fixed allocation of household pastureland. As a re-sult, households attempt to constrain sand dunes and to prevent dune sand from moving onto their pastureland.

Despite the human efforts to secure the landscape with fences, natural processes continue: sand moves across fences. When sand control is not successful, movement of sand across fences creates conflicts between house-holds, such as was the case for Ererdun and his neighbor. Ererdun has made tremendous efforts to reclaim most of his sandy land, but his neighbor in the windward northwest direction has not controlled his sand in the enclosure adjacent to Ererdun's. A moving sand dune, previously 20 m away, has now crossed the fence and covered Ererdun's pasture, consequently creating a new piece of pastureland within the neighbor's fence. With pastureland distribution fixed for thirty years, such neighborly disputes cannot be truly mitigated. Like wildlife that cannot be restrained within conservation boundaries (Naughton-Treves et al. 2003), moving sand does not obey the authority of fences.

Sandy land reclamation inside fences has come with an unintended environmental cost across the landscape. The rapid increase in trees and irrigation has lowered groundwater levels, a problem that has only been exacerbated by the recent drought from 1999 to 2001. This has caused the reduction in interdune vegetation and enabled the expansion of sand dunes. Although tremendous effort is devoted to keeping the sandy land "fixed," it can only succeed in limited areas, and given the connection of the landscape to groundwater, this fixation of landscape in some locations has come at the cost of more moving sand in other locations. With groundwater remaining a "common" resource, household-based pastureland use produces ecological externalities that have not been considered in the privatization of land use. As a result, despite—and partly because of—an upswing in land reclamation, bare sandy areas have also increased. A new pattern on the landscape, a landscape of polarization, has been created, with the expansion and spatial congregation of both planted trees and moving sand dunes. Remote sensing analysis shows that from 1973 to 1997, vegetation with high biomass (mostly planted) in Uxin Ju increased from 102,000 to 428,700 *mu*, and moving sand also increased from 645,000 to over 1.2 million *mu* (Jiang 2004). The bounded condition produced by fencing, while increasing the incentives for investment and reclamation, is leading to environmental problems on a broader scale.

While the increases in planted trees and moving sand may somehow counterbalance each other in terms of the total production of the sandy land, lowland pasture remains victimized by sandy land reclamation. The lowering of groundwater directly worsens the lowland water conditions. Groundwater reduction has been confirmed by remote sensing analysis and interviews. Remote sensing analysis shows that from 1973 to 1997, the area of surface water decreased from 79,890 to 36,500 *mu*. Since lowland surface water is linked with the shallow layer of groundwater, a reduction in surface water bodies serves to indicate a lowered groundwater table. Interviewees pointed out a 2 to 4 m lowering of the groundwater table in their irrigation wells. The increased number of livestock and lack of grazing mobility have only worsened the deterioration of lowland pasture. One interviewee commented that "in the past sheep could hide in *Carex* pasture, but now you can see rabbits running in it." The degradation of Uxin Ju's jewel lowland pasture reduces the capacity for livestock grazing, and has, in part, resulted in the need for sandy land reclamation. An informant from Buridu *gacha*, Baole, put it aptly: "In the past livestock could eat their fill on the pasture, but now, they have to rely on planted fodder and feed as supplement." The forage and feed provided by tree leaves and planted shrubs, grass, and crops, however, have come at a greater cost in terms of human labor. A cycle of increased use, reclamation, and degradation has been set off by decentralization.

Newly Created Inequality, and a New Need for Conservation
The sandy land environment, highly variable in time and space and highly influenced by human interventions, is better described by disequilibrium dynamics (Behnke, Scoones, and Kerven 1994; Wu and Loucks 1995; Zimmerer 1994). This requires that we understand environmental dynamics beyond the aggregate notion of degradation or improvement (Leach, Mearns, and Scoones 1999). Ecological processes connect neighbors and communities, and uneven changes on the landscape call for the sharing of ecological and economic risks. But privatization has effectively destroyed the culture and economy of sharing, and the fixing of pastureland to individual households and fences has created winners and losers.

Inequality among households has emerged. Three kinds of households are poor: those with disabilities, those lacking labor, and those having poor-quality pastureland. It is probably common in any society that the first two kinds of households suffer from poverty; the last kind of poverty has to do with fences. As mentioned earlier, pastureland is distributed according to where people live. People with lowland and less sandy pastures have not only better grazing conditions but also more favorable land to improve upon. People living on sandy land have sandy pastures, and fences prohibit them from grazing elsewhere. Although pasture areas allocated to households are equalized through their conversion to standard pasture, much of the sandy pasture has become even sandier due to the overall pattern of landscape change, as explained earlier. Households living on more sandy areas disproportionately bear more of the negative ecological consequences. A clear illustration of such conditions is provided by Labai and her two adult sons, one of the poor households the government supported in recent years. They have 2,400 *mu* of poor-quality sandy pastureland, equivalent to 134 *mu* in standard pasture. With limits on sheep grazing, they did not have sufficient cash to invest in irrigation wells and equipment, so they lagged behind other households in diversifying land use. Their lack of capital in reclamation only exacerbated the degradation of their allocated pastureland. In 1997, they took a government loan to dig an irrigation well for ten *mu* of cropland, and the government supported them with cash for seeds and fertilizers. They had only 750 trees, compared to the average of 2,800 trees from my sample households. Although they did have enough food, their poverty clearly showed in their almost bare house when I visited in 1999.

The spatial association of sandy pasture and poverty is also found over large areas. Buridu *gacha,* for example, is more sandy, and reclamation efforts there have not achieved anticipated results. Planting did draw sand dunes lower than before, but the total area of sandy land has increased. An analysis of remote sensing images shows that from 1973 to 1997, moving sand in Buridu more than doubled in area, increasing from 316,000 *mu* to 645,000 *mu.* In comparison, during the same period, Chahanmiao *gacha* had only a 37 percent increase of moving sand from 123,000 *mu* to 168,000 *mu.* Given the rapid expansion of moving sand, people in Buridu are much less sanguine about human efforts to improve sandy land than those in Chahanmiao *gacha.*

In 2001, in an attempt to recover seriously degraded pastureland and to help the poor households living on it, the Inner Mongolian regional government launched an "ecological migration" program that funds the relocation of people from the most degraded areas. The program was initiated as a response to a new central agenda to develop the western regions. In 1999, to promote economic development in China's west, which had lagged behind the eastern region, the beneficiary of economic reform policies, the Chinese government put forth various policies and financial incentives under the umbrella of a Western Region Development (WRD) strategy. Since much of the "western region" suffers from ecological degradation, funding and programs for ecological improvement became an important part of WRD. In Inner Mongolia, funds for tree planting have increased, including rewards for people who convert rain-fed cropland to the planting of trees (Feng 2000). The "ecological migration" program emerged in this context. In Ih Ju league, 10,000 people were on the move list in 2001. The league plans to move 68,000 people by 2005. In Uxin Ju, 100 households in the most sandy Buridu *gacha* were on the move list, and each household was to be subsidized by 20,000 Yuan ($2,400) to settle in Chahanmiao, a local town. Use of their allocated pastureland will be banned for five years, by which time, if the pastureland recovers, people can then move back. Since life in the town has been an attractive alternative to pastoral life, more households desire to join the program than the government funds can assist. We have yet to see the success of such temporary relocation. The idea of permanent closure of degraded areas has also been contemplated by the regional and local governments, and development of small towns is seen as the primary way to absorb the relocated population.

Other than strict conservation, programs that restrain use have also been tested. One program experimented with by Ih-Ju league is a move toward grazing bans on all pastures during the entire warm season, so that pasture-land degradation can be alleviated. The pastoralists in Uxin Ju reacted with serious doubts. During interviews in 2001, my informants voiced concerns about not having enough labor to care for livestock raised in stalls; they also questioned whether it is suitable to keep sheep in stalls and whether that practice would deteriorate the quality of sheep wool, a main market product of the local economy. Moreover, they feared that the grazing ban would un-dermine the household economy. The local government officials realize the danger of economic compromise. As one administrator in Ih-Ju League put it, the grazing ban "uses the local people's economic interests in exchange for the nation's ecological benefit; the [central] government should lend us fi-nancial support." The support that has come has been sorely insufficient, leaving the local government scrambling to meet both economic and eco-logical goals.

Conclusion

In the previous two chapters in this volume, Sundberg and Gray have each ex-plored problems of decentralized resource use at the community level that originated in political processes. This paper examines household-based pas-tureland management in China from ecological and cultural perspectives. Echoing Sundberg and Gray, I show that decentralization in Inner Mongolia has not delivered the desired outcome of sustainable use and economic em-powerment for all. After the pastureland was distributed to households and fenced, more intense use has followed, leading to pastureland degradation. Means's (1993) warning about the detrimental environmental impacts of pas-tureland privatization has not been heeded. Even more important, efforts to fix the mobile sand have created their own set of unintended problems. While the sandy land has been more aggressively reclaimed through the planting of trees, shrubs, grass, and even crops, sand continues to move across fences, creating conflicts between neighbors. Moreover, ecological pro-cesses operate at the landscape level, and improvements in certain areas, which draw on and thus lower the groundwater level, only serve to exacer-bate degradation at other locations that are more vulnerable (i.e., more sandy locations or the lowland). Thus, uneven landscape change has occurred. Al-though initial distribution of pastureland was based upon equality among all individuals, postdistribution variability in landscape change has led some to bear more than their fair share of pastureland degradation, since household-based pastureland use in the reform era has effectively eliminated the sharing of ecological and economic risks in the community. Decentral-ization in Uxin Ju has created winners and losers, not through struggle to access, as in the cases documented by Sundberg and Gray, but because of the mismatch between dynamic ecological changes and the bounded decentralized resource use at the household level.

Another important consequence of pastureland privatization and sandy land reclamation concerns changes in lifestyle. As pastures are fenced and trees, shrubs, and crops planted, the nomadic lifestyle becomes impossible. Uxin Ju is going through rapid transformations, and most people are eager to accept these changes. Still, reclamation with fences leaves little room for traditional grazing practices, thereby placing an additional burden on the pastoralists to accept change as a new way of life. This is not to say that these changes are entirely detrimental to the society. While nomadic mobility is disappearing, the Mongols have gained other kinds of mobility, including economic mobility and increased opportunity to adopt an expanded array of economic and land-use practices. However, the alternative mobility cannot replace the important ecological and cultural value of nomadic mobility.

Gray (chap. 12 in this volume) questions whether community is the right scale for resource management; my study casts doubts on households as an appropriate level of use due to the inconsistency of household- and fence-based grassland use with local ecological and cultural processes. Policy implications of this study are threefold. First, to consider ecological processes at the landscape level, household resource use must be assisted with collective management, taking ecological science seriously. Second, the community must build socioeconomic mechanisms to alleviate economic inequalities; only then would sound collective management be possible. Finally, elements of traditional nomadism should be considered, especially for the seriously degraded areas that call for stricter conservation. This study shows that decentralization has offered some hope for economic growth and grassland improvement, but it offers more challenges for long-term sustainable use.

References

Alm, J., and R. M. Buckley. 1994. Decentralization, privatization, and the solvency of local governments in reforming economies: The case of Budapest. *Environment and Planning C: Government and Policy* 12:333–46.

Archibald, S. O., L. E. Banu, and Z. Bochniarz. 2004. Market liberalisation and sustainability in transition: Turning points and trends in Central and Eastern Europe. *Environmental Politics* 13 (1): 266–89.

Assetto, V. J., E. Hajba, and S. P. Mumme. 2003. Democratization, decentralization, and local environmental policy capacity: Hungary and Mexico. *Social Science Journal* 40 (2): 249–68.

Bakir, H. A. 2001. Sustainable wastewater management for small communities in the Middle East and North Africa. *Journal of Environmental Management* 61 (4): 319–28.

Banks, T. 2001. Property rights and the environment in pastoral China: Evidence from the field. *Development and Change* 32 (4): 717–40.

Barfield, T. J. 1989. *The perilous frontier: Nomadic empires and China, 221 BC to AD 1757*. Cambridge, MA: Blackwell.

Behnke, R. H., I. Scoones, and C. Kerven. 1994. *Range ecology at disequilibrium: New models of natural variability and pastoral adaptation in African savannas.* Boulder, CO: Westview.

Bennett, R. J., ed. 1990. *Decentralization, local governments and markets.* Oxford: Clarendon Press.

Brandon, K., K. H. Redford, and S. Sanderson. 1998. *Parks in peril: People, politics, and protected areas.* Washington, DC: Island.

Chen, Y. 2001. Meng gu zu wen hua de sheng tai xue si kao [Ecological implication of Mongolian culture]. *Nei Meng Gu She Hui Ke Xue* [Social Science of Inner Mongolia], no. 5: 34–37.

Feng, G. 2000. Tui geng huan lin zheng ce fen xi yu jian yi [Policy analysis and recommendations for returning cultivation to trees]. *Lin Ye Jing Ji* [Forestry Economics], no. 5: 6–13.

Gutner, T. L. 2002. *Banking on the environment: Multilateral development banks and their environmental performance in Central and Eastern Europe.* Cambridge, MA: MIT Press.

Huang, Z., and B. Song. 1986. Nei meng gu yi ke zhao meng tu di sha mo hua ji qi fang zhi [Desertification and its control in Ih Ju league, Inner Mongolia]. *Zhong Guo Ke Xue Yuan Lan Zhou Shan Mo Yan Jiu Suo Ji Kan* [Memoirs of Institute of Desert, Academia Sinica], no. 3: 35–47.

Humphrey, C., and D. Sneath, eds. 1999. *End of nomadism? Society, state and the environment in Inner Asia.* Durham, NC: Duke University Press.

Jia, H., and Z. Lin. 1994. *Changing central-local relations in China.* Boulder, CO: Westview.

Jiang, H. 1999. *The Ordos Plateau of China: An endangered environment.* Tokyo: United Nations University Press.

———. 2004. Cooperation, land use, and the environment in Uxin Ju: A changing landscape of a Mongolian-Chinese borderland in China. *Annals of the Association of American Geographers* 94 (1): 117–39.

———. 2006. Decentralization, ecological construction, and the environment in post-reform China: Case study from Uxin banner, Inner Mongolia. *World Development*, forthcoming.

Kung, J. K., and S. Liu. 1997. Farmers' preferences regarding ownership and land tenure in post-Mao China: Unexpected evidence from eight counties. *The China Journal*, no. 38: 33–63.

Lane, M. B. 2003. Decentralization or privatization of environmental governance? Forest conflict and bioregional assessment in Australia. *Journal of Rural Studies* 19 (3): 283–94.

Lavigne, M. 1999. *The economics of transition: From socialist economy to market economy.* New York: St. Martin's.

Leach, M., R. Mearns, and I. Scoones. 1999. Environmental entitlements: Dynamics and institutions in community-based natural resource management. *World Development* 27 (2): 225–47.

Lin, G. 1999. State policy and spatial restructuring in post-reform China, 1978–95. *International Journal of Urban Research* 23 (4): 670–97.

Liu, M., and Y. Wang. 1998. Yi ke zhao meng cao ku lun jian she he li yong diao cha bao gao [Investigation report on the construction and use of pasture enclosure in Ih Ju League]. *Nei Meng Gu Cao Ye* [Inner Mongolia Grass Industry], no. 2: 17–19.

Lutz, E., and J. Caldecott, eds. 1996. *Decentralization and biodiversity conservation.* Washington, DC: World Bank.

McMillan, J., and B. Naughton. 1996. *Reforming Asian socialism: The growth of market institutions.* Ann Arbor: University of Michigan Press.

Means, R. 1993. Territoriality and land tenure among Mongolian pastoralists: Variation, continuity, and change. *Nomadic Peoples* 33:73–103.

Naughton-Treves, L., V. C. Radeloff, J. L. Mena, A. Treves, and N. Alvarez. 2003. Wildlife survival beyond park boundaries: The impact of slash-and-burn agriculture and hunting on mammals in Tambopata, Peru. *Conservation Biology* 17 (4): 1106–17.

Nemarundwe, N. 2004. Social charters and organisation for access to woodlands: Institutional implications for devolving responsibilities for resource management to the local level in Chivi district, Zimbabwe. *Society and Natural Resources* 17 (4): 279–91.

Oliveria, J. A. P. de. 2002. Implementing environmental policies in developing countries through decentralization: The case of protected areas in Bahia, Brazil. *World Development* 30 (10): 1713–36.

Pavlínek, P., and J. Pickles. 2000. *Environmental transition: Transformation and ecological defense in Central and Eastern Europe.* London and New York: Routledge.

Pickvance, K. 1998. Democracy and grassroots opposition in Eastern Europe: Hungary and Russia compared. *Sociological Review* 46 (2): 187–207.

Ribot, J. C. 1999. Accountable representation and power in participatory and decentralized environmental management. *Unasylva* 50 (199): 18–22.

Schroeder, R. A. 1999. Geographies of environmental intervention in Africa. *Progress in Human Geography* 23 (3): 359–78.

Sun, J. 1990. Er er duo si gao yuan sheng tai huan jing zheng zh de zhan lue yan jin [Strategic analysis of eco-environmental control of the Ordos Plateau]. *Gan Han Qu Zi Yuan Yu Huan Jing* [Journal of Arid Land Resources and Environment] 4 (4): 45–50.

Sundberg, J. 1999. *Conservation encounters: NGOs, local people, and changing cultural landscapes.* PhD diss., University of Texas, Austin.

Thwaites, R, T. de Lacy, Y. H. Li, and X. H. Liu. 1998. Property rights, social change, and grassland degradation in Xilingol Biosphere Reserve, Inner Mongolia, China. *Society and Natural Resources* 11 (4): 319–38.

Turner, M. 1999. Central-local relations: Themes and issues. In *Central-local relations in Asia-Pacific: Convergence or divergence,* ed. M. Turner, 1–28. Hampshire, UK: Macmillan.

Turner, M. D. 2002. Moral dimensions of "resource struggles" and their treatment in political ecology. Paper presented at the annual meeting of the Association of American Geographers, Los Angeles, California, 22 March.

Unger, J. 2002. *The transformation of rural China.* New York: M. E. Sharpe.

Weber, E. P. 2000. A new vanguard for the environment: Grass-roots ecosystem management as a new environmental movement. *Society and Natural Resources* 13 (3): 237–59.

Williams, D. M. 2001. *Beyond the Great Wall: Environment, identity, and development on the Chinese grasslands of Inner Mongolia.* Stanford, CA: Stanford University Press.

Williams, M. D., and M. P. Wells. 1996. Russia. In *Decentralization and biodiversity conservation,* ed. E. Lutz and J. Caldecott, 109–22. Washington, DC: World Bank.

World Bank. 1997. *World development report: The state in a changing world.* New York: Oxford University Press.

Wu, J., and O. Loucks. 1995. From balance of nature to hierarchical patch dynamics: A paradigm shift in ecology. *Quarterly Review of Biology* 70:439–66.

Wushen Qi (Uxin banner). 2001. *Wushen qi jinmu xiumu lunmu zaixing banfa* [Temporary regulations in rotational grazing and closing of pastureland from grazing]. Unpublished government document.

Zhang, T. 2000. Land market forces and government's role in sprawl: The case of China. *Cities* 17 (2): 123–35.

Zimmerer, K. S. 1994. Human geography and the "new ecology": The prospect and promise of integration. *Annals of the Association of American Geographers* 84 (1): 108–25.

Conclusion: Rethinking the Compatibility, Consequences, and Geographic Strategies of Conservation and Development

KARL S. ZIMMERER

Recognizing the Prospects and Limits of Compatibility

The so-called third wave of conservation is forging close ties to livelihood activities, including agriculture and resource use, that are foundations of sustainability. Recent developments generate effects that span a variety of places, scales, and social groups (Paulson and Gezon 2005; Zimmerer 2006). The emerging effects of third-generation global conservation have included both notable successes and failures, while many other projects are works in progress, and still others are yet to come. This book has demonstrated that mixed results are apparent when third-generation conservation is judged in aggregate, but even specific projects have produced a mix of outcomes if gauged in social and environmental terms. Concluding the book thus begins with a synopsis of the one dozen case studies in which I reflect on how these findings show the promise, as well as the limits, of the linkages of global conservation goals to livelihood-based development (such as agriculture). My synopsis is based on the view that conservation and development activities, even when combined, are often distinct pursuits that frequently entail separate goals and spatial arrangements.[1] This perspective joins forces with those analysts who look both critically and constructively at global conservation and development in order to search for compatibilities and, at the same time, to recognize that sustainability in one domain is not automatically conferred to the other (Hall 2000; Leach, Mearns, and Scoones 1999; Reardon and Vosti 1995; Sen 1999; Sneddon 2000; Zerner 2000).

The aims of biological conservation—to maintain and restore biodiversity in healthy ecosystems consisting of dynamic, interacting biological and phys-

1. Adopting this perspective does not imply that conservation activities are necessarily antagonistic to development, nor does it not mean to adopt the perspectives of either anti-conservation development or anti-development conservation.

ical components—have guided and generated much of the latest phase of environmental globalization. As seen in several chapters, global conservation organizations during recent decades have prioritized tropical rain forest and tropical mountain ecosystems, as well as finer subdivisions recognized at the level of ecoregions (see especially chaps. 9 and 10; also Dinerstein 1995; Mittermeier et al. 1998; Olson and Dinerstein 1998; Olson et al. 2001; Rodríguez and Young 2000). The general goals of designing protected areas are focused on the spatial parameters that maintain the diversity of wild plants and animals through supporting the viability of populations (Goldammer 1992; Soulé and Orians 2001; Young and León 1995; Young and Zimmerer 1998). The concern of global conservation for agricultural and range resources is based on similar yet distinct criteria, namely the maintenance and restoration of biodiversity, productivity, and the ecological health of human-utilized ecosystems (see especially chaps. 2, 6, and 7). Targeting the ecological health of these humanized landscapes has involved components that are clearly shaped by human action, such as domesticated plants and animals and managed soils.

The spatial parameters of conservation areas have been a primary concern in the global expansion of conservation as seen through the eyes of conservation ecologists, biogeographers, resource managers, and protected-area planners (Mooney 1998; Soulé and Terborgh 1999). Protected-area design is infused with a number of spatial-environmental criteria that typically prioritize spatial extent, the coverage of environmental heterogeneity and key habitats, degree of land use and degradation threats, and prescribed management practices of protected areas (Bissonette 1997; Omernik 1995; Terborgh 2002). Of course, the spatiality of conservation territories is inevitably a human choice, often involving disputes and jurisdictional matters that have become increasingly apparent in the growing experience of implementing these territories worldwide. The actual shaping of conservation areas is also based on the interpretation of both human desires and threats, no matter how closely the natural ideal may seem to be matched by the ecological unit, or how well the territorial design appears to reflect the input of expert ecological knowledge (chap. 1).

One well-known spatial design in global conservation is the biosphere reserve (chaps. 1, 4, 6, 8, 9, 10, and 11 in this volume). Numbering more than 400 worldwide, these reserves, coordinated and sanctioned globally by the United Nations (UN), are based on the ideal of a circular design that minimizes negative edge effects (Batisse 1997; Hadley 2002). In reality, a combination of environmental and broadly social and political factors, including the location of nearby settlements, influence the size and shape of global biosphere reserves (Neumann 2004; Simonian 1995). In the case of agricultural and other utilized resources, the spatial design of a growing number of global conservation efforts is predicated on natural hydrologic units, such as watersheds and river basins (chap. 8; Wescoat 1984; Zimmerer 2000). Alternatively, these resource conservation and management efforts, especially common in developing countries, may be designed to correspond to the spaces of target groups, such as villages or rural communities (Leach, Mearns, and Scoones 1999). The latter scale of management is typical of global community-based conservation and decentralization programs, which this volume has described in the context of Latin America, Africa, and Asia (chaps. 11, 12, and 13; see also Leach, Mearns, and Scoones 1999; Roe and Jack 2001).

The general objective of development—to improve the quality of life of poor or disadvantaged people, sometimes referred to as the "development subjects"—is the raison d'être of major global and international organizations. The term "development subjects" has connotations of the social and spatial control that, in many cases, is exercised from above and afar in the globalization of development (Escobar 1995; Ferguson 1990; Mitchell 2002). Raising standards of health, nutrition, education, income, and wealth is generally a goal of development efforts, as is the amelioration of social inequality, including the disadvantages based on gender, ethnicity, class, and corresponding spatial factors. Pursuing equity goals typically involves capacity-building strategies, such as job skills, literacy, and business training; small business and credit (e.g., microcredit) programs; and marketing assistance. An emphasis in rural areas is often placed on community-based development (Brosius, Tsing, and Zerner 1998). Community-based approaches to rural development differ widely in their areas of particular activity and action plans, as well as their philosophical underpinnings; the most politically progressive tend to share an orientation that prioritizes participatory approaches, accessible technologies, culturally appropriate practices, equitable benefits, resource entitlements, and sustained long-term improvements and empowerment (Healy 2001; Leach, Mearns, and Scoones 1999).

Spatial designs of the programs, policies, and projects for development are usually delimited as "action areas," which certain global organizations, such as the World Bank and the Food and Agriculture Organization (FAO) of the UN, delimit through such spatially based procedures as "geographical targeting" and "poverty mapping" (Bigman and Fofack 2000). Food, health, literacy, education, and job programs in urban areas, for example, are likely to show a style of spatiality that takes shape around the residences and livelihood activities of participants from a neighborhood or part of a city (Friedmann 1988; Friedmann and Rangan 1993; Rocheleau, Thomas-Slayter, and Wangari 1996; Porter and Sheppard 1998). Many of these designs are not, however, neatly demarcated into well-defined territories, since considerable spatial heterogeneity of class, ethnicity, gender, and other factors are likely to exist within the urban area that is designated for a program or policy action (Sandercock 1998). Similarly, the spatialities of rural development are differentiated at multiple levels that include the individual person or household (e.g., land privatization in Chinese anti-desertification campaigns, chap. 13); the individual community (e.g., beekeeping groups, chap. 4); socially defined subgroups within one or more communities (e.g., women's artisanal groups, chap. 11); economically defined subgroups within one or more communities (e.g., "green market" forestry producers, chap. 3); producer cooperatives that involve a majority of members of one or more communities (e.g., the organic coffee cooperatives examined in chap. 2); and village-level groups (e.g., the decentralized resource management and development design in West Africa, chap. 12). In a majority of these cases, rural development shows a spatiality that is not only territorial but also networklike, with the latter characteristics multiplied and reinforced through globalization. An understanding of new spatialities that either arise from or interact with globalization contributes to a critique of one main tenet of many initiatives for conservation with development: namely, a tendency to overestimate the extent of territorial integrity.

These findings on global conservation and development lead to our first area of general conclusions. Frequent incongruity between the spatialities needed for conservation and those most appropriate for development is fairly self-evident, yet widespread assumptions exist about the presupposed similarities of conservation areas and the spatial forms of development (chap. 8; see also Brown and Wyckoff-Baird 1992). The contributions to this volume reflect how many conservation organizations and project personnel frequently design the location of projects in order to encourage the support of local residents who live and work at sites that are located in or near to conservation territories. Yet such sites may not be the most appropriate places for a particular socioeconomic intervention from the viewpoint of development processes. Practical examples abound of conservation-with-development projects located along the borders of protected areas, as global conservation organizations seek to obtain the support and lessen the resource-use pressures of local residents. Such sites may not be viable, however, for a particular intervention from the viewpoint of the socioeconomic processes of sound development.[2] Indeed, ill-designed development plans that have been drawn

---

2. For example, the development efforts of conservation that is aimed at the increased control of the grazing of livestock, often cattle, frequently attempt to intensify the use of rangeland or pasture use (via improved pastures, rotational grazing, and other techniques). Since livestock raising in the rural areas that bound protected areas is often highly varied among large ranchers and small farmers, the intensification programs and policies would need to vary considerably depending on this variation. Equally important, this siting of intensified grazing pastures at the edge of the protected area may only lead to the unintended consequence of larger herd sizes that ultimately rely on non-pasture areas.

on spatial blueprints derived from strictly conservation prerogatives may be worse than no intervention at all. (Such unintended consequences, and their frequency in the development projects of conservation-driven initiatives, are discussed further in the next section.)

Globalization has played an influential role in promoting the presumed congruence of conservation and development. Global organizations often assume the congruence of conservation and development for both practical and philosophical reasons (Brown and Wyckoff-Baird 1992). To practitioners, funding agencies, and project personnel, their objective of integrating conservation and development is often premised on a congruent type of territory. Conjoined spatial design is thus often seen as the logical first step that exerted particular appeal to the many planners who heeded calls to sustainable development in the onset of the "third wave" of conservation that has occurred amid globalization since the 1980s (Sneddon 2000). Philosophical grounds for the idea of spatial congruence are also deeply rooted in the organizations and activities of environmental globalization. Here the assumption of a spatial congruence of local conservation and development supplies an important small-area frame of reference for insertion into many global-scale environmental planning agendas (e.g., Ravenga et al. 1998). The appealing simplicity of this spatially congruent vision is at odds, however, with the reality of multiple discordant spatialities, which, as this volume amply attests, are characteristic of environmental globalization.

The second conclusion reached here is that conservation and development are prone to produce a growing multiplicity of spatial forms under ongoing globalization. Environmental globalization has increased the prominence of both contiguous-style conservation territories and networks. This observation raises a major point of general reflection: namely, that environmental globalization and the formation of conservation areas can be fruitfully engaged with extensive investigations and theoretical discussions about the spatialities of globalization (Amin 2002; Massey 1999a, 1999b, 2002; Swyngedouw 1997). One pivot of these debates, which until now have focused primarily on the spatial dynamics of economic and political globalization, is the emphasis on either the territorial forms of globalization in economic systems (e.g., free-trade zones, multicountry trade or customs pacts) or the contrasting, network-style spatialities (e.g., transnational corporate structures and business communities, global activist networks). Prior to this volume this geographic lens has not been turned, however, on the role of globalization in human-environment interactions and, more specifically, on the main environmental issues within the realm of environmental conservation and development. Our volume has broken new ground in this area by identifying in conceptual and empirical terms the importance, functioning, and interaction of both territories and social-environmental networks that occur in global conservation. It is to be hoped that these contributions will spur further studies on the combined territorial and network styles of spatial organization of the continued global "conservation boom" (Zimmerer 2000).

Our third conclusion is focused on practical recommendations in the areas of policymaking and analysis. This contribution is elaborated in the three sections of the conclusion that follow. It begins with a discussion of the unintended consequences of conservation projects, policies, and programs that are covered in the chapters of this volume. These unintended consequences signify not only unpredicted outcomes but also the commonness of multiple results, both positive and negative, that stem from the spatial dynamics of conservation and development programs.[3] Next, the conclusion turns to the evidence that the chapters have presented on the diverse practices of the resource users and rural people that inhabit the expanding interface of global conservation and environmental management in developing countries. These farm- and local-level practices constitute several of the most immediate driving forces leading to unintended consequences. "Micrologics" based on the operating principles of family farms, ranches, and small companies help provide an analytical compass that is necessary to understand more fully the unfolding spatialities of global conservation and development that involve local people and their livelihoods (Bardhan and Udry 1999; Godoy et al. 2000; McSweeney 2002; Mertens, Lambin, and Sunderlin 2000).

Finally, the recommendation of particular "spaces of hope" for global conservation—spatial, environmental, cultural, and social—is the basis of the third section of the conclusion. The term is chosen to signify the proposed framing of conservation approaches that we believe are pivotal to the future of global conservation and its expanding entanglements with the livelihoods and politics of farmers, herders, and other resource users. Such spaces of hope comprise new spatialities that the contributors and I believe can lead to benefits for the broad base of these people, especially the currently least powerful, as well as bringing sustained health to the environment.[4] These entanglements extend also to the environmental scientists, activists, and "global citizens" that actively engage via the consumption and production of ideas, resources, information, and opinions with spatially extensive (fre-

3. The term "unintended consequences" is chosen as a means of reflecting on the outcomes of conservation interventions and related development effects, which now are becoming apparent although they were seldom predicted, even partially, in advance. "Unintended consequences" is also meant to signify the commonplace multiplicity of outcomes—unintended consequences in one type of result may coincide with predicted ones elsewhere.

4. The centrality of both the geographical concerns of space (and with it the idea of place; Amin 2002) and a progressive politics in this environmental globalization is similar in spirit to the political and economic analysis of spaces of hope that are the basis of a well-known treatise in geography and the related social sciences by David Harvey (2000).

quently global) policy and political organizations. "Spaces of hope" describes a particular group of spatialities based on social-environmental networks that, it is argued, must inform the next phase of policymaking and analysis for environmental conservation in order to account more fully for globalization, development, and the expanded interfaces with agriculture, resource use, and other livelihood concerns. In particular, the results of this volume's studies are used to sketch a set of four proposed spaces of hope based on the formal and informal networks of agriculturalists and land users that are crucial to new directions in conservation concerns that are focused on sustainability.

Unintended Consequences of Global Conservation and Sustainability Efforts

Unintended consequences are a frequent outcome of the efforts to address the combination of conservation and development goals. Indeed, unintended consequences have become a common part of global conservation and sustainability efforts that will increase as these activities expand in size, scope, and influence (Dove 1996). Each chapter of this volume can be seen as contributing to an assessment of the unintended consequences of recent conservation initiatives and the unfolding interactions with development change. These unintended consequences are either apparent in existing outcomes or anticipated in the current set of changes that are detailed in the chapters. The thematic nature of these unintended consequences is grouped according to the main divisions of the book, namely the emphasis on spatiality (part 1), scale and livelihood analysis (part 2), transnational and border issues (part 3), and decentralization and environmental governance (part 4).

Part 1 begins with the examination of the global "green market" certification programs for organic coffee production that have been adopted among the small farmers of peasant and indigenous communities in Oaxaca, Mexico (chap. 2). These programs have been successful in promoting the protection of tropical deciduous shade trees in the fields of organic coffee that are spreading through a range of tropical and subtropical environments that grade from the semi-arid to the humid. The agroforest landscapes of shade-grown coffee are fragmented, however, since the decisions about adopting organic coffee cultivation along with the canopy of diverse shading species, in addition to other sustainable techniques, are made at the level of single-farm families or households and their individual field parcels. As a result, and without intending to do so, the global green-market programs have generated the need to address landscape connectivity via the coordination of organic coffee production at the landscape-level scales of the rural cooperative, the community, and the multicommunity level. One response that we recommend in the face of this challenge is to understand better the rationales that lead certain families or households, but not others that are nearby, to adopt the production of organic coffee with its attached landscape elements.

The use of high-resolution satellite imaging via remote sensing is examined in chapter 3 in order to address its possible support to sustainable forest management and global green-market certification programs such as those of the growing Forest Stewardship Council (FSC). These spatial imaging technologies are new, rapidly evolving, and subject to a plethora of potential applications to environment, conservation, and development projects. Potentially widespread use as an environmental certification device is pronounced in many tropical regions and developing countries in general, where many sites are distant from the certifying organizations and agents. One key issue is the social and economic accessibility of these costly technological networks that look to become integral to global conservation (Liverman et al. 1998; Turner 2003). Concerns about equitable social development bring attention to how the costs, technical expertise, and hoped-for benefits of green-market tropical forestry based on remote sensing technologies can be made accessible to a wide network of tropical forest owners and managers. Of particular interest is whether, or to what extent, the global techno-environmental network that is envisioned can be extended to benefit the small-scale units of forest-using individuals and communities that reside in the rural areas of developing countries.

Chapter 4 evaluates the unintended consequences of global conservation's network of beekeeping for locally sustainable ecodevelopment. As discussed in the chapter, beekeeping is especially common in and near the protected areas of tropical forests—both rain forests and tropical deciduous forests. Honeybees favor heavily flowering trees, many of which are characteristic of the sunlight-rich environments of edge habitats, newly opened areas, and disturbed sites, where many woody species produce abundant flowers and relatively large amounts of nectar. As a result, one of the unintended consequences of the many globally networked projects for beekeeping and honey production is to ensure an abundant supply of pollinators for those types of trees that, from the perspective of ecology, behave as early successional or invasive species (i.e., "weedy" trees). In addition to possibly slowing the succession of forest species and vegetative communities (due to the heavy pollination of weedy trees), the honeybee populations may even steer or deflect succession toward a more weedy style of vegetation assemblage than would be present otherwise. Given this unexpected consequence, the various other activities of beekeepers that involve land use and vegetation management (e.g., agriculture, forest extraction, range or pasture use) must be seen as equally important as the focus of beekeeping itself.

Part 2, which is focused on scale and issues of changing livelihoods, begins with the analysis of urban house-lot gardens (chap. 5). Agrobiodiversity and food production in these sites are dependent on a scale of flows created through the transfer of planting material along the networks of social ties that extend to the rural places that surround urban areas. In the case of Santarém, Brazil, these flows of seeds and cuttings from rural Amazonia are the basis for the supply of a large variety of herbs, shrubs, and trees for productive house-lot gardens within the city that furnish a valuable source of food, seasonings, medicine, and fiber to the often-poor households of urban and periurban neighborhoods. Programs, projects, and policies for urban food supply and security would be well advised to support these urban house-lot gardens, an argument that is presented in chapter 5. Yet one unintended consequence of such support efforts would be to overlook or even to sever the urban-rural ties that are so crucial to the support of these flows of seeds and plant cuttings. Households with urban house-lot gardens, who invest work in actively renewing the social networks connecting them to rural areas, need to be seen as maintaining the networks at this scale in conjunction with labor migration, work exchanges, and accessing land and other resources. These economic activities continue to be either directly or indirectly a part of family- and household-level strategies in networks that inscribe a scale traversing the overly simplified urban-rural dichotomy.

Chapter 6 examines the scaling of seed supply in social-environmental networks that support the agrobiodiversity of the Andean potato crop. It finds that key flows occur within and among households of single communities as well as connecting those of neighboring communities and development institutions. This chapter's findings highlight the potential for negative unintended consequences if projects for the community-based conservation (CBC) of agrobiodiversity were to delimit the social and environmental boundaries of single rural communities as the primary demarcations for initiatives addressing the functions of seed supply and availability. In highlighting this concern the chapter can be seen as pointing implicitly to the importance of the diverse rationales and socioeconomic circumstances of the individual farmers, especially women, and households who conduct, choose, and maintain these agrobiodiversity-rich seed flows. In the case of the farm-level production of agrobiodiverse Andean food plants, with emphasis on the Andean potatoes in particular, a disproportionate share of cultivation is known to occur among the more locally well-to-do and poorer farm families (see chap. 6).[5] Similar influences and personal relations may exist among farm families and households in the case of seed flow preferences and capacities. Knowing these determinants of seed flows would aid in avoiding the unintended

5. Farm families and households among the group that is in the middle of the local socioeconomic spectrum, by contrast, are less able to find land and labor resources for agrobiodiversity production.

consequences of programs or projects with good intentions but insufficient understanding of how the key nodes operate in such potential conservation networks.

Various unintended consequences that surround the scientific research and representations of the complex cross-scale linkages of desertification, climate change, and land use in the Sahelian countries of Africa are raised in chapter 7. Without directly intending to do so, the global environmental sciences have tended to widen a chasm of disparate scales in the analytical frameworks and project initiatives designed to address these problems. This chasm has been opened through support for the opposite-tending scales of remote sensing approaches and community-based conservation. Remote sensing approaches are prone to place emphasis on larger areas of environments and land use. In fact, as chapter 7 points out, the use of remote sensing approaches to investigate and guide management in the Sahelian countries is iconic in the sense of symbolizing the technology-based, large-area research that is prevalent among the global environmental change sciences. By contrast, CBC has prioritized the scale of single rural communities in prescribing the interventions that are designed to combat desertification and to contribute to sustainability. Unintentionally, then, these reference scales of remote sensing and community-based conservation have led some to overlook the intermediate scale of land use that is described in the chapter and in other key works on the topic (Turner 2003; see also Lambin et al. 2001). This intermediate scale, which consists of multiple communities and noncommunity lands across a variety of environments, is a key to understanding the impacts of the pastoralism of Sahelian peoples. Furthermore, the intermediate scale requires analysis of the land use rationales across these midsize areas that Sahelian herders employ at the level of individual households and their varied institutions, including the social and environment networks that mediate land use cooperation and conflict with primarily agricultural peoples (Bassett 2001; Turner 1999; see also Gray, chap. 12 in this volume).

Part 3, which is focused on transnational and border issues, begins with an account of the unintended consequences driven through the joint projects of major transnational development and global conservation organizations (chap. 8). The contradictory objectives in the case of the Mekong Delta project in Southeast Asia are hydrologic-based development and, more specifically, the prospect of big-dam projects, on the one hand, and the broad-based arena of biodiversity conservation, on the other hand. Even at first glance, major aspects of this combination are likely to be incompatible under the most probable scenarios. Heavy-impact industry and industrial agriculture that are fueled by big-dam development would incur the high levels of environmental destruction that are antithetical to biodiversity conservation. Still, the type and extent of impact will depend on the potential inroads of sustainability policies and politics, as seems to be happening (at least in the planning stages) with many new hydro-development projects. If successful, even if only partly, the sustainability emphasis would require a finer analytical framework in order to examine the interaction effects of biodiversity conservation with the specific features of a more environmentally sound type of hydrologic-based development.[6]

Analysis in chapter 9 suggests that a sequence of unintended consequences will be unfolding in conjunction with called-for transnational designs for ecoregion-based protected areas in the Andean countries. The prediction of this outcome is based on the sort of reasoning that is discussed in the opening section of this chapter. Here, the ecoregion-based templates would serve as primary or exclusive guide to the design of transnational conservation territories. These conservation areas would need to include a variety of sustainability and development initiatives both in and near the protected areas, a point that is emphasized in chapter 9. Nonetheless, the plans for the socioeconomic side of these transnational projects would probably be dictated by the ecoregion-based spatial design of the protected area. One potential consequence, albeit an unintentional one, would be the treatment of the socioeconomic dimension of these projects in terms of an ideal natural type. The assumptions contained in the idea of a "rain forest community," for example, are likely to conceal the diverse socioeconomic dynamics that characterize most such places and may therefore serve poorly as a general designation. Understanding the potential for transnational project design in this case will depend on socioeconomic analysis that extends beyond the spatially oversimplified assumptions into the variety of farming and land and resource use rationales that exist within and among these communities that are critical to conservation (Coomes, Barham, and Takasaki 2004; Coomes, Grimard, and Burt 2000; Godoy et al. 2000; McSweeney 2004).

Chapter 10 is a carefully constructed plea for closing the gap between the networks of global conservation and environmental science, on the one hand, and these sciences (and scientists) that function primarily within the context of the national level within the country of Peru, on the other hand. In calling for the much-needed engagement and potential convergence of global and Peruvian scientific networks, this chapter also raises the prospect of an unexpected consequence that is positive in this case. This unexpected consequence would be the greater dialogue across conservation networks, both global networks and the Peruvian national arena, with respect to the

6. For example, the sustainable-style intensification of irrigated agriculture would potentially offer compatibilities with a greater level of protection of uncultivated lands. In that case, adequate frameworks for understanding sustainability are dependent on several levels of analysis, including the micro level of the actions of farmers and other resource users.

agriculture and resource use that relate to conservation. Leading figures in Peruvian conservation have been luminaries in recognizing the importance of agriculture and resource use in the relation of people to the environment. Expanding the message of chapter 10, then, it is hoped that this broadening engagement with Peruvian national conservation networks can increase the dialogue among those working in the country, including global conservation and environmental scientists, that are concerned with the *combination* of conservation and sustainability.

Part 4, which examines decentralization and environmental governance, opens with the analysis of a potentially major benefit of the unintentional consequences of the broadening political dimension of global conservation (chap. 11). Global conservation activities and locally based sustainability projects, exemplified in the various efforts undertaken in conjunction with the Maya Biosphere Reserve (MBR) in Guatemala, carry the potential of becoming sites of democratization. This consequence in terms of democratization would be unintended, since it is not the general goal of most conservation organizations, whose main interest still lies in the effectiveness of protected-area management conservation measured as the absence of human impacts. Chapter 11 suggests that global conservation organizations, as well as their national and local counterparts (especially nongovernmental organizations, or NGOs), would be well advised to design and support their projects, to the extent possible, as "sites for democratization" (see also Forsyth 1999; Sundberg 2002). Recognizing and understanding this type of unintended consequences suggests that understanding the microdynamics of these projects will be needed in order to evaluate them as potentially democratizing activities within their larger national arenas (Bryant and Bailey 1997; Carney 2004).

A consequence that is revealed as unintended in chapter 12 offers an example that adds further to the preceding chapter by showing how the local-level decentralization of global environmental governance raises the renewed importance of national political circumstances. New closeness of the links of the local and national in environmental politics is clearly claiming a prominent place as a result of globalization activities. In the case of the village-level decentralization (*gestion des terroirs*) being conducted throughout Burkina Faso, the call for zonification planning at the local level is unexpectedly reinforcing the importance of national governance. Consequently, national-level environmental governance, and particularly nationwide agrarian reform, is a prerequisite for the sort of decentralization that is being planned in Burkina Faso. As elucidated in chapter 12, the effects of this sort of national-level environmental governance would then take shape through the local microdynamics of resource access and land use.

Chapter 13 is centered on a consequence that is unexpected, especially to the planners of the anti-desertification campaigns of Inner Mongolia who chose as their primary approach the combination of the privatization of individual property and vegetation planting (referred to among Chinese environmental scientists as a sort of ecological restoration, with emphasis on the restoration or, in some cases, the creation of economically productive resource systems, such as irrigated farming). This approach, as chapter 13 points out, is similar to the one that guides the privatization of community and cooperative land and water resources through many of the activities of global development networks worldwide. Yet the case study shows the outcome of unintended consequences, namely the improvement of water and vegetation quality on most private parcels while these conditions have worsened in the pasturelands that remain under communal control. This sort of unexpected outcome, with a new mix of improvement (on private lands) and degradation (on community lands), must be seen as an increasing possibility for other areas undergoing these changes in land tenure and environmental management.

"Micrologics" of Agriculture and Resource Use in Conservation and Sustainability Policies

The rationales ("micrologics") of farmers and other resource users, both as individuals and small groups, are one of the most useful perspectives for understanding the immediate or "on-the-ground" interface of global conservation efforts with agricultural activities. Examining these rationales is an important means of understanding the consequences, whether intended or unintended, of the interaction of conservation efforts with activities aimed at sustainability. The term *micrologics* refers to the suite of rationales and culturally conditioned responses that influence behaviors related to resources, conservation, and development. In general, these logics are structured by such conditions as limited economic endowments and poor public infrastructure (health, education, transportation, etc.). More specifically, micrologics of land and resource use encompass responses to economic criteria, such as profit maximization and price signals. Similarly important are several other micrologics that range from human- and cultural-ecological rationales (e.g., risk aversion, seasonal labor scheduling) to political, institutional, and cultural factors, such as stewardship beliefs, culinary preferences, political practices, and aesthetic choices (Blaikie and Brookfield 1987; Kanbur 2002; Mayer and Glave 1999; Scott 1998; White 2002).

The main units of decision making about resource use include the large farm, ranch, and land-user group as well as the small farm or smallholder family, often characterized as a peasant household, and the single resource-using individual (Bever 2002; Creed 2000; Netting 1993). Households and communities are not homogeneous, but their micrologics are shaped by within-group differences based on categories such as age, gender, and ethnicity (Agarwal 1994; Carney 1993; Folbre 1986; Godoy et al. 2000). Much recent research explores the relation of these micrologics to the actual activities of resource use and conservation impacts (Barham and Coomes 1994; Coomes and Barham 1997; Coomes, Grimard, and Burt 2000; Harris 2002; Mertens, Lambin, and Sunderlin 2000; Turner 2000; Turner et al. 2001; Vance and Geoghegan 2002; Zimmerer 2004). These advances are the backbone of this section, and they are used to reflect further on a number of the unintended consequences of conservation that are described in the previous one.

The asset portfolio is a core element of the micrologics employed by peasant households and other decision-making units. The concept of the asset portfolio refers to the access to local physical, environmental, human or social, and financial capital (Bardhan and Udry 1999; Reardon and Vosti 1995). Asset portfolios, in the context of rural household economies in developing countries, consist of such items as land and labor access, nonland resources, and house, vehicle, and livestock ownership. Asset portfolios are a powerful influence on the micrologics of agriculture, resource use, and the type, magnitude, and frequency of environmental impacts (Bulte and van Soest 2001; Coomes, Barham, and Takasaki 2004; Coomes, Grimard, and Burt 2000; Pichón 1997; Rudel, Bates, and Machinguiashi 2002; Takasaki, Barham, and Coomes 2001; Walker, Moran, and Anselin 2000). In this context, the asset portfolio of a household is of equal or greater influence than annual income or political identity as a shaper of several of the main socioeconomic dynamics of resource use. Recognizing the significance of these influences, and examining how they operate, offers the capacity of probing globalization effects in conservation and agriculture in the terms of existing livelihoods.[7]

Diversification of socioeconomic activities at the family or household level, as well as the level of individuals, is a defining trend of the livelihoods of rural people worldwide (Reardon, Berdegué, and Escobar 2001; Reinhardt 2000). Diversification refers to the expanding array of economic activities at the micro level. It typically results in a complex web of ways in which household portfolio assets are used. Diversification is driven in part by the agricultural crisis of declining staple food markets that has deepened due to globalization. For example, diversification is common among maize growers of Latin America and Africa, potato producers of the Andean countries, and the wheat growers of Canada and the United States (Bellon 1991; Bebbington

---

7. The analysis and insights that can be gained by examining asset portfolios at the micro level are being newly advanced in various ways; for example, these advances account more fully for asset variation within and among households due to factors such as gender, ethnicity, age, and duration of settlement (Godoy et al. 2000; McSweeney 2002, 2004; Perz and Walker 2002; Sunderlin and Pokam 2002; Turner 1999; R. T. Walker 2003).

2001). Since diversification is a common response at the farm level, it is an important change process determining how a family or household adjusts the array of activities related to environments and, where applicable, conservation initiatives. Reflecting on the examples of this book, it is possible to see diversification at work in the immediate farm-level contexts of organic coffee farming, sustainable forest management, beekeeping, seed production of agrobiodiverse food plants, tree and shrub planting, and irrigated farming with a development emphasis. In these cases, the promotion of a targeted activity often leads to unintended consequences, both positive and negative, in other aspects of the diversified household-level portfolio of activities and investments. For example, organic coffee production has tended to support the interplanting of nutritious local food plants (chap. 2); less favorable impacts result from reducing the herd movements of Sahelian pastoralists, which undermines soil fertilization processes that are important to the farming activities of these same families (chap. 7).

Both the choice of resource use activity (portfolio choice) and resource allocation level (portfolio expenditure) are important in the environmental outcome of micro-level management. Since certain activities are common to the diversified economies of many farm families, it is often the level of resource allocation and the style of use that matter most to environmental impact. For example, irrigated rice farming in basin-development initiatives for the Mekong Delta (chap. 8) could lead either to destructive impacts or to more sustainable uses, depending on the sorts of cropping systems and soil, pest, and weed management that are used. The process of intensification is important in determining these effects. Intensification refers to the increased allocation of the factors of capital, technological inputs, labor, and other resources per unit area (Bassett 2001; Bassett and Zimmerer 2003; Netting 1993; Smith 1996; Sonneveld and Keyzer 2003; Turner et al. 2001). Intensification can result in more sustainable agricultural and resource use practices, such as certain types of Southeast Asian paddy rice cultivation and the organic coffee production in Mexico that relies on farm terracing, soil management, and a carefully tended canopy of shade trees (chap. 2). Alternatively, intensification can produce environmental consequences that are highly damaging, which is the outcome that chapter 8 concludes is currently most likely as a result of the still-unfolding transnational river basin planning and management in the Mekong countries.[8] The well-informed environmental analysis of agricultural intensification is critical to evaluating properly the building interest in so-called ecoagriculture as a strategy for increasing agri-

---

8. A focus on the intensification of agriculture and consequences for land use is increasingly taken into consideration in the planning and management of global conservation projects (Shriar 2001; Smith 1996).

cultural production that is being advocated by global conservation proponents (Green et al. 2005; McNeely and Scherr 2003).

Disintensification is a micrologic that rivals the prevalence of intensification in many rural places amid global conservation and development projects. Disintensification refers to decreased investment of inputs (capital, technological inputs, labor, built infrastructure) per unit area. The frequency of disintensification has grown with the decline of the economic circumstances of the majority of land users of rural societies worldwide. Predominant urbanization and food overproduction are associated with the characterization of rural societies not as backwater relics waiting to be modernized but instead as shaped or constructed through active, ongoing processes of uneven development. Production activities, such as farming, have become increasingly marginal so that they are now rivaled in many places by "consumption" activities in rural economies (e.g., tourism). Disintensification is often evident in the increased shift from agriculture and food production to livestock raising, and also within the types of livestock, typically from cattle to goats and sheep in semi-arid areas (see chap. 7). While the shifts to or within livestock raising may require fewer labor inputs, the disintensified land use that results frequently worsens pressure on vegetation and soil resources as an unintended consequence.[9]

The micrologic and environmental impacts of disintensification are highly varied. In many cases the results tend in the other direction from the above illustration. For example, disintensification may lead to the reduction of grazing or forest-use activities, and thus a shift to more environmentally sound practices. Micrologics also guide the connections among activities within a farm- or household-level economy, including the ones that may be of greatest interest to conservation. Tropical forest extraction and rights to forest products, for example, are tied to the landholding levels and incomes of small-farmer households (Coomes 2004; Vosti et al. 2003). Such connections are involved in the adjustments of a certain activity (such as off-farm migration) that must be coordinated with other, even opposite-tending changes (such as the disintensification of food plant agriculture). These links at the micro level are commonplace given the multiple economic activities that characteristically exist within the farm family or household. These

9. A related example is withdrawing of labor from agrobiodiversity farming and soil conservation devices (terrace and rock wall fences) that had made for sustainable landscapes in the Andes (see chap. 5; Zimmerer 1993). It is important to note in these examples that disintensification is assessed relative to preceding land use and that the earlier use, if it had represented a long-term arrangement, may have demonstrated a fairly sustainable style of resource management (Denevan 2001; Sneddon 2000).

within-unit logics help to explain the links between the shifting places and scales of intensification and disintensification, which, as seen in the chapters, have produced environmental consequences that are either positive or negative, or both, depending on local spatial differences.[10]

Migration is one of the most important micrologics in resource use to affect global conservation efforts since it is so common and is increasing in magnitude and frequency among rural people in developing countries (Barham and Boucher 1998; Jokisch 2002; Mutersbaugh 2002; Taylor, Rozelle, and de Brauw 2003). The increased role of migration is the result, in large part, of urban growth that ranges from small and medium-size towns and villages to large metropolitan areas and global megacities. Migration and its environmental effects are widely regarded as an integral feature of globalization (Brah, Hickman, and Mac an Ghaill 1999). Migration from the rural areas of developing countries takes the form of a dizzying array of permanent, seasonal, and cyclical strategies. These forms of migration, alone or combined, contribute to the micrologics of farm families and households. Family- or household-level diversification that incorporates migration is producing fairly predictable consequences for land use, such as disintensification in "sending communities" and frontier-style intensification (chaps. 3 and 11). Other outcomes are more unexpected, such as rural-urban networks owed to migration (chap. 5).

Education adds another dimension to the micrologic of farm families and households that interface with global conservation issues in developing countries. Education is widely regarded as a major feature of the asset portfolio that influences economic choices and environmental impacts (Turner et al. 2001). In general, education is thought to reduce land use pressures, at least in the immediate context, since more-educated persons are likely to pursue non-resource-dependent livelihoods (Pichón 1997; Young 1999). Education is also important as a major driver of diversification since the increased schooling of children is a common goal of many families. This goal often requires that the rural households of developing countries pay in order to send their children to schools in towns and cities. The resulting move-

---

10. One vivid example of the shifting places and scales of intensitification and disintensification is organic coffee as a high-value crop among Oaxaca farmers (chap. 2), which may reduce forest use or grazing pressures in the mix of other activities that are undertaken among these families. Simple predictions do not apply to such transfers, since increased earnings without the possibility for other investments might well be channeled into expanded livestock raising (Pichón 1997). Another example is the parcel-level intensification of tree and shrub planting in Inner Mongolia (chap. 13), which is spatially displacing the environmental degradation of lowered water tables and diminishing water supplies to nearby pasture lands, and thus reinforcing the shift away from livestock raising to a greater extent than planners may have expected.

ments connect the countryside to villages or cities via the social networks among children and the adult family members that support their education, which adds another example of increasing rural-urban linkages (chap. 5).

Education in the broad sense of environmental education and awareness is potentially a fertile ground for policy that is informed through the micrologics of farm and resource use. This promise is rooted in the examples of how different groups of farm and resource users are guided by distinct micrologics. As shown in this volume, such different resource-use groups commonly exist within local places or communities (see also Bulte and van Soest 2001; Carter and May 1999; Coomes, Barham, and Takasaki 2004). Distinctions are commonly drawn among the landless, near-landless, and small-, medium-, and large-size landholders, and they may exist in conjunction with differences of gender, race, and ethnicity as well. With the vital importance granted to environmental education and educational reforms associated with democratization (chap. 11), it is important that new learning communities incorporate the awareness of micrologics into the design of projects as well as curriculum development and other innovations.[11]

Spaces of Hope?

The analysis of unintended consequences and the micrologics of resource use leads to a domain of practical recommendations about various new or rethought spatial designs—optimistically termed "spaces of hope"—as introduced in the first section of this chapter. As a result of our studies, the contributors and I believe that such new geographies must be taken into account in global conservation and sustainability. These practical recommendations require the integration of both environmental expertise and development perspectives, which our volume has sought to advance. An inveterate weaving of disciplinary perspectives, lately much advanced in examining human-environment interactions, is now essential at the interface of the spatial design dimensions of global conservation, on the one hand, and the issues involving livelihoods, agriculture, and development change, on the other hand (Raynor and Malone 1998; Schoenberger 2001; Westley and Miller 2003; Zimmerer and Bassett 2003). This volume has applied the perspective of geography to the twin phenomena of territorialization and network building that

---

11. Such micrologics are likely to be equally or more important in shaping the viability of land use alternatives than are isolated factors whose functional impact is often minimal. A typical example of the latter, as discussed above, is whether communities or land users are found on the boundaries of protected areas (chap. 9), which frequently functions as the sole or major design criterion for existing sustainability programs, which results in the categorization of socioeconomic groups that are so varied that the importance of micrologics is overlooked.

are proliferating as a result of the linkages of global conservation with agriculture, land, and resource use.

Multiple, overlapping areas of conservation, management, and land use already exist in numerous protected areas, with many of them currently in the planning phase. The variety of coexisting spaces often combines the conventional territorial form of conservation with the areas that are attached to nearby residents, local land users, government agencies and institutions, activist groups, experts, and citizen coalitions. The future of many conservation areas is likely to depend on the unfolding of these spaces as they are negotiated among these diverse interests that will likely show a growing interface with resource use and livelihood issues. Even with the territorial forms of conservation, various subtypes are shaped by the powers of globalization (Mutersbaugh, chap. 2 in this volume; see also Morehouse 1996; Neumann 1998, 2004). At the end of the spectrum that puts emphasis on less-humanized templates are the "natural" designs such as ecoregions, ecological life zones, and ecosystems. Spatial designs that combine natural blueprints with the ideas of humanized landscapes are varied, from more abstract or political designations (buffer zones) to explicitly utilitarian configurations ("foodsheds") to those that are forged according to deeply cultural models of the landscape.

Examples of the more abstract type of conservation territory are biosphere reserves and conservation corridors, in which process-based and abstract design parameters (concentric circles, connecting polygons) tend to figure as prominently as habitat location and the recognition of human settlement and land use activities. The transnational protected area (TPA), whose expansion is evaluated in chapter 1, is a territorial type that often originates in geopolitical concerns that are then extended to natural habitats and landscapes. Explicitly utilitarian designs are territories based on particular management or use characteristics, such as the extractive reserves of rubber tappers in the Brazilian Amazon. Watersheds and river basins, which are deeply cultural as well as biogeophysical models, are two territorial designs that have become increasingly popular as spatial blueprints for the combination of conservation with agriculture and resource use planning.

The idea of a social-environmental network encapsulates several of the main policy-related findings of this volume. The term *social-environmental networks* refers to the assemblages of linked persons and their activities across areas of resource use. Examples of social-environmental networks, which are summarized below, include farmer, resource cooperative, and land use organizations; seed exchange circuits within and among farm households and multiple communities; pastoralist-farmer groups; and migrant networks that are connected to conservation and resource issues. The links in these social-environmental networks are formed through various sorts of exchanges, transfers, personal relations, and physical movements. These links include the exchange of information, goods, and direct labor inputs that have to do with the environment and resources together with the relations of trust and shared identity that form through kinship, community, working conditions, and political and social mobilization. Social-environmental networks are often sustained through the laboring activities of people in formal organizations (like coffee cooperatives) and informal resource-use networks (like seed-exchange networks). Such work activities and the networks sustaining them are responsible for the support of the real costs of much environmental sustainability, although the basis of such contributions is regularly overlooked.

Social-environmental networks may involve conventional territorial arrangements, such as the attempts to create more complete coverage of shade trees among the organic coffee growers belonging to cooperatives in Mexico. Still more often, the spatial design of social-environmental networks is at least somewhat discontinuous, as described below, and thus differs from the tendency of current conservation to rely on contiguous-tending territories. Social-environmental networks may be based on well-defined environment-centered social movements, such as groups who are protecting their useful forests (Chipko in India, Brazilian rubber tappers; see Barkin 2001; Peet and Watts 2004; Zerner 2000). Still, social-environmental networks also refer here to a broader realm of experience and work activities that involves various commonplace or everyday activities. Mostly quotidian networks of human-environment interaction, such as seed exchange, migration, and herd management, are not only environmentally significant but in many cases are imbued with deep-seated cultural, social, political, or personal meanings (Croll and Parkin 1992; Meagher 2004).

The first type of social-environmental network to be recommended comprises farmer, resource cooperative, and land use organizations. These groups are common throughout many rural areas of developing countries as well as most developed ones. Two of the better-known examples are coffee cooperatives, exemplified in Mexico, and beekeeping groups, such as those in Brazil (chaps. 2 and 4). The potential for groups of sustainable forestry managers is another example (chap. 3). The importance of this type of social-environmental network is rooted in two features that potentially present an advantage to the network members as well as global conservation organizers and organizations. First, the majority of these groups are already in existence, which means that they bring established mechanisms and capacities for social organization, information exchange, and political mobilization that are related to resource management (Healy 2001; Keck 1995). Second, most farmer organizations, resource cooperatives, and land user groups are concerned with issues that are both "productivist" (e.g., increasing milk production in dairy herds) as well as "postproductivist" (e.g., provisioning of non-farm goods and services to tourists and other consumers of rural amenities; Evans, Morris, and Winter 2002). This combination of productivist and post-productivist activities is evident in the characteristic diversification of family- or household-level economies that was discussed in the previous section. Consequently, the organizations of farmers, resource cooperatives, and land users that form in developing countries are often able to address the new opportunities or dilemmas of conservation (e.g., ecotourism, wildlife damage to crops, and disease spread to domesticated livestock). At the same time, these groups are committed to productivist concerns (crop growing, livestock raising) that remain crucial to their rural livelihoods.

In advocating for the increased role of these existing farmer organizations and resource groups it must be mentioned that they have been bypassed sometimes in the third wave of conservation in favor of newly formed institutions (Zerner 2000). One likely reason for overlooking the existing groups is the difficulties that can emerge when national political issues, such as the realities of contentious agrarian reform politics, are raised via the politics surrounding local conservation or resource activities. Rather than viewing the presence of this broader political arena as a drawback, the contributors and I see it as a necessary dimension of conservation and development politics (chaps. 11 and 12). Indeed, the contributions of conservation can be strengthened, rather than diluted, through articulation with national issues like democratization initiatives and agrarian reform.

The next three types of social-environmental networks, which refer to a more informal type of networking, are resource-, social-, and place-based groupings of people and the places where they live and work. Resource-specific networks comprise persons whose activities connect them via flows of resources, labor, and information. These networks tend to exist at scales larger than individual communities. Multicommunity scaling confers an advantage in those issues of resource conservation whose organizational extent requires cooperation, coordination, and capacity building among multiple communities (or, in many cases, subgroups within multiple communities). Seed-acquisition and seed-exchange networks are a prime example of these social-environmental networks (chap. 6), where the seeds of agrobiodiverse Andean food plants are purchased, bartered, and gifted among friends, neighbors, acquaintances, and seed growers who live across multiple communities. The spatiality of these seed flows often connects contiguous communities, although the seed networks do at times connect places and people in noncontiguous locations. Additional examples of social-environmental networks involving labor and information flows are the work-swapping customs that support Andean wetland agriculture and farm terrace building, which have been identified as mainstays of sustainable approaches to agriculture, resource use, and conservation in the tropical mountains of western South America (Denevan 2001; Erickson 1992; Healy 2001).

The third type of social-environmental network is that of the resource, conservation, and land use practices of persons that undertake migration activities. Migrant networks obviously have distinct and varied spatialities, which share the common and defining thread of connecting separate places (Jokisch 2002; Radcliffe 1993; Rudel, Bates, and Machinguiashi 2002). In the rural areas of developing countries one of the most common forms of migration is a rural-rural network that connects the established settlement of "sending communities" to frontier or colonization sites. These networks have long existed in particular places, although the frequency and scope of rural-rural migration and its potential consequences for conservation have expanded during recent decades, with such high-profile issues as rain forest clearing on tropical settlement frontiers. The expansion of rural-rural migration is attributed to farm- or household-level diversification, widespread and chronic farm crises, and the impact of environmental and climate change effects that have weakened the capacities of local livelihoods. The network-style spatiality of migration is found at the farm family or household level, since this unit is coordinating the movements of one or more of its members along with resource use and environmental management decisions, information flows, and investment and marketing choices (Taylor, Rozelle, and de Brauw 2003). At the same time, rural-rural migration networks may also involve subgroups within and among communities who exchange resources and information in making these adjustments.

The environmental stakes in rural-rural migration are elevated, for this mode of movement may link impacts on more than one crucial environment. Prominent examples include the rural-rural migration from areas of tropical mountain environments to the tropical forest lowlands, and vice versa, in world-renowned centers of biodiversity and fragile landscapes that are located in Latin America and South and Southeast Asia (Denevan 2001; Durham 1995; Coxhead and Jayasuriya 2004; Painter 1995). One particularly well-known and important example is the rural-rural migration connecting crucial areas of agriculture, resource use, and conservation in the Andes highlands and the Amazonian tropical forests (chaps. 9 and 10). Our interest in drawing additional attention to the social-environmental networks of these migrants is redoubled as a result of some of the politics of interpreting migration in certain conservation circles (chap. 11). Even now these people are often portrayed in an a priori fashion as environmentally destructive intruders or eco-villains. Such portrayals have occurred via a series of accounts that may be naïve, romantic, or self-serving in depicting the frontier migrants as flocking destroyers of nature in a place where they are not native (Slater 2003).

Rural-urban spatialities are a fourth type of social-environmental network that is recommended as the design basis for the future of conservation and development. Rural-urban migration is apparent in the teeming cities of developing countries, which vary from provincial capitals to major metropolitan centers. It is equally apparent that rural out-migration to urban areas is not a unidirectional or one-way process, but rather is frequently based on periodic return migration as well. At the family or household level of resource management, both in rural areas and urban ones, these ties are apparent in the flows of goods, ideas, labor, technologies, and virtually every sort of possibly transferable item. One vivid example is that of the urban house-lot gardens in cities of the Brazilian Amazon, such as Santarém, that are provided with planting material of trees, shrubs, and herbs through rural-urban social networks (chap. 5). The ties between the city and countryside are increasing in frequency and scope as a result of expanded urbanization. Yet the growing urbanization is frequently based on low-paying employment and difficult living conditions. Migrants find that these conditions can be at least slightly improved through continued connections to rural and periurban people and places through their networks of kin and acquaintances. In addition to urban house-lot agrobiodiversity and food supply, the influence of urban-rural migration is felt strongly in land use and environmental change pressures such as the development of settlements in periurban areas (Jokisch 2002). In general, the significance of these urban-rural linkages for conservation points out a main counterargument to the entrenched dichotomy of conservation into the so-called Green Agenda of rural areas and the artificially decoupled Brown Agenda of urban environmentalism.

The recommendation for conservation to focus on these four types of social-environmental networks is also a call to engage in a wide-ranging consideration of the issues of agriculture and food as central to a broader, more environmentally encompassing approach to global conservation. Each of the social-environmental networks that we are recommending includes an emphasis on farming and food, so it is a logical extension to see these issues as adding to the core of conservation. Agriculture and food issues bring development concerns into extensive direct contact with the realm of the environment. Indeed, worldwide development policy and politics are reaching an unprecedented degree of concern over the security and sovereignty of food supply and agricultural ecosystems, with emphasis on environmental quality, health and sustainability, and affordability, resilience, and reducing vulnerability (Friedmann 2006; Marsden 1997; Marsden and Smith 2005; Orlove and Brush 1996; Thrupp 2000). At the same time, the world of conservation is ever more engaged with agriculture and land use. Global conservation must remain involved with agriculture and land use due to the importance of protected areas located in rural lands of developing countries, which includes the prevalence of new designs such as conservation corridors that promise to place an unprecedented level of emphasis on agriculture and land use.

All the social-environmental networks that we have discussed embrace an environmental view of farming since they combine persons and environments that collectively encompass both the productivist view (i.e., as agricultural production that is sustainable) with consumption-based perspectives (i.e., as food, including diverse and healthy food). These agriculture and food issues are central to many of the organizations, citizens, and individual persons with whom the book's contributors and I have worked. Across a variety of places and people the issues of the quality of food, along with access to high-quality food, are increasingly familiar as popular touchstones of current consumer-based environmentalism in the countries of North America, Europe, and the Global South (Friedmann 2006; Gottlieb 2001). For global conservation organizations, this broader view that encompasses food quality is derived also from the fact that the environmental dimension of agriculture has attained a newfound personal relevance to many of their headquartered personnel and planners. The personal importance of food quality issues at home holds the potential of extending the relevance of these issues to the conservation areas that are far from the centers of their global headquarters.

Our recommendation of these four types of social-environmental networks is also resonant with the recent trend of policy and public interest toward placing increased emphasis on urban environmental issues (Gottlieb 1993, 2001). Urban environmental issues and growing citizen movements for environmental justice have expanded beyond pollution, toxin, and contamination problems to also embrace a broader concern with the issues of food rights, water resources, and environmental health. This more expansive view of urban environmental issues suggests a greater awareness of the vital, ongoing connections of the nature of cities to periurban, suburban, and rural areas (P. A. Walker 2003; Walker and Fortmann 2003; Zimmerer and Bassett 2003). As this volume demonstrates, the social-environmental networks of importance to conservation and development are frequently based on linkages that connect rural people and places to those in nonrural settings, which can include the same people as well as their counterparts. Farmer organizations in developing countries, for example, are typically headquartered in villages and cities, while many of the involved individuals also maintain an active presence in urban affairs with a secondary residence in the city as well. Migration and multicommunity resource networks similarly tend to involve connections to urban or periurban areas even when they are described as rural-rural, since this designation refers merely to the end points of migratory activities. Such connections may occur either directly as part of the migratory travel or marketing activities, for example, or indirectly in that one or more members of the family or household are living or working in these nonrural settings.

These recommendations regarding social-environment networks must be considered with respect to the mainstream ideas and policies that guide global conservation and protected areas. Expanding worldwide interest in conservation corridors offers one area of particular promise since these initiatives are premised on the idea of combining nature protection and concerns for land use sustainability. Conservation corridors can potentially bring a heightened awareness of the interface with agriculture and land use as an integral feature of conservation rather than as an unwanted complexity. Similarly, the social-environment networks recommended here should be of interest to such spatial designs as buffer zones and transition zones that encircle the core "nuclear zones" of protected areas; while the buffer zones and transition zones have become principal design elements of conservation plans, the impetus and ideas for sustainable land use are often lacking. This book's findings urge the immediate investment of expertise in sustainability issues in such areas as buffer zones and transition zones, rather than downplaying through simplification the importance of livelihood ecologies. The latter prospect is not an empty threat since it is advocated currently in the "fortress conservation" approach, which has gained some momentum in seeking to place nearly exclusive emphasis on strictly protected areas by claiming that "The upshot of this new emphasis on sustainability [the third wave of conservation] rather than protection has been a reduced commitment to protected areas" (Kramer, van Schaik, and Johnson 1997, 4).

Also, the social-environmental networks that are highlighted here bring new attention to potentially beneficial partners for environmental conservation that may be offered through the much-expanded emphasis on agroecology and agrobiodiversity within international development. In particular, these networks suggest the potential contributions of programs such as the sustainability groups within the global agricultural centers (such as programs within the Centro Internacional de Agricultura Tropical [CIAT] in Cali, Colombia, and the CONDESAN group that is housed within the Centro Internacional de la Papa [CIP] in Lima, Peru), international development organizations, national agricultural programs, and agricultural NGOs. These linkages can aid in bringing agrarian and livelihood issues into a more central place in global conservation. Environmental globalization that weds both food and conservation issues is potentially united to contend with the strong anti-environmental tendencies of much of the economic globalization that has occurred thus far in the form of global trade agreements, corporate resource raiding at the global scale, and siting preferences for environmental de- or unregulated locations globally (Gwynne and Kay 2004; Roberts and Thanos 2003; Speth 2003). Faced with these realities, the findings of this volume urge global conservation organizations to be open to the consideration of projects, programs, and policies that take the shape of the new spatialities. Such new spatialities require the informed and comprehensive analysis of *both* conventional conservation territories *and* innovative networks, which may occur in multiple, overlapping configurations. The combination of conservation territories and networks can help provide an innovative suite of design capacities for the future of environmental conservation, both global and local, as well as the intermediate scales that must also be valued in order to create viable pathways to sustainability.

Acknowledgments

I gratefully acknowledge the many contributions that my colleague Bradford Barham has made to this chapter through a large number of fruitful discussions, our cotaught graduate seminar, and his commentary at the "Spaces of Hope" conference.

References

Agarwal, B. 1994. *A field of one's own: Gender and land rights in South Asia.* Cambridge: Cambridge University Press.

Amin, A. 2002. Spatialities of globalisation. *Environment and Planning A* 34:385–99.

Bailey, R. 1998. *Ecoregions: The ecosystem geography of oceans and continents.* New York: Springer.

Bardhan, P., and C. Udry. 1999: *Development economics.* Oxford: Oxford University Press.

Barham, B. L., and S. Boucher. 1998. Migration, remittances and inequality: Estimating the net effects of migration on income distribution. *Journal of Development Economics* 55:307–31.

Barham, B. L., and O. T. Coomes. 1994. Reinterpreting the Amazon rubber boom: Investment, the role of the state, and Dutch Disease. *Latin American Research Review* 29 (2): 73–109.

Barkin, D. 2001. Neoliberalism and sustainable popular development. In *Transcending neoliberalism: Community-based development in Latin America,* ed. H. Veltmeyer and A. O'Malley, 184–204. Bloomfield, CT: Kumarian.

Bassett, T. J. 2001. *The peasant cotton revolution in West Africa: Côte d'Ivoire, 1880–1995.* Cambridge: Cambridge University Press.

Bassett, T. J., and K. S. Zimmerer. 2003: Cultural ecology. In *Geography in America at the dawn of the new millenium,* ed. G. L. Gaile and C. J. Wilmott, 97–112. Oxford: Oxford University Press.

Batisse, M. 1997. Biosphere reserves: A challenge for biodiversity conservation and regional development. *Environment* 39 (5): 7–33.

Bebbington, A. 2001. Globalized Andes? Livelihoods, landscapes, and development. *Ecumene* 8 (4): 414–36.

Bellon, M. 1991. The ethnoecology of maize variety management: A case study from Mexico. *Human Ecology* 19 (3): 389–418.

Bever, S. W. 2002: Migration and the transformation of gender roles and hierarchies in Yucatan. *Urban Anthropology* 31:199–230.

Bigman, D., and H. Fofack, eds. 2000. *Geographical targeting for poverty alleviation: Methodology and application.* Washington, DC: World Bank.

Bissonette, J. A., ed. 1997. *Wildlife and landscape ecology: Effects of pattern and scale*. New York: Springer.

Blaikie, P., and H. Brookfield, eds. 1987. *Land degradation and society*. London: Methuen.

Brah, A., M. J. Hickman, and M. Mac an Ghaill, eds. 1999. *Global futures: Migration, environment, and globalization*. London: Macmillan.

Brosius, J. P., A. L. Tsing, and C. Zerner. 1998. Representing communities: Histories and politics of community-based natural resource management. *Society and Natural Resources* 11 (2): 157–69.

Brown, M., and B. Wyckoff-Baird. 1992. *Designing integrated conservation and development projects*. Washington, DC: Biodiversity Support Program.

Bryant, R. L., and S. Bailey. 1997. *Third world political ecology*. London: Routledge.

Bulte, E. H., and D. P. van Soest. 2001. Environmental degradation in developing countries: Households and the (reverse) environmental Kuznets Curve. *Journal of Development Economics* 65 (1): 225–35.

Carney, J. 1993. Converting the wetlands, engendering the environment: The intersection of gender with agrarian change in the Gambia. *Economic Geography* 69:329–48.

———. 2004. Gender conflicts in Gambian wetlands. In *Liberation ecologies: Environment, development, social movements*, 2nd ed., ed. R. Peet and M. Watts, 316–36. London: Routledge.

Carter, M., and J. May. 1999. Poverty and class in rural South Africa. *World Development* 27 (1): 1–20.

Coomes, O. T. 2004. Rain forest "conservation-through-use"? Chambira palm fibre extraction and handicraft production in a land-constrained community, Peruvian Amazon. *Biodiversity and Conservation* 13:351–60.

Coomes, O. T., and B. L. Barham. 1997. Rain forest extraction and conservation in Amazonia. *Geographical Journal* 163 (2): 51–64.

Coomes, O. T., B. L. Barham, and Y. Takasaki. 2004. Targeting conservation-development initiatives in tropical forests: Insights from analyses of rain forest use and economic reliance among Amazonian peasants. *Ecological Economics* 51 (1–2): 47–64.

Coomes, O. T., F. Grimard, and G. J. Burt. 2000. Tropical forests and shifting cultivation: Secondary forest fallow dynamics among traditional farmers of the Peruvian Amazon. *Ecological Economics* 32:109–24.

Coxhead, I., and S. Jayasuriya. 2004. Development strategy and trade liberalization: Implications for poverty and environment in the Philippines. *Environment and Development Economics* 9:613–44.

Creed, G. W. 2000. "Family values" and domestic economies. *Annual Review of Anthropology* 29:329–55.

Croll, E., and D. Parkin, eds. 1992. *Bush base: Forest farm; Culture, environment, and development*. London: Routledge.

Denevan, W. M. 2001. *Cultivated landscapes of Native Amazonia and the Andes*. Oxford: Oxford University Press.

Dinerstein, E. 1995. *A conservation assessment of the terrestrial ecoregions of Latin America and the Caribbean*. Washington, DC: World Bank.

Dove, M. R. 1996. Center, periphery, and biodiversity: A paradox of governance and a development challenge. In *Valuing local knowledge: Indigenous people and intellectual property rights*, ed. S. B. Brush and D. Stabinsky, 41–67. Washington, DC: Island.

Durham, W. 1995. Political ecology and environmental destruction in Latin America. In *The social causes of environmental destruction in Latin America*, ed. M. Painter and W. Durham, 249–56. Ann Arbor: University of Michigan Press.

Erickson, C. L. 1992. Prehistoric landscape management in the Andean highlands: Raised field agriculture and its environmental impact. *Population and Environment* 13 (4): 285–300.

Escobar, A. 1995. *Encountering development: The making and unmaking of the third world*. Princeton, NJ: Princeton University Press.

Evans, N., C. Morris, and M. Winter. 2002. Conceptualizing agriculture: A critique of post-productivism as the new orthodoxy. *Progress in Human Geography* 26:313–32.

Ferguson, J. 1990. *The anti-politics machine: "Development," depoliticization, and bureaucratic power in Lesotho*. Cambridge: Cambridge University Press.

Folbre, N. 1986. Hearts and spades: Paradigms of household economics. *World Development* 14:245–55.

Forsyth, T. 1999. Environmental activism and the construction of risk: Implications for NGO alliances. *Journal of International Development* 11 (5): 687–700.

Friedmann, H. 2006. From colonialism to green capitalism: Social movements and the emergence of food regimes. In *New directions in the sociology of global development,* ed. F. H. Buttel and P. D. McMichael. Amsterdam: Elsevier/JAI, forthcoming.

Friedmann, J. 1988. *Life space and economic space: Essays in third world planning.* New Brunswick, NJ: Transaction Books.

Friedmann, J., and H. Rangan, eds. 1993. *In defense of livelihood.* West Hartford, CT: Kumarian.

Godoy, R., K. O'Neill, K. McSweeney, D. Wilkie, V. Flores, D. Bravo, P. Kotishack, and A. Cubas. 2000. Human capital, wealth, property rights, and the adoption of new farm technologies: The Tawahka Indians of Honduras. *Human Organization* 59 (2): 222–33.

Goldammer, J. G., ed. 1992. *Tropical forests in transition: Ecology of natural and anthropogenic disturbance processes.* Boston: Birkhäuser Verlag.

Gottlieb, R. 1993. *Forcing the spring.* Washington, DC: Island.

———. 2001. *Environmentalism unbound: Exploring new pathways for change.* Cambridge, MA: MIT Press.

Green, R., S. J. Cornell, J. P. W. Scharlemann, and A. Balmford. 2005. Farming and the fate of wild nature. *Science* 307:550–55.

Gwynne, R., and C. Kay. 2004. *Latin America transformed: Globalization and modernity.* 2nd ed. London: Arnold.

Hadley, M., ed. 2002. *Biosphere reserves: Special places for people and nature.* Paris: UNESCO.

Hall, A., ed. 2000. *Amazonia at the crossroads: The challenges of sustainable development.* London: Institute of Latin American Studies.

Harris, J. 2002: The case for cross-disciplinary approaches in international development. *World Development* 30:487–96.

Harvey, D. 2000. *Spaces of hope.* Berkeley and Los Angeles: University of California Press.

Healy, K. 2001. *Llamas, weavings, and organic chocolate: Multicultrual grassroots development in the Andes and Amazon of Bolivia.* South Bend, IN: University of Notre Dame Press.

Jokisch, B. D. 2002. Migration and agricultural change: The case of smallholder agriculture in highland Ecuador. *Human Ecology* 30:523–50.

Kanbur, R. 2002. Economics, social science, and development. *World Development* 30 477–86.

Keck, M. E. 1995. Parks, people, and power: The shifting terrain of environmentalism. *NACLA Review* 28 (5): 36–41.

Kramer, R., C. van Schaik, and J. Johnson, eds. 1997. *Last stand: Protected areas and the defense of tropical biodiversity.* Oxford: Oxford University Press.

Lambin, E. F., B. L. Turner, H. J. Geist, S. B. Agbola, A. Angelsen, J. W. Bruca, O. T. Coomes, R. Dirzo, G. Fischer, and C. Folke. 2001. The causes of land-use and land-cover change: Moving beyond the myths. *Global Environmental Change* 11:261–69.

Leach, M., R. Mearns, and I. Scoones. 1999. Environmental entitlements: Dynamics and institutions in community-based natural resource management. *World Development* 27 (2): 225–47.

Liverman, D., E. Moran, R. R. Rindfuss, and P. C. Stern, eds. 1998. *People and pixels: Linking remote sensing and social science.* Washington, DC: National Academy Press.

Marsden, T. 1997. Creating space for food. In *Globalising food,* ed. D. Goodman and M. Watts, 169–91. New York: Routledge.

Marsden, T., and E. Smith. 2005. Ecological entrepreneurship: Sustainable development in local communities through quality food production and local branding. *Geoforum* 36 (4): 440–51.

Massey, D. 1999a. Geography matters in a globalised world. *Geography* 84 (3): 261–65.

———. 1999b. Philosophy and politics of spatiality: Some considerations. The Hettner-lecture in human geography. *Geographische Zeitschrift* 87 (1): 1–12.

———. 2002. Globalisation: What does it mean for geography? *Geography* 87 (4): 293–96.

Mayer, E., and M. Glave. 1999. Alguito para ganar (a little something to earn): Profits and losses in peasant economies. *American Ethnologist* 26 (2): 344–69.

McNeely, J. A., and S. J. Scherr. 2003. *Ecoagriculture: Strategies to feed the world and save biodiversity*. Washington, DC: Island.

McSweeney, K. 2002. Who is "forest-dependent"? Capturing local variation in forest-product sale, Eastern Honduras. *Professional Geographer* 54:158–74.

———. 2004. Forest product sale as natural insurance: The effects of household characteristics and the nature of shock in eastern Honduras. *Society and Natural Resources* 17:1–18.

Meagher, K. 2004. Social capital or analytical liability: Social networks and African informal economics. Paper presented to the Program in Agrarian Studies, Yale University, New Haven, Connecticut, 19 November.

Mertens, B., E. F. Lambin, and W. D. Sunderlin, W. D. 2000: Impact of macroeconomic change on deforestation in South Cameroon: Integration of household survey and remotely-sensed data. *World Development* 28:983–99.

Mitchell, T. 2002. *Rule of experts: Egypt, techno-politics, modernity*. Berkeley: University of California Press.

Mittermeier, R., N. Myers, J. Thomsen, G. da Fonseca, and S. Oliveieri. 1998. Biodiversity hotspots and major tropical wilderness areas: Approaches to setting conservation priorities. *Conservation Biology* 12:516–20.

Mooney, H. A. 1998. *The globalization of ecological thought*. Oldendorf, Germany: Ecology Institute.

Morehouse, B. J. 1996. *A place called the Grand Canyon: Contested geographies*. Tucson: University of Arizona Press.

Mutersbaugh, T. 2002: Migration, common property, and communal labor: Cultural politics and agency in a Mexican village. *Political Geography* 21:473–94.

Netting, R. C. 1993: *Smallholders, householders: Farm families and the ecology of intensive, sustainable agriculture*. Palo Alto, CA: Stanford University Press.

Neumann, R. P. 1998. *Imposing wilderness: Struggles over livelihood and nature preservation in Africa*. Berkeley: University of California Press.

———. 2004. Nature-state-territory: Toward a critical theorization of conservation enclosures. In *Liberation ecologies: Environment, development, social movement*, 2nd ed., ed. R. Peet and M. Watts, 195–217. London: Routledge.

Olson, D., and E. Dinerstein. 1998. The global 2000: A representation approach to conserving the earth's most biologically valuable ecoregions. *Conservation Biology* 12 (3): 502–15.

Olson, M., E. Dinerstein, E. Wikramanayake, N. Burgess, G. Powell, E. Underwood, E. D'amico, et al. 2001. Terrestrial ecoregions of the world: A new map of life on earth. *BioScience* 51:933–38.

Omernik, J. 1995. Ecoregion: A framework for managing ecosystems. *George Wright Society Forum* 12:35–50.

Orlove, B. S., and S. B. Brush. 1996. Anthropology and the conservation of biodiversity. *Annual Review of Anthropology* 25:329–52.

Painter, M. 1995. Upland-lowland production linkages and land degradation in Bolivia. In *The social causes of environmental destruction in Latin America*, ed. M. Painter and W. Durham, 133–68. Ann Arbor: University of Michigan Press.

Paulson, S., and L. L. Gezon, eds. 2005. *Political ecology across spaces, scales, and social groups*. New Brunswick, NJ: Rutgers University Press.

Peet, R., and M. Watts, eds. 2004. *Liberation ecologies: Environment, development, social movements*. 2nd ed. London: Routledge.

Perz, S. G., and R. T. Walker. 2002: Household life cycles and secondary forest cover among small farm colonists in the Amazon. *World Development* 30:1009–27.

Pichón, F. 1997. Colonist land-allocation decisions, land use, and deforestation in the Ecuadorian Amazon frontier. *Economic Development and Cultural Change* 45 (4): 707–44.

Porter, P., and E. Sheppard. 1998. *A world of difference: Society, nature, development.* New York: Guilford.

Radcliffe, S. 1993. The role of gender in peasant migration: Conceptual issues from the Peruvian Andes. In *Different places, different voices: Gender and development in Africa, Asia, and Latin America,* ed. J. H. Momsen and V. Kinnaird, 278–87. London: Routledge.

Ravenga, C., S. Murray, J. Abramovitz, and A. Hammond. 1998. *Watersheds of the world: Ecological value and vulnerability.* Washington, DC: World Resources Institute and Worldwatch Institute.

Raynor, S., and E. Malone, eds. 1998. *Human choice and climate change.* Columbus, OH: Battelle Press.

Reardon, T., J. Berdegué, and G. Escobar. 2001. Rural nonfarm employment and incomes in Latin America: Overview and policy implications. *World Development* 29 (3): 395–409.

Reardon, T., and S. Vosti. 1995. Links between rural poverty and the environment in developing countries: Asset categories and investment poverty. *World Development* 23 (9): 1495–1506.

Reinhardt, N. 2000. Latin America's new economic model: Micro-responses and economic restructuring. *World Development* 28 (9): 1543–66.

Roberts, J. T., and N. D. Thanos. 2003. *Trouble in paradise: Globalization and environmental crises in Latin America.* London: Routledge.

Rocheleau, D., B. Thomas-Slayter, and E. Wangari. 1996. *Feminist political ecology: Global issues and local experience.* London: Routledge.

Rodríguez, L. O., and K. R. Young. 2000. Biological diversity of Peru: Determining priority areas for conservation. *Ambio* 29:329–37.

Roe, D., and M. Jack. 2001. *Stories from Eden: Case studies of community-based wildlife management.* London: Biodiversity and Livelihoods Group, International Institute for Environment and Development.

Rudel, T. K., D. Bates, and R. Machinguiashi. 2002. A tropical forest transition? Agricultural change, out-migration, and secondary forests in the Ecuadorian Amazon. *Annals of the Association of American Geographers* 93:87–102.

Sandercock, L. 1998. *Making the invisible visible: A multicultural planning history.* Berkeley: University of California Press.

Schoenberger, E. 2001. Interdisciplinarity and social power. *Progress in Human Geography* 25 (3): 365–82.

Scott, J. C. 1998. *Seeing like a state: How certain schemes to improve the human condition have failed.* New Haven, CT: Yale University Press.

Sen, A. 1999. *Development as freedom.* New York: Anchor.

Shriar, A. 2001. The dynamics of agricultural intensification and resource conservation in the buffer zone of the Maya Biosphere Reserve, Petén, Guatemala. *Human Ecology* 29 (1): 27–47.

Simonian, L. 1995. *Defending the land of the jaguar: A history of conservation in Mexico.* Austin: University of Texas Press.

Slater, C., ed. 2003. *In search of the rain forest.* Durham: Duke University Press.

Smith, N. J. 1996. *Biodiversity and Agricultural Intensification.* Washington, DC: The World Bank.

Sneddon, C. S. 2000. "Sustainability" in ecological economics, ecology, and livelihoods: A review. *Progress in Human Geography* 24 (4): 521–549.

Sonneveld, B. G. J. S., and M. A. Keyzer. 2003. Land under pressure: Soil conservation concerns and opportunities for Ethiopia. *Land Degradation and Development* 14 (1): 5–23.

Soulé, M. E., and G. H. Orians, eds. 2001. *Conservation biology: Research priorities for the next decade.* Washington, DC: Island.

Soulé, M. E., and J. Terborgh. 1999. *Continental conservation: Scientific foundations of regional reserve networks.* Washington, DC: Island.

Speth, J. G. 2003. *Worlds apart: Globalization and the environment.* Washington, DC: Island.

Sundberg, J. 2002. Conservation as a site for democratization in Latin America: Exploring the contradictions in Guatemala. *Canadian Journal of Latin American and Caribbean Studies* 27 (53): 73–103.

Sunderlin, W. D., and J. Pokam. 2002: Economic crisis and forest cover change in Cameroon: The roles of migration, crop diversification, and gender division of labor. *Economic Development and Cultural Change* 50:581–606.

Swyngedouw, E. 1997. Neither global nor local: "Glocalization" and the politics of scale. In *Spaces of globalization: Reasserting the power of the local,* ed. K. R. Cox, 137–66. New York: Guilford.

Takasaki, Y., B. L. Barham, and O. T. Coomes. 2001. Amazonian peasants, rain forest use, and income generation: The role of wealth and geographical factors. *Society and Natural Resources* 14:291–308.

Taylor, J. E., S. Rozelle, and A. de Brauw. 2003. Migration and incomes in source communities: A new economics of migration perspective from China. *Economic Development and Cultural Change* 52 (1): 75–101.

Terborgh, J., ed. 2002. *Making parks work: Strategies for preserving tropical nature.* Washington, DC: Island.

Thrupp, L. 2000. Linking agricultural biodiversity and food security: The valuable role of agrobiodiversity sustainable agriculture. *International Affairs* 76 (2): 265.

Turner, B. L. II, et al. 2001. Deforestation in the southern Yucatán peninsula: An integrative approach. *Forest Ecology and Management* 154:355–70.

Turner, M. D. 1999. Merging local and regional analyses of land-use change: The case of livestock in the Sahel. *Annals of the Association of American Geographers* 89 (2): 191–219.

———. 2000. Drought, domestic budgeting and wealth distribution in Sahelian households. *Development and Change* 31:1009–35.

———. 2003. Methodological reflections on the use of remote sensing and geographic information science in human ecological research. *Human Ecology* 31 (2): 255–79.

Vance, C., and J. Geoghegan. 2002. Temporal and spatial modeling of tropical deforestation: A survival analysis linking satellite and household survey data. *Agricultural Economics* 27:317–32.

Vosti, S. A., E. M. Braz, C. L. Carpentier, M. V. N. D'Oliveira, and J. Witcover. 2003. Rights to forest products, deforestation, and smallholder income: Evidence from the Western Brazilian Amazon. *World Development* 31:1889–1901.

Walker, P. A. 2003. Reconsidering "regional" political ecologies: Toward a political ecology of the rural American West. *Progress in Human Geography* 27 (1): 7–24.

Walker, P. A., and L. Fortmann. 2003. Whose landscape? A political ecology of the "exurban" Sierra. *Cultural Geographies* 10 (4): 469–91.

Walker, R., E. Moran, and L. Anselin. 2000. Deforestation and cattle ranching in the Brazilian Amazon: External capital and household processes. *World Development* 26:683–99.

Walker, R. T. 2003. Mapping process to pattern in the landscape change of the Amazonian frontier. *Annals of the Association of American Geographers* 93:376–98.

Wescoat, J. L., Jr. 1984. *Integrated water development: Water use and conservation practice in western Colorado.* Chicago: University of Chicago Department of Geography.

Westley, F. R., and P. S. Miller. 2003. *Experiments in consilience: Integrating social and scientific responses to save endangered species.* Washington, DC: Island.

White, H. 2002. Combining quantitative and qualitative approaches in poverty analysis. *World Development* 30:511–22.

Young, E. 1999. Balancing conservation with development in small-scale fisheries: Is ecotourism an empty promise? *Human Ecology* 27 (4): 581–620.

Young, K. R., and B. León. 1995. Connectivity, social actors, and conservation policies in the central Andes. In *Biodiversity and conservation of neotropical montane forests,* ed. S. P. Churchill, 653–61. New York: New York Botanical Garden.

Young, K. R., and K. S. Zimmerer. 1998. Conclusion: Biological conservation and new geographical approaches. In *Nature's geography: New lessons for conservation in developing countries,* ed. K. S. Zimmerer and K. R. Young, 301–27. Madison: University of Wisconsin Press.

Zerner, C., ed. 2000. *People, plants, and justice: The politics of nature conservation.* New York: Columbia University Press.

Zimmerer, K. S. 1993. Soil erosion and labor shortages in the Andes with special reference to Bolivia, 1953–1991: Implications for conservation-with-development. *World Development* 21 (10): 1659–75.

———. 2000. The reworking of conservation geographies: Nonequilibrium landscapes and nature-society hybrids. *Annals of the Association of American. Geographers* 90 (2): 356–69.

———. 2004. Cultural ecology: Placing households in human-environment studies; The cases of tropical forest transitions and agrobiodiversity change. *Progress in Human Geography* 28 (6): 795–806.

———. 2006. Cultural ecology: At the interface with political ecology—the new geographies of environmental conservation and globalization. *Progress in Human Geography* 32 (1): 63–78.

Zimmerer, K. S., and T. J. Bassett. 2003. Future directions in political ecology: Nature-society fusions and scales of interaction. In *Political ecology: An integrative approach to geography and environment-development studies,* ed. K. S. Zimmerer and T. J. Bassett, 275–296. New York: Guilford.

Contributors

J. CHRISTOPHER BROWN
Department of Geography
1475 Jayhawk Blvd
213 Lindley Hall
University of Kansas
Lawrence, KS 66045-7613
jcbrown2@ku.edu

LESLIE C. GRAY
Environmental Studies Institute
Santa Clara University
500 El Camino Real
Santa Clara, CA 95053
lcgray@scu.edu

HONG JIANG
Department of Geography
550 North Park St.
University of Wisconsin
Madison, WI 53706
hjiang@geography.wisc.edu

TAD MUTERSBAUGH
Department of Geography
University of Kentucky
1331 Patterson Office Tower
Lexington KY 40506-0027
mutersba@uky.edu

JANE M. READ
Department of Geography
144 Eggers Hall
Syracuse University
Syracuse, NY 13244-1020
jaread@maxwell.syr.edu

LILY O. RODRÍGUEZ
Programa Parque Nacional
    Cordillera Azul
San Fernando 537
Miraflores, Lima 18
Peru
lilyrodriguez2@terra.com.pe

RODRIGO SIERRA
Department of Geography and the
    Environment
University of Texas at Austin
210 West 24th St., 334
Austin, TX 78712-1098
rsierra@mail.utexas.edu

CHRIS SNEDDON
Department of Geography
6017 Fairchild
Dartmouth College
Hanover, NH 03755
sneddon@dartmouth.edu

JUANITA SUNDBERG
Department of Geography
University of British Columbia
1984 West Mall
Vancouver, BC V6T 1Z2
Canada
sundberg@geog.ubc.ca

MATTHEW D. TURNER
Department of Geography
550 North Park St.
University of Wisconsin
Madison, WI 53706
turner@geography.wisc.edu

ANTOINETTE M. G. A.
   WINKLERPRINS
Department of Geography
314 Natural Sciences Building
Michigan State University
East Lansing, MI 48824-1115
antoinet@msu.edu

KENNETH R. YOUNG
Department of Geography and the
   Environment
University of Texas at Austin
210 West 24th St., 334
Austin, TX 78712-1098
kryoung@mail.utexas.edu

KARL S. ZIMMERER
Department of Geography
150 Science Hall
550 North Park St.
University of Wisconsin
Madison, WI 53706
zimmerer@wisc.edu

Index

activism. *See* environmental justice

Africa: agriculture, 22, 281, 282, 286–90; agro-biodiversity, 159; beekeeping, 101–3, 108, 109, 111; decentralization, 200, 277, 293, 296; environmental science, 166–83; sustainable forestry, 74; urban gardening, 123, 130, 133, 137; world-region overview, 16, 17, 317, 318, 324, 328. *See also* Sahel

African Sahel. *See* Sahel

agrarian reform. *See* land reform

agribusiness, 10n3, 14, 142, 143

agriculture: in Africa, 22, 281, 282, 286–90; in the Andean countries, 17, 141–60; in Brazil, 15, 16, 17, 121–37, 322; in China, 296–97, 301, 303, 305; development, 336, 337; disintensification, 52, 154–55, 330; environmental conservation and, 3, 5, 6, 7, 9, 12, 14, 45, 49–53, 66, 117–19, 156–60, 315, 316, 320–24, 325–29, 330–33; globalization, 5, 6, 12, 13, 41, 328; intensification, 57, 155, 328, 329; near Manu Biosphere Reserve (Peru), 141–60; in Maya Biosphere Reserve (Guatemala), 21, 264, 266, 269; in Mekong Delta, 328; in Mexico, 12, 13, 45–47, 49–67; in mountainous areas, 17, 49–67, 141–60; in Peru, 141–60; protected areas and, 32, 34, 146, 159; tropical ecosystems and, 12, 13

agrobiodiversity: in Andean countries, 141–43, 154, 158–60; in Brazil, 121, 122, 127–28; and environmental conservation, 316, 322–23, 336, 339; as global conservation issue, 9, 16, 17, 118

agrodiversity, 16, 121, 122, 127, 136

Amazonia. *See* Amazon region

Amazon region, 14, 15, 71–88, 92–113, 121–37, 216, 233, 236, 241, 243, 323, 333, 336

Andean countries, 17, 19–20, 141, 146, 158–60, 189, 212–27, 229–48, 325–26, 328. *See also* Bolivia; Colombia; Ecuador; Peru; Venezuela

Andean potatoes, 17, 118, 148, 153, 159, 323, 328

Andes mountains, 17, 18–20, 150–56, 216, 218, 230, 233, 234, 236, 240, 243, 336

Asia, 32, 74, 123, 200, 317

Asian Development Bank, 18, 192, 195, 203, 205n15, 207, 208

asset portfolio of resource users, 148, 328

audit trail, 56, 57. *See also* environmental certification

Audubon Society, 233

Australia, 32, 109, 197

Baltic ecological networks, 32

beekeeping, 14, 15, 45, 47, 92–94, 100–113, 318, 322, 334

berm cultivation, 122

big dam (large-scale hydroelectric) development, 203–5, 324

biodiversity: certification and sustainability projects, 50, 51, 64, 92, 121, 127, 136, 142, 166, 167; and decentralization, 256, 263, 315; definition, 212; importance in global conservation, 1, 2, 5, 11, 14, 18–20; "third wave" conservation approaches, 316, 324, 336; in transnational conservation, 187–89, 191–92, 193, 197, 201, 202, 212–13, 217, 222, 225, 226, 229, 230, 239, 241, 244

biological corridors. *See* conservation corridors

biosphere reserves: in global conservation, 8, 15, 18, 21, 31; and sustainability projects, 45, 51, 67, 146, 149; "third wave" conservation approaches, 256, 262, 265, 316, 333, 339

birds. *See* wildlife

desertification, 17, 22, 23, 119, 173, 257, 298, 301, 318, 324, 327

development: and decentralization, 273, 277, 286; globalization, 2, 3, 15; and sustainability projects, 47, 94–97, 106, 108, 111, 137, 143, 166; and "third wave" conservation approaches, 315–21, 324, 325, 327–32, 336; transnational, 192, 247

devolution, 20, 178, 256, 259, 278. *See also* decentralization

discourse analysis, 5, 6, 23, 92, 94, 100–108, 208, 260, 265–66, 268, 336

disintensification, 52, 154–55, 330

diversification, of household-level economies, 15, 155, 328, 329, 330, 331

"Earth Summit," 14, 29

East Asia, 32

Eastern Europe, 297

ecoagriculture, 329

ecodevelopment, 15, 45, 47, 322

ecohydrology, 193

ecological analysis: agrobiodiversity, 146, 148, 155; desertification, 166–80, 297–98; in global conservation, 16, 17, 18, 310, 333; protected-area design, 213, 214, 217, 225; and sustainability projects, 51, 66, 72, 74, 76–77, 82, 84, 109–11, 117–19, 127; water resources, 201–2

ecological disturbance, 72, 83, 322

ecological restoration, 230, 296, 301–4, 327

ecological theory, 16, 17, 18, 51, 67, 156–57, 172–74, 308

economy: and decentralization, 257, 297, 301; and environmental conservation history, 229; globalization, 2, 9, 15–17, 21, 27; sustainability projects, 46, 50–57, 65, 73, 85–87, 92, 95, 124; and "third wave" conservation approaches, 318, 327–32; transnational, 203, 216. *See also* development; household-level economies

ecoregions, 18–20, 23, 27, 187–90, 195, 199, 208, 212–27, 325–26, 333; definition, 216

ecotourism, 11, 27, 330, 335

Ecuador, 17, 19, 144, 158, 189, 212, 213, 218, 221

education, 51, 53, 64, 97, 106, 112, 169, 232, 269, 286, 331, 332

ejido communities (Mexico), 21, 52–56

environmental certification, 3, 12–14, 45, 46, 49–67, 71–75, 84–87, 100, 144, 318, 321, 322

environmental conservation: agrobiodiversity, 141; and decentralization, 259, 262, 277, 281, 308, 324–26, 340; and globalization, 3–7, 15, 17, 23–34; semi-arid environments, 166, 167, 181; in sustainability projects, 52, 57, 63, 87, 92, 105, 108, 112, 117; transnational, 193–98, 212–27; in urban gardens, 121–22. *See also* protected areas

environmental degradation: and decentralization, 280, 282, 298, 301, 306, 308; and environmental science, 166, 168, 174; and global conservation, 316, 324, 327, 331n10; and globalization, 2, 17, 18; and sustainability projects, 59, 71; transnational, 200

environmental globalization: and decentralization, 255–57, 266, 278; definition, 1, 2; and environmental science, 166, 180; global overview, 1–11, 18, 23–34, 35; and sustainability projects, 45–47, 112, 119; in "third wave" conservation approaches, 315, 316, 319, 332; transnational, 192, 212

environmental governance, 9–11, 18–19, 21, 64, 181, 255–57, 259, 266, 273–74, 277, 291, 293, 321, 326

environmental interdisciplinarity. *See* interdisciplinary knowledge

environmentalism, 5, 272, 338

environmental justice: and decentralization, 270, 272, 274, 278; in global conservation, 3, 4, 5, 6, 11, 21, 22, 27, 47, 320; and sustainability projects, 95, 96, 106; in transnational conservation, 199, 229

environmental management. *See* management of resources

environmental planning and policy: agrobiodiversity, 156–59; beekeeping, 93; and decentralization, 260, 262, 277, 280, 291–93; in global conservation, 1, 2, 3, 7, 45; semi-arid environments, 176; and "third wave" conservation approaches, 320, 321, 333; transnational, 213

environmental politics. *See* politics

environmental risk. *See* risk

environmental science: in conservation and sustainability projects, 45, 51, 62, 66, 79, 117–19, 122, 142, 143, 156–58; in environmental globalization, 1, 7, 8, 9, 15–18, 34, 35; and global environmental change, 167–74; and "third wave" conservation, 316, 320, 324, 325, 326, 332; and transnational conservation, 189, 190, 201, 225, 233, 234, 236

ethnic groups, 4, 243, 255–56, 279, 291, 304, 332

ethnographic techniques, 21, 22, 259, 260, 266

Europe, 21, 32, 73, 74, 97, 133, 338

European Union (EU), 27, 63, 97, 218

feminist political ecology, 260

fences, 23, 296, 301, 304, 305, 310–11